Praise for Andrea Wulf's

THE BROTHER GARDENERS

"A fascinating and beautifully researched story."
—*The Philadelphia Tribune*

"Engaging. . . . Lavishly researched. . . . Wulf never allows her material to overwhelm a vivid sense of the big picture, which keenly informs her sparkling narrative."
—*Bookforum*

"*The Brother Gardeners* is beautifully researched and equally well written." —*American Scientist*

"As Wulf triumphantly shows, plants and gardens reveal a wider view of the forces that shape society. . . . Rarely has the story of English plants been told with such vigour, and such fun."
—*The Times Literary Supplement* (London)

"The best book this year is *The Brother Gardeners: Botany, Empire and the Birth of an Obsession*."
—*The Independent on Sunday* (London)

"Well written. . . . Andrea Wulf brings this formative period of plant history to life." —*American Gardener*

"*The Brother Gardeners* is an excellent, hugely entertaining and instructive tale, and Wulf tells it well."
—*The Guardian* (London)

Andrea Wulf

THE BROTHER GARDENERS

Andrea Wulf was born in India and moved to Germany as a child. She trained as a design historian at London's Royal College of Art and is coauthor (with Emma Gieben-Gamal) of *This Other Eden: Seven Great Gardens and 300 Years of English History*. She has written for *The Sunday Times* (London), *The Wall Street Journal*, and the *Financial Times*, and she reviews for numerous newspapers, including *The Guardian*, *The Times Literary Supplement*, and the *Mail on Sunday*. She appears regularly on BBC television and radio. *The Brother Gardeners* was longlisted for the Samuel Johnson Prize.

www.andreawulf.com

THE
BROTHER
GARDENERS

BOTANY, EMPIRE AND THE
BIRTH OF AN OBSESSION

Andrea Wulf

VINTAGE BOOKS
A Division of Random House, Inc.
New York

FIRST VINTAGE BOOKS EDITION, MARCH 2010

Copyright © 2008 by Andrea Wulf

All rights reserved. Published in the United States by Vintage Books,
a division of Random House, Inc., New York. Originally published in
Great Britain in slightly different form by William Heinemann,
a division of Random House Group Ltd., London, in 2008, and
subsequently published in hardcover by Alfred A. Knopf, a division of
Random House, Inc., New York, in 2009.

Vintage and colophon are registered trademarks of Random House, Inc.

A portion of this work originally appeared in
Early American Life magazine.

The Library of Congress has cataloged the Knopf edition as follows:
Wulf, Andrea.
The brother gardeners : botany, empire and the birth of an obsession /
Andrea Wulf.—1st American ed.
p. cm.
Originally published: London : William Heinemann, 2008.
Includes bibliographical references and index.
1. Horticulturists—Great Britain—History—18th century.
2. Plant collectors—Great Britain—History—18th century.
3. Gardening—Great Britain—History—18th century. I. Title.
SB61.W85 2008b
635.0941—dc22   2008055080

Vintage ISBN: 978-0-307-45475-1

Author photograph © Dick Bartling

Printed in the United States of America
10   9   8   7

To Brigitte and Herbert

Our England is a garden, and such gardens are not made
By singing:—"Oh, how beautiful!" and sitting in the shade,
While better men than we go out and start their working lives
At grubbing weeds from gravel-paths with broken dinner-knives.

RUDYARD KIPLING, *"The Glory of Gardens"*

ontents

CONTENTS

A NOTE ON THE PLANT NAMES

In order to avoid the unwieldy use in the text of both the common and Latin names of plants, I have used either one or the other depending on the name by which a plant is most likely to be known. However, every plant is listed in the index under its common name (with the Latin name in brackets) and under its Latin name (with its common name in brackets).

When eighteenth-century correspondents only used common names of a genus such as Solomon's seal and it has been impossible to identify the exact plant species, this name is used throughout the book.

Additional information on the plants that played an important role in the eighteenth-century garden and their introduction to Britain can be found in the Glossary at the end of the book.

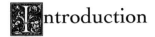# Introduction

But here it is worth noting a minor English trait which is extremely well marked though not often commented on, and that is a love of flowers. This is one of the first things that one notices when one reaches England from abroad, especially if one is coming from southern Europe.

GEORGE ORWELL, *The Lion and the Unicorn:*
Socialism and the English Genius, 1941

When I left my home town of Hamburg more than a decade ago, few of my friends possessed a garden. Even now, they mostly still prefer to live in huge apartments, putting off buying houses with outside space for their retirement. In Germany, gardening is often considered an occupation for pensioners. I certainly thought of it like that. It meant growing a few pansies in almost bare flowerbeds, planting out your allotment in regimented rows, or obsessively trimming your hedge with a ruler (on one occasion a neighbour sent a policeman to our house to inform my parents that our unruly hedge was at least four inches too close to the pavement).

I was therefore amazed, when I moved to London in the mid nineties, to find a nation obsessed with gardening. The shelves of my local newsagent groaned beneath lavish displays of gardening magazines; everywhere I went I saw signs for garden centres, and my new friends all seemed to think that the best way to spend a weekend was to visit the grounds of a stately home (unless they had an allotment, in which case the thrill of digging and weeding could not be surpassed). It seemed that everybody had a garden, and if they didn't, they wanted one: young and old, fashionable and staid, middle-class or working-class. At parties I listened for hours to twenty-something Londoners rave about plants or moan about horticultural disasters. I went with a trendy graphic designer to a nightclub, only to listen for half the evening to the minute details of the yield of his vegetable plot. On one occasion, a beautiful and perfectly groomed young woman told me that it was pointless throwing snails over the garden wall on to the railway embankment because they would return—she knew for certain because she had marked their shells with white paint.

Before long, I had my own garden—a tiny patch of green at the back of a typical London terrace. The previous owners had lovingly tended their flowers and herbs, and before I moved in they gave me a little tour of the garden. Only half listening, because I was much more interested in the house, I carelessly nodded when they asked me to care for the plants.

Yet by the time the house was renovated almost a year later, the garden had been destroyed by neglect and building work. Walking out of my back door, I was confronted by a jungle of thistles and dandelions, endless metres of bindweed, and knee-high grass. Guiltily I set to work, weeding and digging, and found beneath the weeds strangled shrubs and the remains of the flowers. Squashed and suffocated as they were, I hoped to make them flourish again.

But I had a problem: how was I supposed to care for these plants if I didn't know anything about them? My horticultural journey began with an excursion to a bookshop and the purchase of a glossy plant dictionary. Identifying my plants proved to be more complicated than I had anticipated and I took to sending startled visitors into the garden in case they could help. Soon I learnt that the large bushy plant in the corner, with a blossom resembling an elderly woman's flowery bathing cap, was *Hydrangea macrophylla*, with the more revealing common name, mophead. There was also a jasmine which perfumed the air in the summer, and a clematis that had taken over most of the garden wall. It seemed that I also owned a red camellia and an old fuchsia that leaned against the shed. Nothing special, I thought, until I began to discover the extraordinary history behind these seemingly ordinary plants.

It was neither a gardener nor a botanist who gave me a glimpse of these hidden stories but a mathematics professor with a penchant for eighteenth-century prints, hieroglyphs and languages. It was not that he had an extensive knowledge of plants, but what he did know, he was passionate about. When

I asked his wife how to prune my straggly fuchsia, he launched into a lecture that lasted the whole of dinner. "Fooks-ia," he chided, was the correct pronunciation, not "fjoosha," because it was named after the German Leonard Fuchs, a Renaissance botanist. He made a point of explaining that, in the first half of the sixteenth century, Fuchs had been one of the first scientists to shake off deep-rooted beliefs in mysticism in favour of direct observation of nature.

Later, I learned that Fuchs had never even seen a fuchsia because he was long dead when the first specimen arrived in Europe from South America. But I also found out that Joseph Banks—the naturalist on Captain Cook's circumnavigation of the globe—had thought the fuchsia so valuable that, in 1788, he had carried one into the greenhouse at Kew on his head, unwilling to trust such a treasure to any of the gardeners (incidentally, it was also in 1788 that Banks acquired from the Far East the ancestor of my *Hydrangea macrophylla*, the first Asian hydrangea to put down roots in English soil).

A few weeks after the "fooks-ia" conversation I received a parcel containing three books written by the Swedish botanist Carl Linné, also known as Linnaeus. It was the continuation of my botanical lecture. This time I was being educated about my daughter's name, Linnéa, which is an old-fashioned Swedish girl's name as well as belonging to a delicate pink forest flower (the reason why I had chosen it). This flower, I now discovered, had itself been named after Carl Linnaeus—the botanist who had classified the natural world and invented a standardised botanical nomenclature. I also learned that Linnaeus had played a central role in popularising botany as a genteel pastime for the middle classes of eighteenth-century England.

As my interest grew, I visited the Botanic Garden at Kew and the Chelsea Physic Garden to learn more. I joined the armies of garden fanatics who spent weekends wandering along neatly laid-out flowerbeds, staring into tree canopies or hiding from the rain in steamy greenhouses. I had never quite realised before how different leaves could look—glossy, fleshy, heart-shaped, sword-like, paper-thin, serrated, creased ... Equally, blossom presented a never-ending panoply of colour and shape. After countless days at Kew and Chelsea and hours of questioning gardening friends, I began to be able to tell some plants apart and knew about their cultivation. Penstemon had to be cut back in spring, for instance, whereas tradescantia preferred an autumn trim; and my hydrangea should be pruned less severely than the rose. I divided my delphinium clumps and harvested my first tomatoes. My kitchen became a greenhouse as I tried to grow some annuals from seed. I was incredibly proud when I succeeded in taking cuttings from dahlia tubers— quadrupling my plants. I showed them off to everybody who came round to my house. And whenever I was working in a library, I would use my lunch breaks to search out plant books. Then one day I opened a volume that fundamentally changed the way I looked at my plants.

This was the *Gardeners Dictionary*, first published in 1731 by Philip Miller, the head gardener of the Chelsea Physic Garden. Listing all known plants in cultivation in the England of his day, Miller provided thousands of entries, giving each plant's country of origin, sometimes the year of introduction and advice on cultivation. He suggested planting-patterns for climbers, evergreens or perennials for open borders, and plants suitable for shaded parts of the garden. He even advised on pest control, soils, propagation and the dangers of wind and frost. The book seemed to be the ancestor of the modern dictionary I had bought—just less glossy and without illustrations.

What most surprised me was how much of what I had learned about gardening during the previous months had been developed during the eighteenth century, and how many of my plants had been introduced into Britain at that time. As I delved further, it became clear that Miller was part of a much larger plant-collecting and botanical network that stretched to every known corner of the globe. When I then discovered the correspondence between Peter Collinson, a wealthy English merchant, and the American farmer John Bartram, it ushered me into a world in which flowers, trees and shrubs took precedence over war and politics. I realised that the English landscape garden had its roots in America because it was Bartram who had dispatched from Philadelphia to London hundreds of boxes filled with seeds and plants, furnishing the groves and shrubberies that would later be imitated everywhere in Europe and in America herself. Over four decades Bartram and Collinson transported America's evergreens, magnificent trees and colorful shrubs to Britain, transforming the parkland into a new "natural" landscape. Like a slowly developing photograph, a picture began to emerge of a horticultural and botanical revolution which had laid the foundations of the English garden.

Soon I too was in the grip of an obsession. I saw how letters sent between Collinson and Bartram, but also between famous Enlightenment men like Carl Linnaeus, Hans Sloane, Benjamin Franklin and Joseph Banks, fostered an international community where plants and ideas could be exchanged across vast distances. By the mid eighteenth century American trees were flooding into Britain, changing the landscape for ever, and by the end of the century Banks had added thousands of plants from Africa, Australia and the Far East. In tandem with the expanding empire the numbers of plants that arrived in Britain increased, providing gardeners with an ever wider choice. As the nineteenth century dawned, even the most humble garden could boast exotic flowers and shrubs. The gardens had undergone such seismic changes that England, which only a hundred years earlier had been largely parochial and insular, had emerged as the garden of the world. Now, when I walked out into my little plot, I saw it not as a chaos of unidentifiable plants but as the ordered result of pioneering work by an extraordinary and dedicated group of men who turned their fellow countrymen into a nation of gardeners. This is their story.

Prologue

The Fairchild Mule

Yet nature is made better by no mean
But nature makes that mean: so, over that art
Which you say adds to nature, is an art
That nature makes. You see, sweet maid, we marry
A gentler scion to the wildest stock,
And make conceive a bark of baser kind
By bud of nobler race: this is an art
Which does mend nature, change it rather, but
The art itself is nature.

WILLIAM SHAKESPEARE, *The Winter's Tale,* Act 4, Scene 4

On an early summer's day in 1716, Thomas Fairchild went into his Hoxton garden, closed the door of his potting shed and set in motion a chain of events so momentous that in time no gardener would ever think about plants in the same way again. At the same time, it led Fairchild, a devout Christian, to live in fear of God's wrath for the rest of his life.

The leading nurseryman of his day, Fairchild was famed for his rare plants, his talent for coaxing the most recalcitrant bulbs into bloom, and his exquisite grapevines. At forty-nine, he was still a bachelor, having dedicated his life to horticulture, and his bulging belly was testimony to his success. No day went by without his working at the nursery in Hoxton, a suburb known for his market gardens some two and a half miles from the City of London, but on this particular occasion Fairchild was not merely sowing or pruning. Alone in his potting shed, he placed a collection of flowerpots on the workbench. Some were planted with sweet Williams, others with carnations. Taking a feather, Fairchild drew its tip across the stamen of a sweet William—his hands, roughened by their daily contact with soil and water, were nevertheless delicate and steady. Then he gently brushed the stigma of a carnation with the feather, so that the dust-like pollen crossed from one flower to the other. The work was quickly accomplished. Now he had to wait.

6

" Fairchild Mule"
1717

Over the following months the seeds ripened inside the carnation, each containing its parents' traits. But it was only when spring came and those seeds grew and then flowered that Fairchild could see his creation: a pink flower that melded the carnation's double blossom with the clustering flower heads of the sweet William. It was the world's first man-made hybrid.

At a time when most botanists were still refuting the idea that plants reproduced sexually like animals, the Fairchild Mule, as the hybrid came to be known, was tangible proof of this proposition. Fairchild was aware of the sexual theory, and knew that the male pollen from the sweet William had fertilised the female stigma of the carnation. Yet, instead of rejoicing in his achievement, he was anxious. His act was an incendiary one because it contradicted the universally held belief that God had given birth to all plant species on the third day of Creation, while changes in plants' appearances were explained as "accidents" and "varieties." So what had driven Fairchild to undertake this blasphemous experiment?

For today's gardeners—spoilt by tens of thousands of plant varieties that provide a plethora of choice for every season—it is perhaps difficult to imagine just how dull and dreary the late seventeenth- and early eighteenth-century garden looked for at least five months of the year. At the beginning of the eighteenth century, England's biggest nursery, Brompton Park, was only able to supply plants that flowered between February and the end of September. In spring, regiments of tulip, narcissi and hyacinth lined flowerbeds, and in summer, the different heights of lavender, lily, snapdragon and periwinkle created waves that rose and fell between the topiary, but in autumn and winter there was hardly any colour at all. No fiery autumn foliage, no frost-dusted blooms. Only rows of yew and holly sheared into immaculate pyramids, cones and balls, and the dark lines of thin ribbon-like flowerbeds twisting in swirling curves through gardens, all empty.

The English garden of Fairchild's day was known for its turf and topiary, not its large collection of flowers or shrubs. It consisted, as one French visitor observed, "principally of extremely smooth lawns ... divided into plots and squares by long wide paths" as well as "holly, yew, laurel, and cypress cut in all sorts of shapes and figures." These gardens sought beauty in geometrical splendour rather than movement and colour. Intricate patterns in the turf provided the variety, not drifts of blossom. Gardeners laid out what plants there were, carefully choosing their positions in the overall composition, all trimmed and manicured. Not even a strong wind could rustle the leaves of the stiff green sculptures—the only noise in this stillness came from the gurgling water of the fountains.

England had never led the fashion; instead, gardeners had imitated Continental fashions such as cascading Italian Renaissance water terraces or marching avenues lined by rows of stately trees. They had copied French grand baroque gardens such as Versailles, in which trees were shaped into rigid walls of hedges and symmetry reigned over nature, and learned from the Dutch the theatrical display of jewel-like individual plants. Throughout the seventeenth century English gardeners had been so behind their European colleagues that two decades before Fairchild's experiment, when William and Mary had arrived from the Netherlands to claim the British throne, they had been frustrated by the paucity of flowers available in London nurseries. To furnish the gardens and greenhouses of Hampton Court, they had imported plants from the Netherlands instead. Dutch gardeners had benefited considerably from the increased trade that came with the expansion of the empire. As Dutch merchant ships criss-crossed the globe, they picked up floral treasures along the way, and brought them back to the nurseries of Holland and Flanders.

The Dutch East India Company had even founded a botanic garden at Cape Town in which South African plants were tended, as well as some from India, Sri Lanka and the West Indies. It became a horticultural clearing-house for the Dutch colonies and provided copious bulbs and seeds for the gardens of the Netherlands.

Thomas Fairchild shared William and Mary's disappointment with England's poor selection of plants, and had bought some of his stock from Dutch nurseries. At the same time he had worked hard to expand his network of contacts, so that he was ready to exploit any opportunity to procure rare plants from abroad. As a result, his little half-acre nursery in Hoxton was bursting with plants never before seen in an English garden.

Exiguus spatio, variis sed fertilis herbis.

Engrav'd by Henry Fletcher London 1729.

This engraving illustrates the prevailing taste for formal French gardens at the time that Fairchild experimented with his hybrid. In a garden that is ruled by geometry, tall hedges are clipped into green walls and the centre path is lined with topiary. Exotics are displayed in vases as valuable specimen plants.

He particularly adored the colourful shrubs, flowers and trees to be found in the American colonies. But these were difficult to procure. Since the foundation of the first English settlement, Jamestown in Virginia, in 1607, few North American plants had made it to England, most having perished during the long journey—often not even one in 500 seeds arrived in a healthy enough condition to germinate. There were some successes: John Tradescant the Younger had been the first professional gardener to travel to the new colonies in the late 1630s for a plant-hunting expedition. Mainly exploring the forest and swamps around Jamestown, he had returned with the first of the North American Michaelmas daisies, the American sycamore and the tulip poplar. But there were still very few American plants available in commercial nurseries.

Fairchild, though, sold Tradescant's sycamores as well as Virginian sunflowers, asters, goldenrod and rudbeckia. *Aesculus pavia* from North America, a small tree which was introduced as "scarlet flower'd horse chestnut" in 1711, blossomed for the first time in Fairchild's garden, and he was so successful at cultivating tulip poplars that he was able to supply "the chief Gardens abroad." In the Hoxton nursery there were plants from across the world. Fairchild sold lilies of all sorts, including a new species with white and purple stripes that he had "purchas'd from abroad." In 1710 bladder senna had come from the Caucasus, and Fairchild's customers were delighted by its strangely shaped petals—which looked like pouting lips— soon it would adorn Georgian shrubberies.

Some exotics were only available from Fairchild: hellebores, for example, one horticultural publication explained, "are so rare, that they are scarcely to be seen, unless at Mr. Fairchild's at Hoxton." Reputedly, he was the first to grow evergreen Portugal laurel and the Oriental poppy, as well as the first to make *Nerine sarniensis* bloom repeatedly every autumn. Some thirty years after he had established his nursery, Fairchild could boast that more than twenty species blossomed in his garden each December. His collection was so exceptional that his nursery became as important a landmark on visiting gardeners' itineraries of London as the royal garden at Hampton Court.

Yet Fairchild wasn't satisfied. Well-stocked as his nursery was, he still had to rely on others to increase his selection, and his correspondents were often as unreliable as the weather, which frequently caused the loss of cargo in storms at sea. Over the years Fairchild grew so frustrated about the small number of plants he could get hold of that he decided to take matters into his own hands and meddle with nature herself.

The creation of the Fairchild Mule was a watershed event not only because it revealed nature's workings—the sexual reproduction of plants—but because it demonstrated the power of an empirical approach to gardening based on observation and experimentation. In order to make his discovery, Fairchild had relied on the most recent philosophical and scientific methods, casting aside the vernacular myths and superstition that still ruled his profession.

The few botanical books that were published at the dawn of the eighteenth century repeated the flawed advice from ancient Greek and Roman horticultural treatises. Many gardeners, for example, believed in the moon's influence over plants, not because it had been scientifically proven but because "[f]rom Pliny to this day most Authors have been of that opinion." As late as 1693 one writer instructed "[w]hen you sow to have double Flowers, do it in the Full of the Moon." Writers also believed that vines should be pruned only when the full moon was in line with certain star constellations, or that lemons could be turned red by grafting them on to pomegranate and mulberry trees, or that apples could be made sweeter by watering the tree with urine. Some even insisted that white lilies would turn purple if their stems were dipped in red wine, or that flowers became more scented when soaked in cinnamon-infused water. And even the pragmatic horticultural writer John Evelyn, who published the first monthly to-do list in 1664, still instructed his gardeners to transplant "when the wind is south or west."

Although at the same time men like Isaac Newton explained the workings of the universe through a set of physical laws, most people still held deeply rooted and archaic worldviews. In this old framework, forces of nature such as thunderstorms, earthquakes and droughts were interpreted as an expression of God's almighty power, making the comets that shot through the skies in the closing years of the seventeenth century signs of God's anger and heralds of political unrest. In this God-centred view of the world, each plant and animal was like a cog in the divine machine and each had its place in the hierarchy dependent on their relationship to and use for man.

This belief that man's destiny was intertwined with nature nurtured a great number of superstitions and myths about the meaning and uses of plants. Bay trees and beeches, for example, were often planted near houses to ward off lightning, while pregnant women were thought at risk of miscarriage if they stepped over cyclamen. Similarly, any apple tree which fruited and flowered at the same time was viewed with particular foreboding. The names given to plants often reflected the myths associated with them: Motherdee (red campion), for example, was feared because it was believed that the parents of any child who picked it would die. Meanwhile, in medicinal use of plants, apothecaries and physicians insisted that by matching the shape and colour of plants to different organs, illnesses could be cured—a method called the "Doctrine of Signatures." The oil from the walnut, for instance, with its likeness to the human brain, was a remedy for head wounds; the lung-like leaves of lungwort were used to treat respiratory diseases; and jaundice was treated with yellow herbs.

Fairchild, however, was not convinced of this close correlation between man and nature. The experiments he had made with grafting plants on to different rootstocks and his observation of the movement of sap had taught him that leaves, fruits and roots all fulfilled functions that were necessary for the plant's *own* nutrition and reproduction. It was this reliance on observation rather than myth that allowed him to make the innovations he did. He would have agreed with John Locke, who had written some twenty years earlier, "[T]here is noe thing constantly observable in nature, which will not always bring some light with it and lead us farther into the knowledge of her ways of workeing."

Fairchild's hybrid was so revolutionary that the Royal Society itself invited him to present a dried specimen of it at one of their Thursday meetings. For scientifically minded men like Fairchild, the Royal Society was the most important institution in Britain. Founded in the 1660s "for the improvement of naturall knowledge by Experiment," it disseminated rational thought and propagated new ideas. Once a week its like-minded

The Royal Society (1660)

PROLOGUE

members met in the belief that exchanging knowledge would create a more enlightened and therefore better world. Bridging continents, languages, political hostilities and religions, these men were convinced that through reason and observation the principles of nature could be elucidated.

Experiments were the backbone of all enquiry at the Royal Society, ranging from the science of pressure cookers, which tenderised bones into an edible soft paste, to exploding dogs and even blood transfusions carried out between humans and sheep. These scientific "performances" had in fact become so strange that Jonathan Swift mocked them in *Gulliver's Travels* (1726) when he created the fictional Academy of Lagado, in which dusty philosophers extract sunbeams from cucumbers, soften marble into pillows and pump air into dogs' guts— all thinly disguised allusions. Robert Hooke, for example, had been obsessed with air pumps, and having suffocated uncountable dogs and birds in vacuums, then volunteered to experiment on himself. The Society's secretary and future president, Hans Sloane, also seemed to care little for convention or the sentiments of his genteel neighbours in Chelsea when he dissected the decaying corpse of an elephant on the lawn at his house— all in the name of natural philosophy.

Many of the experiments furthered the advancement of knowledge, the progress of technology and an understanding of the natural world. Clocks, telescopes and surveying instruments that improved cartography, astronomy or navigation were first presented at the Thursday meetings, while fellows introduced new mathematical theories and botanical discoveries. Fairchild was not a fellow of the Society himself—probably because of his humble background. But some of the fellows with an interest in botany regularly used Fairchild's horticultural talents to confirm their theories, asking him to perform experiments at his nursery, and he was also an avid reader of the Society's publications. One book in particular had fascinated him: *The Anatomy of Plants* (1682), in which the author and Royal Society fellow, Nehemiah Grew, declared stamens to be a plant's male organs and pollen the spermatic fluid.

Despite being respected by the Society as "the most rational Gardener," with a "true Bent of Genius," Fairchild had never been asked to attend a meeting himself, so it would have been with some trepidation that he made the carriage journey from Hoxton to the headquarters of the Royal Society at Crane Court near Fleet Street, on 4 February 1720, almost three years after the Mule had first flowered. It was the secretary, Hans Sloane, who presided over the meeting that day, instead of the seventy-seven-year-old president, Isaac Newton, who was ill. Sloane sat in a large armchair at the top of a long table, behind him, above the blazing fire, a portrait of the royal patron, George I. Down each side of the table, the fellows were seated on benches upholstered in green, while one man sat at a separate little desk, ready to take the minutes.

Sloane was the physician to the king and ran a successful practice in fashionable Bloomsbury, as well as being one of the most passionate plant-collectors in Britain. As a young man he had been the personal physician to the Duke of Albermarle in Jamaica and when he had returned to England in 1689 with the embalmed body of his employer, his bags were full of pickled animals and hundreds of dried plant specimens that nobody in England had ever seen. Now, aged fifty-nine, Sloane had long given up plant-hunting himself but regularly financed expeditions to augment his collection.

The botanical lecture that Thursday was to be given by Patrick Blair, a fellow of the Society who had worked with Fairchild on a number of theories. It seems that when, a few weeks earlier, Blair had suggested to Fairchild that he present the novel hybrid to the Society, Fairchild's religious scruples had caused him to panic, and he had changed the story of the Mule's conception, claiming that serendipity rather than human agency had produced it. Given that even Isaac Newton insisted that the universal laws he was revealing had initially been devised by God, Fairchild's fears of being regarded as blasphemous were understandable. After all, he also believed that God was the divine architect. For him the complexity of plant reproduction expressed God's omnipotence, and the contemplation of flora revealed "the Power and Wisdom of the Creator." So, when it came to displaying the dried and mounted specimen of the Mule to the fellows, Blair told them that it was "of a middle nature between a sweet William and a carnation" —an "accidental" hybrid which Fairchild had found in his nursery in a plot where the two plants grew so close together that the flowerheads had touched and thereby pollinated each other.

Whether or not the fellows believed Blair's story, only few understood the significance of the hybrid for gardens. One such forward-looking man was the horticultural writer and friend of Fairchild, Richard Bradley, who prophesied that "[a] Curious Person may by this knowledge produce such rare Kinds of Plants as have not yet been heard of." No longer would gardeners have to rely on the God-given flowers and plants that lay at hand. Fairchild heralded a new era, in which England came to lead gardening fashions in Europe and America. Within five decades of his experiment, English gardens were being copied across the world and no other country could boast so many foreign plants in cultivation.

Foreigners toured the country to see the gardens because "[t]his art is very highly thought of in England," while English gardeners were admired for their ability to "paint" with plants and thought to be "the true artists of their nation." In fact, gardens and their content had become so important to the English that sales particulars of houses advertised and high-lighted shrubberies, specimen trees and pleasure grounds "planted with various valuable Exotics." Flower-sellers pushed their carts through the streets selling their wares, women dec-orated their hair with silky petals, and the first gardening magazine was launched in 1787.

A nation of amateur gardeners was born. As the nineteenth century dawned, nurserymen were beginning to mass-produce hybrids in order to fill the beds and new conservatories of the middle classes. Businesses like Fairchild's were no longer a rarity: garden-owners could purchase plants from the more than 200 nurseries which had sprung up everywhere in the country. And it was not only hybrids that these nurseries stocked. Their beds were filled with the thousands of new species that had been introduced to England as her merchants plundered the globe. Cedars, pines and other evergreens provided winter interest, while rhododendrons paraded showy blossoms in late spring. In early summer magnolias and tulip poplars flowered and in autumn the russet foliage of Amer-ican deciduous trees set the landscape alight.

Though Fairchild continued to collect plants from around the world after his creation of the Mule, he did not live to see the full extent of what he had set in motion. He died on 10 October 1729, bequeathing to his parish church the sum of £25: for preaching "annually forever" a sermon on "the wonderfull works of God in the Creation"—a prophylactic measure in case he had provoked God's scorn. Fulfilling Fairchild's wishes, priests would preach that man had "never been able to produce any new species."

To this day, the "Fairchild Sermon" is delivered annually in St. Giles, Cripplegate, the ancient church where John Milton worshipped and which is now hidden in the labyrinth of London's Barbican. Every first Tuesday in Pentecost, the Worshipful Company of Gardeners march towards the altar in pairs, dressed in fur-rimmed gowns and ceremonial chains, carrying their trademark silver spade, while outside, in the concrete jungle, a few lonely flowers creep over balconies and windowsills—pelargoniums, lobelias, roses and many more. They are hybrids, and Fairchild's legacy to the world.

Part I

ROOTS

1

"Forget not Mee & My Garden"

*There ought to be gardens for all the months in the year,
in which severally things of beauty may be then in season.*

FRANCIS BACON, "Of Gardens," 1625

The first three months of the year were always the busiest time for the cloth merchant Peter Collinson, for it was then that the ships from the American colonies arrived in London. But on this January morning in 1734 he was concerned not with the arrival of reels of wool or bales of cotton but with an altogether different cargo. Awaiting him at Custom House, down by the docks, were two boxes of plants that, for Collinson, were the most exciting piece of merchandise he had ever received.

As he hurried towards the Thames from his Gracechurch Street office, in the financial centre of the city, Collinson could see the clusters of tall masts above the rooftops and hear the cries of stevedores as they unloaded precious goods from the holds. The stretch of the river between London Bridge and the Tower was the main harbour of London and more than two thousand vessels—besides barges, wherries and ferries—created "a forest of ships." Moored side-by-side, the vessels left only narrow channels for the barges between them, and the wharves, quays and stairs that lined the river were so crowded it was hard to move. These ships brought tea and silk from China; sugar and coffee from the West Indies; spices from the East Indies and corn and tobacco from the American colonies. The river was, as one visitor said, the "foster-mother" of London, pumping money, goods and life into a city which more than half a million people called their home—the largest metropolis in the world.

Collinson was one of the many merchants benefiting from the huge expansion of trade that had occurred since the accession of King George II. Soon to be forty, he had inherited the cloth business from his father a few years earlier and was involved in shipping cloth all around the globe, with his main market in the American colonies. Between the 1720s and the 1760s exports to the American colonies quadrupled, while those to the West Indies multiplied almost by seven, providing untold wealth to a new class of businessman. As Daniel Defoe wrote, "our merchants are princes, greater and richer, and more powerful than some sovereign princes."

And, as London grew, so too did the trade to be done within the city. In one year alone, Londoners consumed nearly 2 million barrels of beer, 15 million mackerel and 70,000 sheep. London was one vast consumer market and the streets thronged with trade. Not far from Collinson's office were the shops of Cheapside and Fleet Street, whose large windows created the impression that they were "made entirely of glass." Sweets, cakes and fruits were stacked in precarious towers, and even the apothecaries' colourful potions were lavishly displayed to entice the passers-by. Tourists wandered for hours, admiring what one called "the choicest merchandise from the four quarters of the globe." When dusk settled on the city, thousands of candles threw a soft light on to glittering jewellery, polished silverware and framed engravings. Lanterns fastened to the front of each house created a luminous necklace along the streets, giving London a permanently festive atmosphere.

Although Collinson was not one of the richest merchants in the city, he was very comfortably off. He lived with his beloved wife in a "little cottage" in the "pleasant village" of Peckham.* It was "the most Delightfull place to Mee," he said, where he could retreat from the "hurrys of town," and where he could indulge the great passion of his life: gardening. Collinson had been fascinated by the natural world from an early age, when he wandered the garden of his grandmother, full of topiary and "curious plants." His grandmother, who had brought him up, had encouraged his enthusiasm and taken him on many trips to the nurseries around London, in particular Fairchild's little plot in Hoxton where the adolescent Collinson had gazed in

---

* Collinson regarded himself as a family man. He missed his wife whenever they were apart, insisting that just seeing her handwriting "made my heart Leap for Joye."

awe at the exotic world contained in the hothouses. As he got older, his interest in nature only increased. "I must be peeping about," he wrote, confessing that nature haunted him everywhere he went.

During the week, though, Collinson was forced to attend to his business in the City, where he felt like "a Cockney who Lives in a Wilderness of Chimneys." His main trade was with the colonies in North America, in particular with Pennsylvania, which, due to its rapid rise in population, was one of the fastest growing markets for cloth. And it was from the docks of Philadelphia that Collinson's plant boxes had been dispatched.

To get to the Custom House, Collinson would have hailed one of the many porters that carried busy merchants on sedan chairs to their appointments. The porters forged a path through the crowds by pushing and crying, "By your leave," and they were so fast that pedestrians who did not jump out of their way fast enough were knocked to the ground. His office was at the heart of the city, not far from the Royal Exchange, that engine of trade in whose courtyard merchants like Collinson met to sell and buy their stocks—said to be "the wealthiest corner of earth." Nearby were many coffee houses in which more business was conducted, such as Lloyd's Coffee House in Lombard Street, which became the venue for marine insurance, and the Pennsylvania Coffee House in Birchen Lane. In the latter, Collinson traded and negotiated but also heard the latest news from America, since it was frequented by captains, and visitors from the colonies who often had their lodgings there when they first arrived in London. There was hustle and noise everywhere. Shopkeepers advertised their wares, animals were herded to the slaughterhouses and musicians played at the street corners. The cacophony of voices, church bells and rattling coaches was interrupted each hour by the shouts of the watchmen who called out the time and the state of the weather.

The scene in the Custom House was similarly chaotic and teeming. Here, in a room that ran the whole length of the 190-foot-long building, merchants and captains as well as tourists queued for hours along the rows of counters and desks in order to retrieve their goods, trunks and luggage, and to declare their imports. Foreigners found the confusion and crowds in the so-called Long Room daunting and were shocked to see that the customs officers pocketed a share of the duty themselves. Even Collinson, who had "good friends among the Commissioners," often wasted many hours here.

Today, however, his patience was rewarded. Peering into the two wooden cases from Philadelphia, Collinson could hardly believe what he saw. Inside were hundreds of seeds neatly wrapped in paper, a few living plants and, most extraordinary of all, two flourishing kalmia cuttings.* Collinson had admired the kalmia's many hundred puckered pink flowers that opened like mini-umbrellas in drawings but had never seen a real one. Nobody had ever seen them growing in England, since other cuttings had never survived the journey from America, which took between five and twelve weeks.

For a number of years now, Collinson had been using his trading connections to augment his flowerbeds with the horticultural spoils of distant countries. "Forget not Mee & My Garden," he pleaded with his business partners, and occasionally, along with rolls of cloth, he might find some seeds in a little paper bag, or the remains of a shrub that had been severely neglected. Plants suffered greatly on sea voyages. Often the boxes and barrels with trees and shrubs were put on the open deck, exposed to wind, salt water and fluctuating temperatures. Rats and mice feasted on the leaves and often built their nests in the safety of the cosy boxes, while sailors helped themselves to plant water or even the rum in which choice specimens were stored. In addition, since plant boxes were the least valuable cargo on merchant ships, they were often the first freight to be jettisoned during storms or pirate attacks. One of Collinson's correspondents blamed the continuous failure to procure healthy plants on the captains, who were, according to him, so stupid that "[o]ne may as soon tutor a monkey to speak, or a French-woman to hold her tongue, as to bring a skipper to higher flights of reason." Whenever Collinson knew a passenger, he asked them to guard the boxes, thinking "it

---

* In order to avoid confusion, all plants are referred to under their Linnaean name or current common name, even if they were called something different at the time. The kalmia, for example, was in the first half of the eighteenth century either chamaedaphne, chamaerhododendron or azalea. Collinson also often used its common name, laurel.

might be a pretty amusement for them to peep & Look after it," but to no avail—"Hitt or Miss Lucks all," he concluded. Nine out of ten plants perished on the journey, and even fewer went on to flower once planted in English soil.

On this occasion, though, Collinson had been lucky. The captain had been more helpful than others and had stowed the plant boxes underneath his own bed, where they had profited from the warmth of the cabin. At last Collinson would be able to create the kind of garden he longed for, bringing a new world of plants to Britain, and so celebrate God's abundance. For Collinson, every foreign species in his garden testified to God's creation, making it akin to a horticultural bible. "[I] admire them [plants] for the sake of the Great & all Wise Creator of them to Enlarge my Ideas of his Almighty power & Goodness to Mankind," he said. His garden in Peckham was therefore a place of "Sublime Contemplation" through which he might discern a closer understanding of the divine plan that underpinned nature.

Although these ideas were widely held, they were even more pertinent to Collinson because he was a Quaker. William Penn, the founder of the Quaker settlement Pennsylvania, had already in 1693 encouraged his fellow believers to live with nature because "there we see the Works of God." Similarly, Collinson believed that plants were a direct route to a relationship with God, while other material possessions were only a pompous display of wealth—they were "Man's work" and only for "Pride, Folly and Excess," Penn had declared. Thus Collinson, who could have dressed in flamboyant and luxurious silks, favoured modest dark clothes without any of the embroidery, ribbons and ornate buttons that were so fashionable at the time. No lace was applied to his plain white shirts, nor did he wear colourful waistcoats. Equally, in his house he eschewed heavily decorated furniture or lustrous fabrics, adding colour and beauty instead with cut flowers and "nosegays."

As a Quaker, Collinson had been prohibited from going to university, but his enquiring mind meant that he educated himself through observation, books, and a prolific correspondence with men around the globe. In 1728, age thirty-four, he had become a fellow of the Royal Society. Collinson's annotations in the margins of his books reveal a man who engaged

Peter Collinson dressed in plain clothes. The book to the left
is Linnaeus's *Hortus Cliffortianus* (1738), depicting an engraving
of *Collinsonia* (see page 127 for a reproduction of it).

deeply and enthusiastically with their contents. He even used
dinners served at friends' houses to broaden his knowledge, on
one occasion asking Hans Sloane, the president of the Royal
Society, about the classification of a fish he had eaten. He
thought that a "rational pursuit" such as horticulture or natural
history lead to health, pleasure and profit. Though some might
consider dabbling with plants a mere amusement, Collinson
believed they made men virtuous and temperate—traits that he
and his faith held dear.

With his election as a fellow of the Royal Society, Collinson
became part of the so-called "Republic of Letters" in which
theories, experiments and ideas transcended national, political
and religious boundaries. "I think there is no Greater pleasure
then to be Communicative & oblige others," Collinson encouraged one of his correspondents, adding persuasively, "Wee
Brothers of the Spade find it very necessary to share among
us." And to prove his commitment—with the underlying hope
of receiving some American plants in return—Collinson even

agreed to be the unpaid London agent for the subscription library that Benjamin Franklin and a group of friends had established in Philadelphia in 1731. Thereafter, for more than two decades, Collinson would choose, buy and ship books to Philadelphia, enlightening the learned circles in the colonies. The success of the library, Franklin later insisted, was "greatly owing to his [Collinson's] kind Countenance and good Advice."

As if to emphasise his primary interest, the first book Collinson had selected for the library was a horticultural publication, but despite such hints no plants had arrived. Consequently Collinson had asked the secretary of the Library Company and a fellow Quaker merchant, Joseph Breintnall, for some help. Breintnall, however, was unable to collect plants himself—or as Collinson described it, was keen "to get rid of my importunities"—but had recommended a local farmer who was known for his good knowledge of native species and who was willing to send regular floral dispatches. This man was John Bartram, the sender of the two exciting boxes. As soon as Collinson examined them, his spirits rose. He knew he was dealing with someone who understood plants and could further his aims. Bartram had, after all, been clever enough to realise that if a plant was to arrive in Britain in January, it should be of a variety that flourishes in a colder climate.

Like Collinson, John Bartram had adored botany since childhood and called it his "darling study," but had little time to indulge his passion since his farm on the outskirts of Philadelphia occupied almost all his time. Initially the work had involved draining the swampy ground and quarrying stone for his house; later he had to tend to his vegetables and corn. His first wife and one son had died the year he had bought his farm in 1727, but he had remarried within two years and his new wife, Ann, bore him a child roughly every two years.

To a man like Collinson, who spent much of his spare time at the great estates of aristocrats or meeting the intellectual elite at the Royal Society, Bartram's circumstances appeared somewhat lowly. But in fact Bartram's life was fairly privileged by colonial standards. His house, for example, unlike most colonial farm dwellings—which were one-room log cabins—was a fine stone building with four rooms. Only three years earlier, Bartram had completed it by carving a commem-

The Botanist.

John Bartram, as imagined and illustrated by Howard Pyle in his article "Bartram and His Garden" in *Harper's New Monthly Magazine* in 1880.

orative stone which read "God Save" in Greek—an indication of his aspirations and ambitions. Collinson, though, chose to believe that Bartram was a "plain Country Man," who happened to know a lot about plants and who was willing to gather seeds, acorns and cones for him on a regular basis.

Like Collinson, Bartram was a Quaker, his grandfather having followed William Penn to America to avoid persecution and in hope of a better life as a colonial farmer. And as was the ethos of his religion, Bartram was unashamed of his background; he was strong-minded, unwavering in his faith and outspoken in his opinions. Such was his steadfastness that two decades later he was expelled from the Darby meeting by his fellow Quakers for insisting that Jesus was a mortal man. Despite being accused of heresy, Bartram refused to bow.

This episode also reveals a man who relied more on his rational thoughts than on his faith. Bartram was inquisitive and observant. He adored scientific books more than any other, but they were expensive and few were available in the colonies. But even if he had been able to purchase them, he would not have been able to read most of the botanical publications because, he explained, "ye lattin pusels mee." His correspondence with Collinson would allow him to broaden his knowledge in other ways.

The hunger for American plants had been growing in England ever since the plant hunter Mark Catesby had returned from his expeditions in Virginia and Carolina almost ten years previously. But Catesby had brought only a few seeds back to England. Most of them had been for Thomas Fairchild, who had been able to raise a small number, such as "Mr. Catesby's new Virginian Starwort" (*Aster grandiflorus*) and "Virginian Purple Sun Flower" (probably *Echinacea purpurea*). A handful of seeds had also gone to Collinson, a present from Catesby for his help in finding sponsors for the expedition. But largely, Catesby's cargo had consisted of dried specimens for the herbaria of English collectors, and a trunk full of drawings he had made. These he engraved and published in ten parts from 1729 to 1748, with Collinson's financial help, as *Natural History of Carolina, Florida, and the Bahama Islands*. The illustrations were so unique that even gardeners with a small salary bought what was, at twenty guineas, the most expensive botanical book of its age. Plant-lovers were entranced by the pictures of magnolia buds which unfolded into enormous, flawless white petals against a background of shiny green leaves, much like silk appliqué on velvet curtains. They were gripped by Catesby's depiction of the lilac blossom of American wisteria, which seemed to tumble from the branches like bunches of grapes, or by callicarpa, a highly unusual plant which wore its garish purple berries around its branches like bead bracelets. The aromatic bark of Carolina allspice, Catesby tantalisingly promised, was like cinnamon, while the scent of the copper red petals would fill the air with the perfume of pineapples.

One of the reasons why American trees and shrubs were so desired in England was that Catesby believed they could be grown in the open rather than in hothouses, unlike the tender bulbs and flowers that arrived from places like South Africa. Part of the reason why Collinson had found it so difficult to obtain American specimens from his trading correspondents was because, as he later recalled, "what was common with them but rare with us they did not think worth sending." All this was about to change. For the first time, an American-born man who had a strong understanding of natural history promised to send annual seed boxes and living plants to England. If Bartram dispatched his boxes regularly, Collinson was certain he could transform the English garden.

The letters between Peter Collinson and John Bartram reveal how their relationship evolved during the four decades of their correspondence. In the first year, Collinson could barely hide his excitement, and wrote to Bartram every few weeks requesting different seeds or cuttings. "Pray send a root or two" of hellebore, Collinson wrote in one letter—they were so rare that a few years earlier only Thomas Fairchild had grown them—and in another, "[p]lease to remember all your sorts of Lillies as they happen in thy Way," and "don't let escape" the lady's slipper orchid, his favourite plant. A yellow one already flourished in his garden at Peckham, but he wanted a variety because the sculptural blossom "makes my mouth Water." Collinson asked Bartram so often for his favourite plants that he soon adopted the shorthand: "thee knowest what please me."

Collinson also sent precise instructions on how to pack the boxes. The first cargo that Bartram had sent, though full of neatly wrapped seeds, had been of little use as none of the paper bags had been labelled with the name of the plants. Because of this, if a seed failed to germinate, it was impossible for Collinson to reorder it since he couldn't identify it. When Collinson suggested Bartram name the seeds, however, he ran into a problem: though Bartram recognised many plants in the wild, he did not know their proper names, because he had never had a "person or books to instruct me." Collinson's solution was simple: when sending a particular seed, Bartram should press two examples of the plant, dry them for two weeks, then mount each on a separate piece of paper, both numbered the same. One set of dried specimens should be sent to England, where Collinson had arranged for the most eminent botanists to classify them, while the duplicates with the corresponding numbers should be retained by Bartram until Collinson next wrote with the relevant names. Cross-referencing the numbers, the farmer could then learn the names himself and Collinson could easily reorder from these lists. This way, they established an efficient system which would function much like a mail-order catalogue, as well as serving as a useful learning tool for Bartram.

Collinson's instructions were long and elaborate in order to ensure that Bartram understood every detail. His large energetic handwriting often covered many pages, with new paragraphs added whenever he found a spare moment during the working day, or late at night in front of the fire. "I always write in a Hurry," wrote Collinson, excusing his rushed letters, "and write phaps the Same thing over & over again for my Memory is Burdened with a Thousand things." By the time Collinson's letters arrived in Philadelphia, or Bartram's in London, some were unreadable because moisture had blotched the ink or, as Collinson wrote to Bartram, "I have much to do to read thy Letter for some Mischievous Insect has Eaten thy Letter in large holes in four places." Collinson also instructed Bartram to hand-copy each letter, which, though time-consuming, was necessary in order to make sense of the responses which often arrived ten or twelve months later. Collinson did the same; "I keep a regular Account of Letters & by whom Answered," he explained to Bartram, so that they would each know what was being referred to.

Bartram followed Collinson's advice closely and praise followed immediately: "The seeds in No.2 was nicely pack'd," Collinson wrote. Equally, the dried specimens which were so important for the identification of the plants arrived perfectly dry and firmly mounted—"I think thee has Discharged that affair very Elegantly," Collinson applauded. And when Bartram packed some seeds and roots together in the same box, Collinson patiently explained that the moisture of the roots ruined the seeds.

But in these early letters Collinson was also somewhat con-descending to Bartram: in the social hierarchy the American farmer was well below Collinson, who regarded him as humble and uneducated. For his part, Bartram always addressed Collinson with deference, knowing that his access to educa-tion and the wider world depended on this man. When he sent Collinson some observations on rattlesnakes, for example, he apologised that it was "oddly huddled together," to which Collinson was kind yet patronising: "thy style is much beyond what one might expect from a Man of thy Education." Bartram's subservience paid off: Collinson agreed to read the letter to the Royal Society, a forum of scientific enquiry which Bartram would never otherwise have been able to penetrate. There could also be financial advantages, Bartram hoped. Collinson, for example, dispatched clothes for Bartram's family, or books, paper and tools. As postage always had to be paid by the addressee, Collinson packed the presents in the trunks to the Library Company, so that Bartram saved the money.*

Bartram's cargo arrived so regularly that Collinson made a special germinating bed, six foot by three, in Peckham for his new American plants. Whenever a box of living shrubs arrived, Collinson carefully emptied it and spread the remaining earth on to this bed so he would not lose a single seed. "I never trust the Servants with such Curiositys," he assured Bartram when he received the valuable rarities. Keeping the American soil, Collinson was sure, would also ease the seeds' acclimatisation to their new home. In June 1735, after some failures during the first year, Collinson was able to report that one-third of the seeds had germinated, "[t]his year Better Luck attends thy Seeds." There was a new species of hellebore, six sorts of fern "very different from ours," goldenrod and honeysuckle. With the success, however, came also disappointment. Some seed perished quickly or never matured, some lay dormant for two or three years, and some took decades until they blossomed for the first time. Collinson had to draw upon huge reserves of patience. Until a seed grew and flowered, he had often no idea what it was. He wrote of Virginia(n) bluebells, which had arrived as roots without leaves or stem, "How often I survey'd It and anticipated the Time of its growth, by Imagineing what Class & what sort of Flower & Leafe, such an odd Root must produce."

But just as Collinson was beginning to enjoy the fruits of his labours, and to pass some of the seeds and plants on to his friends, the whole enterprise was suddenly endangered by an argument about payment. In October 1735, about two years after he had dispatched the first box to London, Bartram wrote to Collinson accusing him of failing to compensate him properly. Although Collinson shared some of the cargo with his friends and they sometimes paid Bartram in kind for his troubles, the compensation was irregular and according to their own generosity. Bartram argued that he had often neglected the work on his farm in order to collect seeds, acorns and

---

* This arrangement of using the trunks for Bartram's packages lasted for twenty-two years, until the new secretary of the Library Company put a stop to it.

cones. In exchange, he complained, he had only received a few goods and some European seeds and bulbs—but only "one Sixth part" of what he had sent.

Collinson had certainly sent Bartram remuneration in the form of goods. In March 1735 he dispatched a parcel of cloth, writing to Bartram, "Pray give no body a hint how thee or thy Wife came by the Suite of Cloths," worried that some of his other correspondents might expect similar gifts. However, Bartram clearly felt this wasn't enough. When Collinson replied, in February 1736 (once Bartram's letter had crossed the ocean), he blamed Bartram for not being patient enough, but then relented somewhat and explained, "Thee may assure thyself thee shall not fail of Suitable and grateful returns from me. Phaps I may be slow but I am sure." Two weeks later, Collinson—obviously feeling guilty—packed some parcels of silks, woollen cloth for coats for Bartram and his children, a summer gown for his wife, and some seeds. "Receive it in Love as it was sent," Collinson wrote, in the hope that it "will Convince thee I do what I can," but more he could not do because "I have affairs of greater Consequence to Mind"—as if to remind the farmer of his elevated station.

Collinson did, in fact, do more. Clearly worried by Bartram's complaints, he spent the next three weeks drumming up contributions from the other recipients of Bartram's boxes. On 12 March he wrote that he had collected more than £18 for the outstanding debt, and even better, he had devised a regular subscription system that would be fair and profitable, in which the English collectors and gardeners would pay an annual sum for a certain amount of tree and shrub seeds. If Bartram collected and sent them, Collinson suggested, he would distribute the cargo, and then oversee the accounts and chase the payments. Six weeks later Collinson had organised twenty guineas for Bartram's coming autumn collection, a contract that could be extended on an annual basis. The disadvantage was that payment would only be made once the customers received the goods—this was common and reasonable, Collinson explained, as "the gentlemen should first see what they have for their Money." The risk remained firmly with Bartram, as he would receive no compensation if the boxes were lost at sea.

Meanwhile, Bartram, not having yet received Collinson's

reply to his first complaint in October, became increasingly frustrated. Ignorant of Collinson's new scheme, he talked to the members of the Library Company in Philadelphia and other scientifically oriented men about his unfair treatment. It was only a matter of time before the news reached James Logan, America's foremost scientist, amateur botanist and Chief Justice of Pennsylvania. Logan had come as William Penn's secretary to the colony and was famed for his learning and his extensive library—which consisted of more than 3,000 books. A few years earlier, he had adopted Bartram as his protégé and given him his first three botanical books. Furious about Bartram's exploitation, Logan decided to reprimand Collinson, whom he knew personally from a visit to England. As Collinson sat down in June 1736 to explain to Bartram the subscription system he had worked so hard to organise, Logan also took up his quill to write a sharp letter to Collinson. Asking Collinson to procure a botanical book for him in London so he could give it to Bartram as compensation for the wrong he had suffered, he launched his attack:

> I shall make a present of it to a person thou valuest . . . Bartram has a genius perfectly turned for botany . . . but he has a family that depend wholly on his daily labour, spent on a poor narrow spot of ground, that will scarce keep them above the want of the necessaries of life. You, therefore, are robbing them while you take up one hour of his time without making proper compensation for it. Both thyself, at the head of so much business, and thy noble friend, and friends, should know this; no man in these parts is so capable of serving you, and none can better deserve encouragement, or worse bear the loss of his time without a consideration.

The two letters crossed. When Collinson was reading the accusation against him, Bartram was receiving the news that his English friend had, after all, honoured his promise. Bartram's answer to Collinson is lost, but Collinson immediately sent the botanical book, explaining that Logan "has shown a very Tender Regard for Thee In his letter to Mee"—tactfully euphemising the criticism.

With this reply, Collinson acknowledged that Bartram was more than just a servant or supplier. He and his acquaintances longed for Bartram's plants and realised how much they needed him for the plans they had for the English garden. Never before had such numbers of American plants arrived in England, and under no circumstance would Collinson risk losing Bartram's trust and willingness to deliver more. At the same time Bartram understood that he also had some power over Collinson. Slowly their relationship was changing.

2

"The bright beam of gardening"

Science! Thou fair effusive ray
From the great source of mental day,
Free, generous and refined!
Descend with all thy treasures fraught
Illumine each bewilder'd thought
And bless my labouring mind.

MARK AKENSIDE, "Hymn to Science," 1744

If you have a garden and a library,
you have everything you need.

MARCUS TULLIUS CICERO

As Peter Collinson and John Bartram began to import America
to England, a small revolution was taking place amid the
flowerbeds of England: gardening, traditionally the preserve of
the aristocracy, would eventually become a passion for people
in other strata of society as well. For hundreds of years, rich
landowners seeking to demonstrate their power, wealth and
taste had spent vast sums on armies of professional gardeners
to design and tend their parks and parterres. Now, amateur
gardeners began to take an obsessive interest in their smaller
plots, helped by the invaluable advice of gardener Philip Miller.

In 1731, Miller had published a remarkable book. His
Gardeners Dictionary was the first systematic and comprehensive manual about practical gardening, advising on every
aspect of horticulture from how to protect tender flowers against
frost or pests, to which species suited particular areas of the
garden, such as north-facing plots, avenues or groves. It also
included summaries of scientific treatises on subjects such as
plant anatomy and the causes of rain, as well as an introduction to botanical terms and instructions for the construction
of stoves, hotbeds and the "Mercurial Thermometer." Its concept
was so insightful that it became an instant success, influencing
every gardening book that followed it. Indeed, Miller's

Dictionary laid the foundation for much of our modern horticultural knowledge, and provided a template on which all plant encyclopaedias are based today.

Collinson immediately understood the value of the *Dictionary* for the gardener. And when Bartram asked for recommendations as to what books he should study, Collinson could think only of the *Dictionary*, as he believed it to be "a Work of the greatest use and no Lover ought to be without." During the first three years of their plant exchange Bartram had to go to the Library Company to consult the *Dictionary*, until Collinson convinced one of the other subscribers to Bartram's boxes to send a copy of the first edition as a gift. The index of Latin and English plant names, Collinson thought, might be particularly useful, as he would no longer have to list both in his orders to Pennsylvania. Equally, when he sent Bartram English seeds and bulbs, he just advised "Consult Millar [*sic*] on all these" instead of having to write out lengthy explanations about their cultivation. The "noble Present," Bartram assured Collinson, was much appreciated—"I love such kind of bookes that treats of any branch of botany."

With little access to other botanical publications, Bartram might not have been able to appreciate the great gap in the market that Miller filled, but the *Dictionary* was innovative and visionary. Other bestselling contemporary publications such as John James's *Theory and Practice of Gardening* (1712) were for gentleman gardeners who wanted formal parterres below their drawing room windows. These were instructions on geometry and mathematics, explaining how to "draw a Right or straight Line upon the Ground" and how to use protractors, compasses and graphometers to trace a design. They contained little hands-on gardening information. Miller's *Dictionary*, on the other hand, contained a complete catalogue of all plants that were in cultivation in Britain as well as details on their cultivation and propagation. Each genus was given a separate entry—sorted alphabetically—which was further subdivided into the different species.

Under "G," for example, Miller listed more than forty different sorts of geranium. He specified pot sizes, gave recipes for soils, instructed that some species needed more sand for drainage, and explained that geranium seedlings had to be shaded.

Though common practice today, these were revolutionary instructions for the eighteenth century and cleared out the cobwebs of myths and superstition. Miller did not follow the frequently given advice that planting and sowing should take place according to moon cycles. Instead, he explained that geranium seedlings should be transplanted after two months' growth because they needed to have "rooted sufficiently." For the first time a gardening book approached horticulture in a scientific way, or as contemporaries called it "in a natural Way as Mr. Miller."

Another reason for the *Dictionary*'s instant success and its long-lasting legacy was that it was written in plain English instead of the florid style that had been so typical in many other treatises. Each entry was unfussy and easy to understand: pots needed stones or shreds at the bottom for drainage and peach trees should always be pruned in autumn, rather than in spring, as it was better for the plant and more convenient for the gardener. Everything, Miller wrote in the preface, was "in a methodical Order, whereby a Person may readily turn to any particular Article, of which he wants to be informed." The *Dictionary* intended "to inform the Ignorant," Miller insisted, not the learned.

Miller's success came from his long experience as a gardener, and his reliance, like Fairchild, on his own observations. As the son of a family of nurserymen, he had been groomed for the profession from a very young age. His father had ensured that his son received an education as sophisticated as that of a gentleman and much better than that of other boys of his background. Miller had learned several languages and was so fluent in Italian that he would later translate a botanical book from that language; he had also been tutored in arithmetic, drawing and natural history, and had reputedly read every horticultural textbook ever published in England.

No portrait exists of him, but he was later described by one of his son's friends as a man who always dressed in "plain old-fashioned English dress" and who eschewed wigs, ruffles and other fanciful attire. Instead, Miller was passionate about his profession and determined to introduce others to the pleasures of gardening. He believed that daily exercise and an outdoor life cleansed and strengthened the body, leaving it "pure and uncontaminated." Physical work in a garden, so Miller thought, encouraged blood circulation and "expels the morbid Parts through the Pores." And so each morning he rose from what he called "those refreshing Slumbers" to embark on another day of hard but joyous digging.

The scene of his labours was the Chelsea Physic Garden, which since his arrival in the early 1720s he had completely transformed. Originally the four-acre garden had been laid out by the Society of Apothecaries for their apprentices, in order that they might learn to differentiate herbs accurately so as not to poison their patients. As such, the purpose of the garden was to be a giant living *materia medica**—providing a research tool for the apothecaries as well as rare ingredients for some of their medicines.

But when Miller had first seen the garden, the flowerbeds were poorly stocked and the place dilapidated. His predecessors had subsidised their salaries by selling off precious

The Chelsea Physic Garden from the river front, showing two of the cedars of Lebanon and the greenhouse.

---

* Physicians and apothecaries kept plant samples in so-called *materia medica*—boxes and drawers which had many compartments, like a printer's type case, filled with seeds, roots and dried leaves as well as minerals and animal substances.

specimens, and visitors were shocked to see that hardly any exotics remained and that the greenhouse was "wretched, with few plants in it." The drunkenness of some of the gardeners and apothecaries' apprentices was one problem, and as a warning to others the committee of the garden had pinned a note to the greenhouse door saying that "no Wine or other Liquors ... [was] to be brought into or Drank in the sd Garden." But the garden had also run into financial difficulties, and it was only the generosity of Hans Sloane (then the secretary of the Royal Society) that saved it from bankruptcy. As a passionate plant-collector Sloane understood the importance of the Physic Garden, which was part of his Chelsea estate. He gave the Society of Apothecaries use of the land in perpetuity for an annual payment of £5— under the single condition that, every year, exactly fifty new plant specimens be sent to the Royal Society "ffor improving naturall knowledge," he explained, until the count rose to 2000.

When the Society of Apothecaries looked for a new head gardener, they found Miller to be uniquely suited for the task. For a few years he had run his own nursery in Southwark, until his horticultural genius became his downfall: so successfully had he improved the poor soil on his plot that his avaricious landlord calculated it could be rented out at a greater profit to someone else. With eviction looming, Miller had applied for the position at the Chelsea Physic Garden and arrived with recommendations that he was "superior to most of his occupation."

By 1731, when Miller published the *Dictionary*, he had built the Chelsea Physic Garden into a thriving botanic collection. "[P]lants as well as foreign as Domestick," the garden committee reported, were "much augmented," the accounts were for the first time in credit and fellow gardeners enthused that "Botanick art is greatly encreasing at London." This had all been achieved despite an ineffective and stiflingly bureaucratic committee. When Miller had taken up his position they had introduced a long list of rigid rules that set out the different duties for the director and the head gardener. Miller, though, had mostly ignored any instructions from above. Since the director came only occasionally to the garden, Miller gradually took control, determined to run and manage it as he saw fit. He dealt with accounts, correspondence, acquisitions and the general management, and was always at hand when needed. At first he lived in an apartment above the greenhouse, but when he married and had children he moved to a larger house just across the road in Swan Walk, at the east side of the garden. A new little gate in the east wall, just opposite Miller's house, allowed him to enter that way, rather than having to go to the main entrance at the river.

Miller was a careful observer of everything that went on in his garden. Following the lead of Fairchild, whose "Mule" he included in his *Dictionary* along with information about hybrids and the sexual reproduction of plants, he watched nature with a careful eye. He even improved on Fairchild by discovering that insects were the agents of pollination. Miller removed the pollen from half of his flowering tulips in order to investigate if they would still produce fertile seeds. Two days later he saw bees "working on [the] Tulips" which he had left untouched. When the bees emerged from the flowerheads their bodies and legs were dusted in golden pollen, which, as Miller observed with astonishment, they offloaded in the tulips from which he had removed the stamen (which holds the pollen). By carrying the fertile dust between the flowers, the bees became the horticultural cupids in the garden. Miller could now explain why many of the thriving plants in the hothouses of great gardens and nurseries failed to reproduce, as not many bees found their way into the enclosed spaces.

Each evening before going to bed, Miller would sit down to write up the day's events. He filled the pages of his thick leatherbound diary with notes on his successes and failures: when he pruned which species, the varying effects of sheep's and cow's manure, or how he protected tender flowers from wind. These notes were the basis for his advice in the *Dictionary*. And because he had assembled one of the best plant collections in the country, Miller was able to provide gardeners across the country with practical information on almost all the trees, shrubs and flowers available in England.

The amassing of such a vast collection had been possible because of Miller's international network of collectors. He corresponded with horticulturists, plant-collectors and fellow gardeners around the world, swapping observations and plants. Though the Physic Garden committee insisted that "no plants rootes or flowers be sold exchanged or delivered out of the sd Garden without the permission of the Director & one at least of the Committee," Miller ignored the rule and offered cuttings and seeds from Chelsea in return for new ones. He swapped Welsh flowers for seeds from Colombia, and parted with cones of cedar of Lebanon in return for bulbs from the Cape. He even made his contacts complicit when he urged them to be discreet, warning, "pray take no notice to any body what I sent you." So driven was Miller that he even paid for the postage, packaging and travel expenses from his own meagre salary. In time, he too became a subscriber, through Collinson, to Bartram's five-guinea boxes.

In order to further expand his knowledge and contacts, in 1727 Miller had travelled to the Netherlands to meet the owners and head gardeners of the greatest collections in the world. Every gardener, Miller wrote, "should travel to Holland and Flanders for 'Improvement.' " In Leiden he had met Herman Boerhaave, the most eminent professor of medicine and botany in Europe, and the head of one of the most admired and long-established botanic gardens. The rectangular flowerbeds, divided by neat gravel walks, brimmed with rare plants, making it the most complete botanical collection Miller had ever seen. In the greenhouse he had admired, for the first time, a flowering American torch-thistle, twenty feet high, nestling against the ceiling.

Boerhaave's exotics had galvanised Miller to create a similar collection in Chelsea, and within a few years he had achieved his goal. As Miller's international reputation as a gifted gardener, teacher and "Learned botanist" spread, even Boerhaave began to ask for trees and shrubs "to enrich us with his treasures" because "I know of no one in your country who is more capable to identify and distinguish them." By 1732, a decade into his custodianship and five years after his visit to the Netherlands, Miller received a letter from Boerhaave signed off with the wish "Remember, I beg you, my garden"—Dutch gardeners now requested exotics from England.

For Miller, Holland had also been a powerhouse of horticultural innovation. He had, for example, visited Meerburg, a garden near Leiden where the first pineapples in Europe had been produced. For many, pineapples were the proof that man controlled nature and therefore a most sought-after status symbol. They required constant heat and light for two years, even when icicles clung to greenhouse windows and short winter days enveloped the garden in darkness. Miller had been able to get some valuable tips on how to nurture these fickle fruits—instructions which were a few years later available for any gardener who read the *Dictionary*.

Miller also revived the use of tanner's bark to produce heat for tender exotics. William and Mary had brought the invention from the Netherlands to Hampton Court to protect their orange tree collection during the winter months. With its unique heating properties, the bark was a much-needed improvement on horse dung, which English gardeners had used for more than a century to fill their hotbeds—rectangular wooden frames that were closed with a sloping glass lid. It was a by-product of leather manufacture, a rough powder made of oak bark which generated a constant heat of about 25 degrees centigrade. Unlike manure, which produced lower temperatures that were exhausted after a few weeks, tanner's bark kept its warmth for up to six months. This also meant that the beds did not constantly have to be renewed and tender plants could be left undisturbed. Once a hotbed cooled down, Miller only needed to add a little fresh tanner's bark and give it a good toss with a pitchfork, and the heat could be reactivated for another two or three months.

At Chelsea there was a twelve-foot-long and six-foot-wide hotbed of tanner's bark into which Miller plunged pots of exotic flowers. Reports about the new method quickly spread through England, and fellow gardeners travelled to Chelsea to examine it. The West Indian flowers and shrubs in Miller's hotbed were thriving so vigorously, one visitor wrote, that they were actually "in greater Strength than they were observ'd in Jamaica." For the first time, tropical plants could be cultivated on a much larger scale than ever before. And, even more importantly, the fruits of these extraordinary horticultural successes were no longer confined to specialists and professionals, because the *Dictionary* disseminated the new methods among amateur gardeners.

Miller also advised his readers to build greenhouses to protect their plants, like the one at Chelsea, which was a brick construction with large windows to the south. He suggested that wooden shutters be installed in order to protect the plants during frosty nights, while inside, the walls were to be painted white in order to reflect the sun. Miller also advocated the use of "stoves"—which were hotter than a greenhouse. He recommended two different types which could either be installed in a separate building, or be attached to the greenhouse as in Chelsea: one was a dry stove and the other was a tan or bark-stove. In the so-called dry stove, potted plants (mainly succulents) were kept on ascending stone steps—almost like the seats in an auditorium. The tan-stove was built on Miller's success with tanner's bark hotbeds and contained a pit filled with bark into which potted exotics were placed, allowing gardeners to grow even those "which were thought to be impossible before," Miller wrote smugly.

It had been the Dutch who had introduced hothouses to England when William and Mary installed some at Hampton Court in 1690.* The Dutch had been the first to understand the idea that exotic plants needed sunlight, not just heat, and had introduced glass as the most important building material. But in the decades that followed, English gardeners began upgrading the Dutch designs. Because the smoke of open coal fires harmed many plants, Fairchild replaced them with flues that ran below the steps on which the pots were placed. Miller now advised the installation of flues which ran in a criss-cross pattern along the back wall of the greenhouse, covering the surface like a latticework. As these were longer than straight flues, the heat was more efficiently distributed, obviating the need for an additional furnace. Miller had persuaded the committee that his design for the flues and oven was better than the architect's suggestions, even though it increased the costs of the greenhouse and stoves by another £125, bringing the total to almost £1,900.

---

* Many English gardeners had learned from William and Mary's hothouses at Hampton Court. The Duchess of Beaufort, who had been one of the most passionate plant-collectors in the seventeenth century, had sent her servant to write a report and exact description of the construction and technology. Fairchild's hothouse at Hoxton had also been inspired by the Dutch methods used in Hampton.

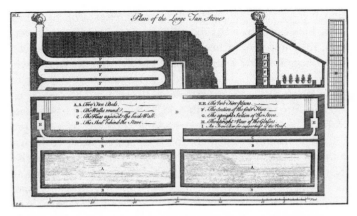

Philip Miller's plan for a large tan-stove. The squares marked as "A" are the beds filled with tan. The flues marked "F" snake up, unlike those in the greenhouse, which ran in criss-cross patterns along the wall.

By the time Miller published his *Dictionary*, Chelsea was better equipped than any other garden in the country to grow plants from any climate. Hotbeds, cold frames, glass stoves in different sizes and temperatures, as well as a sophisticated greenhouse provided a wide range of environments. Colleagues often brought rare seeds to Chelsea in the hope that Miller would be able to raise them. He succeeded, for example, in growing two rare papaw plants from seed. He was also the first to lure coconuts into germinating and to force hyacinths into blossom in glasses filled with water rather than soil, bringing colour and scent indoors during grey winter months, a method still used today.

And as much as he was a diligent gardener, Miller was also a determined self-promoter, making sure that everybody heard about his achievements: raising difficult exotic plants was challenging, but showing off was as much fun. "We have in flower several exotics in vast perfections," Miller wrote to another gardener, adding, "and I doubt not that we shall be able to manage the most tender plants in ye world." It was an exhilarating challenge to coax these delicate plants into life, and the horticultural gossip network spread the news.

Miller was particularly proud of the "firsts" that he had grown. When, for example, he procured from Chile a parcel of seeds that had been almost entirely eaten by insects, he still managed to raise two plants of *Brugmansia arborea*. Their white blossom dangled as six-inch-long trumpet-shaped flowers from their upper branches and was much admired by visitors. The fragrance of "a single tree," Miller later wrote, "will perfume the air of a large garden." He was also the first to grow one of today's most colourful shrubs: *Hibiscus rosa-sinensis*, an evergreen shrub which he procured from Asia. Nobody had seen anything like it, Miller boasted, as it changed colour from white when the buds appeared, to "blush rose colour" and then to dark purple as they withered. He was also the first to receive camphor trees from tropical Asia. In England they remained a rarity for decades, but in Florida its bark would later be harvested for medicine and mothballs. And though the four cedars of Lebanon had already been planted at Chelsea in 1683—the stately tree had been introduced to Britain in 1638—it was only under the care of Miller that they produced cones with ripe seeds. This was such a sensation that Hans Sloane brought a branch of nine cones to one of the Thursday meetings of the Royal Society to show to the fellows.

With these successes Miller proved that he was the most innovative gardener of his generation, bringing together abstract theory with practical methods; curiosity with orderliness and ambition. Miller was unique because his stewardship of a growing plant collection was combined with his broad education, his understanding of the commercial trade, and his talent as experimenter. Never short of self-confidence, he promised himself that Chelsea "will in a short Time outvie all the Gardens of that Kind in Europe."

However, there was one problem: the naming of these new plants was causing great confusion, since gardeners and nurserymen were naming them at will. To Miller's great irritation even his boss, the director of the Chelsea Physic Garden, settled the names "according to his own fancy." The resulting nomenclature was riddled with inconsistencies, repetitions and labyrinthine classifications. For garden owners the confusion meant that cultivation advice could be wrong and that the same plant could be ordered twice under different names. Customers who ended up with duplicates accused their suppliers of being "a Knave or a Blockhead."

The problem was not entirely new. Already in 1629 John

Parkinson, in his bestselling plant catalogue *Paradisi in Sole: Paradisus Terrestris*, declared that "scarce one of twentie of our Nurserie men doe sell the right, but give one for another," and warned not to trust them as they "cheate men of their money." More than sixty years later, Samuel Gilbert, another horticultural writer, remained frustrated by "the Mercenary Flower Catchers" who pretended to sell new exotics but only invented names for "old Flowers to enhance their price." The influx of new American species intensified these conflicts. Even the professor of botany at Oxford was confused when he received Bartram's species, because different authors of botanical books "varies so much in Names," he complained, "that one hardly understands one another."

In the 1720s Miller and Fairchild had gone some way towards rescuing the reputation of their profession. Together with London's best nurserymen and gardeners, they had founded the Society of Gardeners, a body that set out to bring an end to the confusion by regulating the naming of plants. As the most talented of their profession, they believed that between them they would be able to compile a complete catalogue of all foreign species growing in the English garden. And "to compare such Things as should be received from abroad, with those already in the English Gardens, and to discover where the real Difference (if any) lay." Once a month, they each packed a box with specimens and made their way to Newhall's coffee house in Chelsea, near the apothecaries' garden. Here, animated by wine and ale, they compared, discussed and classified the new flowers, trees and shrubs. Once agreed upon, the names were entered into a plant register which was to be published when complete.

It was a slow process, and demanded great patience. To be certain about the similarities and differences between foreign and native plants the members of the society often had to wait patiently for blossom or maturity. Until then, unknown species were listed with descriptive names, such as "Mr. Catesby's new Climber"—later called *Wisteria frutescens* or American wisteria. In the end, the Society of Gardeners was short-lived and did not complete the huge task it had set itself, but Miller copied many of the notes about names to use them in the *Dictionary*. In his turn, he inspired new names when the *Dictionary* became the main reference work in the world of horticulture and gardeners began referring to plants by such shorthand names as "Miller's early dwarf" or "Miller 1st sort."

The *Dictionary* made Miller famous. When it was presented at the Royal Society it was praised as the first publication that pulled together the practical knowledge of a gardener with the writerly abilities of "Men of Letters." The poet and gardener Alexander Pope declared it the gardening book from which "you draw all your knowledge," and even Miller's enemies admitted that the *Dictionary* was "the chief book that is read by the gentlemen who study the art of gardening." It was so successful that the phrase "as Miller directs" or "according to Miller" would become ubiquitous. Even fifty years on, garden writers believed that Miller had "contributed more to the advancement of practical gardening than any individual."

What was particularly remarkable about the success of the *Dictionary* was that it jumped class divisions. Whereas other botanical publications had only aristocratic patrons, even nurserymen and badly paid gardeners bought it. Of course old acquaintances of Miller's like Collinson and Catesby were among the first to sign up for a copy, but many others also subscribed to the first edition, financing the printing of the publication. William Love, the head gardener at Stowe, for example, was one of the 400 subscribers, along with Herman Boerhaave and others from the international league of botanists; Edmund Halley, the famous astronomer; and even the prime minister, Robert Walpole. Merchants and clergymen bought the *Dictionary*, and wealthy landowners procured copies for their head gardeners and themselves, annotating them with their horticultural triumphs and failures—just as they would note the births, deaths and marriages in their family Bibles.

The *Dictionary* and its success reflected the emerging spirit of the time, and in particular of the Enlightenment, in which the traditions of the ancients were replaced by a rational approach to knowledge and nature. As such the *Dictionary* stood at the beginning of an age that fostered monumental reference works which hoped to bring clarity to the world: Denis Diderot would publish the first volume of his *Encyclopaedia* two decades later in 1751, Samuel Johnson would order language in the first comprehensive English dictionary in 1755 and in 1768 the *Encyclopaedia Britannica* would bring knowledge into the homes of the middle classes.

Even after the publication of his *Dictionary,* Miller continued to record his observations and experiments every evening as he had done all his life, integrating these notes and any new plant species that arrived in England into the larger and revised editions that were published throughout his lifetime. In 1735, four years after the first edition, Miller issued the *Abridgement,* which was smaller but still contained much of the same information as the folio edition. More practical in the potting shed or garden because of its size, it was also more affordable. Now, for a few shillings, every gardener could buy the *Abridgement* and partake of Miller's expertise. This moved the craze for exotics beyond the realm of aristocratic collectors with their professional gardeners to a new generation of amateurs. Miller's *Dictionary* was, as one admirer declared, "the bright beam of gardening issuing from the dark cloud of ignorance."

3

"My harmless sexual system"

*When I observe the fate of Botanists, upon my word I doubt
whether to call them sane or mad in their devotion to plants.*

LINNAEUS, *Critica Botanica,* 1737

In the summer of 1736, as Peter Collinson was attempting to
placate John Bartram, and Philip Miller was working on a new
edition of his *Dictionary*, a young Swedish botanist thought it
necessary to travel from the Netherlands to England to meet
all the important gardeners and botanists. Not yet thirty years
old, Carl Linnaeus* was already celebrated by botanists in
Holland, who, he claimed, "sought him as a small oracle."

Such self-promotion was a constant feature of Linnaeus's
personality. He was conceited and arrogant, calling himself the
"prince of botanists" and insisting that he could achieve more
in nine months than the best of his colleagues in several years.
The son of a keen gardener, Linnaeus had been fascinated by
plants for as long as he could remember, and he liked to tell
the story that this love had been instilled through his "mother's
milk," for "whenever he cried . . . she [his mother] put a flower
in his hand, and he was quiet immediately."†

For the past ten months Linnaeus had worked for the wealthy
Anglo-Dutch merchant George Clifford, in Hartekamp near
Haarlem. As the director of the Dutch East India Company,
Clifford had access to the flora from across the world, and his
four hothouses held hundreds of plants from southern Europe,
Asia and Africa. His American collection, however, was small.
Linnaeus, having heard that English gardens "were full of rare

---

* When Carl Linnaeus was knighted in 1761 (antedated 1757) he changed
his name to Carl von Linné. To avoid confusion he remains Linnaeus
throughout the book.
† To ensure his own legacy Linnaeus dictated notes about his life and achievements (using the third person voice) to his pupils for the use of biographers
after his death.

plants" from America, had convinced Clifford to finance the trip to London, promising to procure for his patron all the plants that were unknown on the continent.

Arriving in London that July, the impoverished Linnaeus presented a shabby and unkempt appearance. He lodged with the chaplain of the Swedish church in Princes Square, just to the east of the city, and immediately set to work. Each day, a scrap of paper in his pocket with the address of his lodgings on it in case he got lost, for Linnaeus spoke no English, he sallied forth on his botanical tour. He was a man in a permanent state of hurry, who couldn't abide "slow people." It was as if life was too short to complete the task he had set himself: to bring order to the natural world. "Linnaeus's ambition was limitless," one of his pupils later wrote, and marching through London's streets from one appointment to another, the Swede was convinced that "God himself led him with his own almighty hand" and that one day he would be regarded as the most famous scientist of his age.

To achieve his goals, Linnaeus was willing to cheat, to exaggerate, to undermine his competitors and to be sycophantic. He had, for example, embroidered and inflated the danger of his expedition to Lapland four years previously in order to present himself as an intrepid adventurer. Using the diaries of earlier travellers to the region, he had invented an imaginary itinerary on which he had met with a "wild forest with untamed wolves," a fall into a deep crevasse and a rock that almost crushed him to death. In order to swindle the Royal Society of Sweden, which had financed the trip, Linnaeus had even lied about the mileage—although he was fastidious enough to make sure that it corresponded to the distances covered in other travellers' accounts.

The Lapland "adventure" also played an important role in the courtship of his fiancée, Sara Elisabeth, the daughter of a Swedish physician, whom he had wooed the year before his visit to London (dressed in a full Lapland costume, complete with pointy shoes, fur gloves and a drum). Within two weeks his proposal had been accepted. Having been so successful with her, Linnaeus also wore his costume to impress the Dutch botanists.

The first Englishman that Linnaeus set out to impress was

Peter Collinson. Linnaeus and his employer Clifford had seen the drawings of Collinson's American plant collection which his friend Georg Dionysius Ehret had made in the garden in Peckham during the previous year. Clifford had been so amazed by their rarity and beauty that he had purchased every single one of the pictures. "We are all devoted to the love of exotic plants, especially those from America," Linnaeus declared, and, thanks to Bartram, nowhere outside the American colonies would he be able to see more of them than in Collinson's garden.

Since his visit took place in late July, Linnaeus had missed seeing some of the most beautiful flowers in bloom, such as Collinson's beloved lady's slipper orchids, with their purpley brown twisted petals hanging over the yellow pouch-like part of the blossom. The black locust had also shed the last of its dangling clusters of white flowers, as had the shade-loving perennial Solomon's seal. With some luck, though, Linnaeus would have caught a glimpse and a fragrant waft of *Magnolia virginiana*—one of Collinson's great rarities. It had flowered during May and June, so Linnaeus certainly failed to see its most showy performance, but a few isolated flowers often clung to the branches throughout the summer months. It excited Collinson every year, and Mark Catesby had painted its ivory blossom and purple-coloured seed cones, noting that it spread its exquisite scent through Collinson's garden. Collinson's magnolia was, Miller had written a few years previously, "the largest tree of this kind" in the country. To have such a well-established magnolia was the envy of all gardeners but many of Collinson's other trees were still in their infancy, as Bartram's dispatches were largely composed of seedlings or little saplings, and they had been planted in the Peckham garden for three years or less.

Even if Linnaeus failed to see the flowering of the magnolia, he would have been able to enjoy the vigorous perennial golden-rod, which Collinson ordered regularly from Bartram, and which was about to burst into its blossom—although today's gardeners might be more dubious about its attraction, as it easily becomes invasive. Linneaus would also have admired Collinson's success with *Sassafras albidum,* a columnar tree from the laurel family that Miller described as "one of the most difficult Trees to grow with us." Collinson most certainly showed Linnaeus his new kalmia, as only a few weeks earlier he had informed Bartram that this "ticklish plant" had survived the transport.

Also in Collinson's garden were England's first Iceland poppy, with its petals like creased satin, and the first of the many rhododendrons that would eventually populate England's parks and gardens. Collinson had seen an engraving of the rhododendron in Catesby's *Natural History*, and ordered it from Bartram. There were also many lilies in different hues and shapes, including an eight-foot giant *Lilium superbum*. Towering more than two feet above their heads, it was the tallest Linnaeus had ever seen. Each of its stems carried up to forty speckled orange flowers with sculpted petals that resembled mini-turbans—hence its common name, the Turk's-cap lily. Even today, when it is readily available, gardeners continue to admire how it outgrows their shrubs and young trees. And maybe Linnaeus was also lucky enough to see the first tropical orchid ever to have flowered in England, Collinson's *Bletia purpurea*, which had probably been sent by Catesby when he was in the Bahamas in 1725.

Walking between the flowerbeds, the two men spoke only through the plants, as Collinson had no grasp of Swedish or Latin. Pointing or bending towards a sweet-smelling blossom, picking a flower, twig or leaf to investigate the plants further, Collinson showed off his garden while Linnaeus indicated which species he would like to take back to the Netherlands. He left Peckham with a large collection of seeds and specimens for drying for Clifford.

Linnaeus had been employed by Clifford largely to classify and order the merchant's dried plant collection, a so-called herbarium, so that it could later be published as the *Hortus Cliffortianus*. Linnaeus was the perfect man for the job, in that he was an obsessive classifier who professed himself unable to "understand anything that is not systematically ordered"; he even went so far as to classify botanists (distinguishing between "collectors" and "methodisers," which were then subdivided into subclasses, sub-sub classes and sub-sub-sub classes, such as "compilers," "systematists" or "corollists"). Only seven months before his trip to London, in December 1735, Linnaeus had published a book that he believed would change the world. His revolutionary *Systema Naturae* proposed an entirely new way of classifying plants using a system based on their reproductive organs, and how many stamens and pistils they possessed. His London trip provided the perfect opportunity for him to sell his ideas to the English.

Linnaeus had conceived of this new system because he had found the existing classifications inaccurate and over-complex. Rather than relying on scientific observation of the plant itself, many had been devised from such human perspectives as medical utility or edibility. For many centuries plants had therefore been arranged in such groups as "Hot or sharpe biting Plants" or "Strange and Outlandish Plants," or even "The Unordered Tribe," others just alphabetically.*

In England the most commonly used classification systems were those of the naturalist John Ray and of the French botanist Joseph Pitton de Tournefort. Ray's method—set out in *Methodus Plantarum Nova* (1682) and his *Historia Plantarum*, a vast catalogue of all known plants published in several volumes between 1686 and 1704—required the comparison of a range of characteristics: flower shapes, for instance, or fruit, seeds and habitat. The disadvantage of this approach was that it was very complex. It took years of practice to learn, required expert tuition and access to collections of dried specimens and to expensive, hand-coloured botanical books which often ran to a thousand pages. Whenever a botanist came across a new species, he had to compare it to vast numbers of other specimens in order to identify it—a particularly impractical and laborious approach at a time when more new plants were being discovered than ever before. Yet despite these shortfalls, one

---

* Theophrastus, flora's first classifier, developed a system in 300BC that was actually more sophisticated than what followed over the next 1,800 years. Unlike later botanists, Theophrastus divided all 500 plants he knew into groupings that evolved from the plants themselves, rather than their medicinal or culinary value. Arranging them into trees, shrubs and herbs, the system seemed logical but became imprecise when it came to the definitions of the different groups. A tree, for example, was defined as having one trunk, while shrubs had several, according to which a pomegranate, which was often trained into having multiple stems, would have to be placed as a shrub rather than a tree.

of Ray's greatest achievements was his concept of a species as the basic taxonomic unit. Though botanists had grouped different plants together before, Ray was the first to envisage nature's hierarchy, defining the species as the smallest unit.

In a search for further simplicity, Tournefort had departed from Ray's multi-trait comparative method and instead concentrated on a single characteristic—the formation and shape of the petals in the flower head—the corolla. Linnaeus had initially worked according to this method but had then invented an even easier approach. He followed Ray in retaining species as his taxonomical base unit and Tournefort in using only one characteristic to distinguish classes, but instead of the corolla, Linnaeus had chosen the reproductive organs as determining traits.

The so-called sexual system, Linnaeus insisted, was far superior, as it was much easier to learn and simpler to use than Ray's or Tournefort's method because anybody who was able to count could apply it. With only a few notes and three tables, Linneaus had reordered the entire natural world. Flowering plants were divided into twenty-three classes according to the number of their male organs—the stamens, which he called "husbands." These classes were then further distinguished through the number of pistils at the centre of the flower—the female organs, or "wives."* For those complaining that some flowers were too small for their organs to be seen, Linnaeus advised employing a magnifying glass, while at the same time using the occasion to brag that *he* didn't need to because of his excellent eyesight.

The battle to find the most accurate classification method continues to this day. Over the centuries many systems have been invented and overthrown, but new technologies and discoveries have opened other avenues to be explored.

---

* The system included, in fact, twenty-four groups, though only twenty-three which could be classified according to the number of stamens because the last class included all plants that appeared flowerless, such as mosses, horsetails and ferns. Linnaeus named this class *Cryptogamia*, meaning "hidden marriage," from the Greek *kryptos* (hidden) and *gamia* (marriage). Plants with one stamen were *Monandrias*, with two were *Diandrias*, with three *Triandrias* and so on. In the same manner the number of pistils assigned the plant to an order. The canna, which had one stamen and one pistil, was therefore a *Monandria Monogynia*.

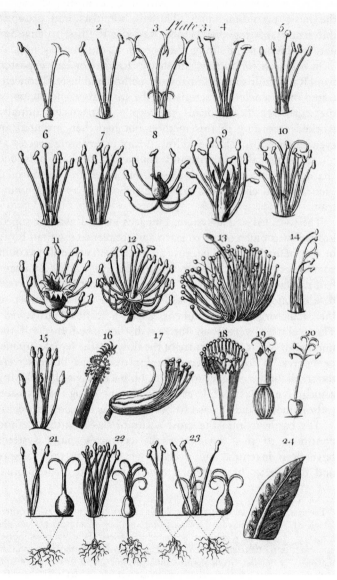

The twenty-four classes of Linnaeus's sexual system according to the number of stamens.

Today's botanists are examining the DNA of plants to discover their relations, a project which once accomplished will present a changed order of the natural world. For the first time we will be able to understand the relationships of plants based on their genes rather than on their appearance (though classifications of DNA-sequences and morphology more often than not corroborate each other), allowing botanists to construct a branching tree of evolution. Papayas and cabbages, for example, have the same ancestor, while the tropical lotus is related to the plane tree.

Just as this new DNA-based method reflects the concerns of our age, so Linnaeus's sexual system reflected the concerns of his. At a time of imperial expansion, when the discovery of new worlds added thousands of previously unknown species to existing plant collections, the sexual system, Linnaeus believed, would be invaluable, as it could be used everywhere, in the wilderness of the new colonies or in the libraries of learned societies in London.

Yet although it made sense to create a common international language, Linnaeus knew it would be difficult to convince other botanists to depart from their traditions. If his system was to succeed, it had to be accepted in England, as the country was becoming the most important horticultural market in Europe. In order to prepare the ground, he and his Dutch friends had sent his *Systema Naturae* to the key people well in advance of his visit to London. Miller, for example, had received a copy a few weeks after its publication, which he had given to the library of the Chelsea Physic Garden, while two copies were dispatched to Hans Sloane, one for the plant-collector himself and one for the Royal Society. Letters eulogising Linnaeus's achievements also preceded his visit: the three tables of the sexual system, a Dutch botanist explained to an English colleague, were "so well made up, that there is nothing existing in nature, which is not to be found there under his classes." Not for a century, he claimed, had there been "one that was so learn'd in all parts of Natural History as he; and that not superficial, but to the bottom."

Linnaeus was so desperate to convince the English botanists and collectors of his method that he even altered a passage of the *Systema Naturae* in which he had criticised Albert Seba, a naturalist who was highly respected in Britain (in particular by Miller and Sloane), in order to please them. The Dutch botanist Johan Frederik Gronovius had advised him in October 1735 that "if you want to make a splash in England, it is vital to get Sloane and Miller on your side." Upsetting Sloane and Miller would be unwise, Gronovius wrote, because then "you can be sure of getting a bad reception in England, something I would very much like to avoid—not least because I have every hope of seeing you elected to membership of the [Royal] Society the moment you turn up in London." Linnaeus's strategy worked only in part, because though he was granted an audience with Sloane, he was not yet made a fellow of the Royal Society.*

It was important for Linnaeus to ingratiate himself with Sloane because, having taken over the presidency of the Royal Society after Newton's death almost a decade previously, Sloane held the most influential position in the scientific world, as well as being the owner of England's largest collection of herbaria—having spent a reputed £50,000 on his dried plants. To examine Sloane's herbaria was essential for Linnaeus as, having created his sexual system, he now planned to publish a catalogue of all known plant species organised according to his new method. Although Sloane was always willing "on proper Notice to admit the Curious to a sight of his museum," Linnaeus took the precaution of bringing a letter of recommendation from the leading Dutch botanist and head of the Leiden botanic garden, Herman Boerhaave, one of Sloane's regular correspondents in the Netherlands—probably in the hope of being received as a scholar not a tourist.

In his note Boerhaave clumsily introduced Linnaeus, who was almost fifty years Sloane's junior, as "to be worthy to be seen with you," and, continuing in the same vein: "who sees you together shall see a couple the like of which the entire world will hardly ever see again." To place the unknown Swede on equal footing with the man whom Boerhaave used to greet as the "FIRST OF THE SCHOLARS in a nation praised before all others for its fame in learning" was certainly flattering to

---

* Linnaeus was not elected for another seventeen years; poignantly he had to wait until May 1753, four months after Sloane's death.

Linnaeus, but perhaps somewhat offensive to Sloane. But even if Sloane was insulted, he sent a note to Linnaeus's lodgings inviting him to consult the herbaria. And so Linnaeus left Princes Square on 28 July to cross town in order to see the seventy-six-year-old collector in his house in Bloomsbury.

The tour of Sloane's museum usually lasted several hours, since it took in eleven rooms and a huge number of objects ranging from a human foetus to elephants' teeth, Egyptian antiquities and 23,000 medals. There were also human remains such as a syphilitic skeleton, kidney stones and a nine-foot bone which was claimed to be from a giant's leg. Curious visitors were drawn to 4,000 insects and butterflies that were kept in glazed boxes, 7,000 different preserved fruits and various stuffed animals, as well as drawers bulging with shells, fossils and precious stones. In low cabinets, jars filled with preserved snakes, fishes and human organs made a macabre display. The tour was usually punctuated by a coffee-break in one of the adjacent rooms, where Sloane showed "all manner of curious books," a habit he probably considered stopping four years after Linnaeus's visit when the musician Georg Friedrich Händel stained a valuable manuscript with a muffin.

Most interesting for Linnaeus, though, were the 230 volumes of Sloane's herbaria. Because the dried specimens had come from so many different owners, they were ordered according to a whole variety of systems. Some were kept in what one visitor described as the "true English fashion in prodigious confusion in one cabinet and in boxes," but most were kept in bound volumes in which several specimens were mounted on each page in order to save space. Though this kept everything neatly together, it created problems when it came to the systematic ordering of the collection. Since many of the specimens were sorted alphabetically, problems arose when a new species had to be added to the volume and there wasn't room in the correct place. Sloane would either paste it in the wrong place (cross-referencing it) or start a new volume. "One must run over the whole classe to find a name," some botanists had complained. Linnaeus doubtless felt superior when he considered how he had avoided this problem by mounting his own and Clifford's specimens individually on loose sheets, then sorting them according to the sexual system, and placing them into cabinets that had twenty-four compartments, one for each of his classes. New species could easily be integrated in between the stacks of paper.*

For years Sloane had tried to sort his collection, but his constant acquisition of new specimens, and the demands of his profession as the most fashionable physician had delayed him. Many specimens were not even properly labelled, such that Sloane had written on some of his volumes "Many not named, chiefly for want of leisure," or "Omitted to be named for want of time." For Linnaeus this rendered the scientific value of the herbaria negligible. Having read Sloane's book on Jamaica, Linnaeus knew it would be difficult to convince him to accept the necessary changes because Sloane had written there that any alterations to botanical names and methods were "a great Obstruction to the Knowledge of natural Things."

Sloane had chosen John Ray's method of classification as laid out in the *Historia Plantarum*. Thus each plant entry in the *Historia Plantarum* that referred to one of Sloane's specimens was annotated in the margins of the book with the shelfmark of its location in the herbaria. For many, this seemed perfectly logical, because Ray had been regarded as the greatest naturalist in England for more than fifty years, but the belligerent Linnaeus disagreed. "What was he?" Linnaeus doubted Ray's achievements, "undoubtedly an indefatigable man in collecting, describing, &c.; but in the knowledge of generic principles, less than nothing." To avoid losing Sloane's support at the Royal Society, Linnaeus did not criticise his methods to his face. Nonetheless he wrote haughtily to a friend in Sweden, "Sloane's collection is in complete disorder."

Maybe Linnaeus was thinking of Sloane's herbaria when he described the following situation two years after his visit to London: "A foreign and unknown plant is brought before two botanists, of which one [is] an empiric, ignorant of systems, but the other a systematist. The empiric tries to divine the

---

* Individual specimens were glued on to loose half sheets, then all species from one genus were placed and held together in a folded full sheet and sorted in the appropriate drawers. Two of the cabinets with twenty-four compartments can still be seen in Linnaeus's summer house in Hammarby, Sweden.

family from the appearance, he recalls every memory—which of this kind of plant he has seen before. Day and night he turns over his herbaria, he reads over all his books, especially those with plates, ignorant in what place among so many thousands of plants or in which author it may be found"—while the botanist who used the Linnaean method just needed to count the stamens and pistils.

Yet Linnaeus encountered even more reluctance than he had anticipated when trying to introduce his system into Britain. Though the classification by stamens and pistils was easy to understand, the problem was, as Collinson later wrote to Bartram, that it "tends but to Embarrass & perplex the study of Botany." Many botanists were scandalised to learn that plants made love in the flower head—the "bridal bed"—which God had "adorned with such precious bedcurtains, and perfumed with so many sweet scents." The calyx (the mostly green petal-like parts at the bottom of the flower head) Linnaeus had named the "bedroom," and in a later publication he called it "the lips of the cunt." Nor did it help that, according to Linnaeus, one wife frolicked with "six husbands" in the lily's flower head, while tulip poplars enjoyed "[t]wenty males or more in the same marriage." Oaks, pines and birch featured husbands and wives that lived in one house but had "different beds," while marigolds and ash trees were even more promiscuous, since their flowers contained husbands that "live with wives and concubines." The whole plant world suddenly seemed to be involved in a horticultural orgy—even the twenty-fourth class of flowerless plants was drawn into "Clandestine Marriages" where the "Nuptials were celebrated privately." Only a few flowers were "virtuous"—like the canna, whose one stamen and pistil were in a monogamous relationship.

After visiting Sloane, Linnaeus went to see Philip Miller at the Chelsea Physic Garden—not only to try to obtain specimens, but in an attempt to explain his method in person. Miller was in some sense even more important than Sloane, because the promulgation of the sexual system in the bestselling *Dictionary* would guarantee success in Britain. But Linnaeus had heard rumours that Miller was "a difficult man" and approached him gingerly. To get the most out of Miller, Linnaeus decided he "did not want to contradict him at all" when he went to see him. And so, as on his visit to Sloane, Linnaeus remained silent while Miller gave him a tour of the Physic Garden in Latin, despite noting with disapproval that the plants were named and classified according to the old methods of Ray and Tournefort. The next day, Linnaeus heard that Miller had gossiped about him in a coffee house, saying "the botanist of Clifford's doesn't know a single plant."

The insult was unbearable, so, the next time he visited Chelsea, Linnaeus altered his tactics. As the two men leafed through the herbarium that was kept in the library above the greenhouse, Linnaeus made sure to show off his knowledge and attempted to demonstrate to Miller the advantages of his system. But the more he said, the angrier Miller grew, until, as Linnaeus recounted, he "scowled at me." Miller, who hated to be contradicted, had taken Linnaeus's remarks as criticism and was at the same time irritated that his guest preferred to sit in the library looking at dried specimens when the flower-beds were ablaze with the rarest blooms and in "greatest perfection." As a practical gardener, Miller simply could not understand why Linnaeus would miss an opportunity to look at living plants.

Offended by Linnaeus's parade of superiority, Miller stormed off, leaving the Swede to worry about how to obtain the rarities he had promised Clifford. But when Miller returned in the evening, Linnaeus invited him to a tavern and "regaled him with a few glasses of wine," after which botanical camaraderie prevailed. In the end, and after a few more visits to Chelsea, Linnaeus received from Miller "a fine parcel for Clifford" made up of many duplicate specimens from the herbarium. But despite the rapprochement, Miller was still dismissive about Linnaeus's sexual system, believing that it was conceived with hubris rather than a quest for knowledge. Linnaeus had discarded the old system, Miller later complained to the professor of botany in Edinburgh, "without the least reason for so doing, but that of having his Method establish'd and made universal." He believed that the mistakes in the system arose from Linnaeus "not having seen the plants." The sexual system, Miller predicted, "will be of very short duration."

As Linnaeus toured Britain canvassing, he continued to meet scorn. In Oxford, the professor of botany Johann Jacob

Dillenius rudely introduced him to a colleague as "the man who has thrown all botany into confusion," thinking Linnaeus wouldn't understand because he spoke no English. On the third day of his visit to Oxford, however, Linnaeus correctly identified a flower he had never seen before by counting the stamens and pistils, thereby demonstrating the simplicity of his method. In Linnaeus's version of the story, Dillenius was so in awe as to beg him "under tears and kisses ... to live and die with him." Apparently he even offered to share his salary with Linnaeus, just to keep him in Oxford. Unsurprisingly, Dillenius described their meeting differently. Immediately after the visit (which was only eight days long, rather than the month that Linnaeus claimed it to be), Dillenius wrote to a fellow botanist that although Linnaeus had "a thorough insight and knowledge of botany," he also thought that "his method wont hold." To make sure that the smug Swede was under no illusion, Dillenius wrote to Linnaeus himself, "I do not doubt that you yourself will, one day, overthrow your own system."*

When Linnaeus left England at the end of August, his bags were full of dried specimens from Miller, Dillenius and Collinson, as well as many living plants, cuttings and seeds from America. He had the coveted *Magnolia virginiana*, a cutting from the southern catalpa which Catesby had introduced in 1722, and from Collinson seeds of the deciduous sassafras. "[A] whole bunch," Linnaeus said.

Although the collecting part of his visit had been successful, he had failed to convince the English botanists to support his sexual system. Most accepted his methodical genius, but not his bold classification. Sloane did not rearrange his herbaria and Miller held out against using Linnaeus's method in the *Dictionary*. Even gardeners who had not met Linnaeus had an opinion. Thomas Knowlton, for example, Lord Burlington's

---

* One criticism was that Linnaeus's system was "artificial," which meant that it did not take the "natural" relationships between plants into account, but instead forced in some cases unrelated plants into one class, just because they had the same number of stamens. Linnaeus was aware that his system was imperfect, but it gave him a basis to start classifying the world's flora. He had stated in the *Systema Naturae* that one day he might replace this artificial system with a more natural one, but until then, he insisted, his sexual system was the best classification.

gardener at Londesborough in Yorkshire wrote that Linnaeus's system was "altogether whimsicall and ridiculous" because a classification of the plant world could not be founded on a single characteristic—nobody, Knowlton insisted, could "go beyond a Ray, a Turnefort." Collinson thought the "Science of Botany is too much perplex'd already . . . and I have not time to make any proficiency in the New Method." The indecent connotations of the sexual system continued to be criticised for years. The professor of botany in Edinburgh (one of Miller's regular correspondents), for example, called Linnaeus's observations "too smutty for British ears," while another botanist wrote to Sloane, "I doubt very much if any Botanist will follow his lewd method." Even thirty years on, the *Encyclopaedia Britannica* still insisted that Linnaeus's words were more licentious than "the most obscene romance-writer."

In America it was different. Unlike the narrow-minded English botanists, the colonists welcomed Linnaeus's unfussy system, as the lack of books and botanical catalogues made Ray or Tournefort's more complex methods impractical. James Logan, Bartram's mentor, for example, praised Linnaeus's accomplishments when he read *Systema Naturae* in the summer of 1736. Since much of America's flora was still unnamed, plant-collectors like Bartram were often confronted with new species that needed to be classified quickly. "The performance is very curious, and at this time worth thy notice," Logan wrote to Bartram, the day after he had read Linnaeus's publication. Bartram would not be able to read it himself because it was written in Latin, but Logan was prepared to translate it for him, because once explained "thou wilt . . . be fully able to deal with it thyself."

Bartram was quick to understand the system and was, as Logan later told Linnaeus, "in awe." To prove his proficiency Bartram sent some plant lists to Logan, all examined for their stamens and pistils. When Collinson heard that Bartram was dissecting flowers, he was most displeased. Clearly still bruised from the accusation of exploitation the previous year, Collinson now suggested that Logan might be profiting from Bartram himself. "[N]o doubt but he [Logan] considers thee for any Time taken up from thy own Affairs . . . in order to satisfy his Enquiries," Collinson wrote to Bartram, as a reminder that it had been Logan who had so worried that the English consortium of collectors were not paying enough. The whole idea of examining plants according to the sexual system, Collinson sneered, might be "a pretty amusement for those that have it [i.e. time] hang upon their hands"—such as wealthy gentlemen scientists like Logan—but "for thee and Mee I think we can't allow it." When Bartram asked repeatedly for a magnifying glass to assist his studies, Collinson remained offended, and replied, "pray make this complaint to J. Logan . . . as thy Enquiries in some Measure to be owing to Him & thee art his pupil." For his part, Bartram was disappointed that Collinson hadn't thought to send him a copy of the *Systema Naturae*, nor even inform him about it. Collinson retorted that he'd considered it useless to Bartram since it was in Latin. Besides, Collinson continued, in England "very few like it" and botanists had not "agreed about it"; personally he preferred Tournefort and Ray.

Back in the Netherlands, Linnaeus continued to be smothered in praise. Boerhaave, for example, hailed his work as the culmination of "unrivalled science," and prophetically claimed that "[a]ges to come will applaud, good men will imitate, and all will be improved." Gronovius, the most adamant advocate of the sexual system, was working on his *Flora Virginica*, the first systematic catalogue of North American flora, and organised it according to Linnaeus's method. Having paid for the printing of *Systema Naturae* in 1735 (because Linnaeus could not afford it), Gronovius now wrote to his English correspondents that the method and tables were so great "that one should be blamed who hath them not hanging in his room."

Linnaeus loved the compliments but was hurt and offended by the reaction of the English botanists—although nothing could entirely shatter his overblown self-confidence. "[W]ithout the system chaos would reign," he declared, and to ensure that his future biographers would be aware of his importance, he later wrote in his autobiographical notes that "no one before . . . had ordered all products of nature with such clarity." Let the pugnacious botanists be all "puffed up," he wrote to a correspondent, "[b]ut when these same people have passed a few years in the field of battle, they become so mild, candid, modest, and civil to every body, that not a word of offence escapes them."

He accused the English for not understanding the superiority of his method, dismissing the whole profession (with the exception of Dillenius): "There is nobody in England who understands or thinks about genera." Linnaeus derogated Miller's botanical achievement to plain plant-collecting, scoffing that it was enough for Miller "if he can procure American plants, living or dried." But Linnaeus—a man so fond of public flattery that he wore his Order of the Polar Star every day when he received it a few years later—found it difficult always to maintain his equilibrium in the face of criticism. "I perceive you take somewhat amiss what I wrote," one fellow botanist responded to a sulking letter, surprised about Linnaeus's lack of humour about his jokes on "the great concourse of husbands to one wife."

When Linnaeus heard rumours, in April 1737, that the celebrated German botanist Albrecht von Haller was composing an attack against him, he immediately wrote to him, presenting eleven arguments as to why he ought not to publish the pamphlet against "my harmless sexual system." Linnaeus's pleadings ranged from plain sycophancy to detailed botanical explanations, and from bragging to modesty. All distinguished botanists, he lied, "have given me their encouragement," and towards the end of the letter he warned "Look over the whole body of controversial writers, and point out one of them who has received any thanks for what he has done in this way." But despite these protestations the disapproval continued.

A month later Dillenius wrote to Linnaeus from Oxford that the sexual system was inadequate because the sexual organs of plants were "mere gewgaws" and "in my judgement, altogether useless, superfluous, and mischievous." To attenuate the tension, Linnaeus dedicated his latest work, *Critica Botanica*, to Dillenius. But publishing etiquette required that authors asked permission from the person they intended to dedicate their book to, and Dillenius made it clear that he did not like to be associated with Linnaeus's work when he replied, "I feel . . . much displeased." And for the future, Dillenius suggested, Linnaeus should not be so irritable when judged, because "you are not very patient under the attacks of adversaries." Indeed, Linnaeus's mood swings and irascibility were his greatest weaknesses. For though he was brilliant, systematic and logical, he was also petulant and childish. Linnaeus euphemised this by calling himself "sensitive," but even one of his admiring pupils admitted how quickly his beloved teacher could "lose his temper," and that he would do "all he could to erase them [his enemies] from the annals of science."

One of the antagonists was Johann Georg Siegesbeck, the demonstrator of the botanic garden in St. Petersburg. In the midst of his quarrel with Dillenius, Linnaeus received the news that Siegesbeck was printing an attack on the sexual system— "The work is very short," his source in Russia reported, "but its brevity is, in my opinion, counterbalanced by spite and arrogance." Shocked by the announcement, Linnaeus's anxious mood turned into fury when, two months later, he read in the critical pamphlet that his classes and orders were a "fictitious matrimony of plants." Under no circumstances would God have allowed such outrageous relationships, Siegesbeck asserted, condemning the whole system as "loathsome harlotry" because it was like "a tour round the bedchambers of prostitutes." Linnaeus complained that "my reputation must suffer," and with typical melodrama described how Siegesbeck had destroyed his credibility in Sweden.

According to Linnaeus his return from his three-year stay in the Netherlands to Stockholm, in June 1738, was insufferable because "all, with one voice, declared that Siegesbeck had annihilated me." His career was in ruins. The prospect of receiving a professorship in botany destroyed, Linnaeus was forced to try working as a doctor because, like many other botanists, he had a medical degree. But to his surprise nobody would send even their servants to him because, thanks to Siegesbeck, he had become "the laughing-stock of every body on account of my botany." But as always Linnaeus's version of the story was exaggerated and over-dramatic, as only two years after his arrival in Sweden he was offered a professorship at the University of Uppsala.

4

"Pray go very Clean, neat & handsomely Dressed to Virginia"

Slave to no sect, who takes no private road,
But looks, through nature, up to nature's God!

ALEXANDER POPE, *"An Essay on Man"* (1733–4),
chiselled above the front door of John Bartram's house

The land around John Bartram's farm and Philadelphia was rich with plant life. Within a day's ride Bartram could find kalmias, tulip poplars or Collinson's favourite lady's slipper orchids, but it was inevitable that, to feed the growing hunger of his transatlantic customers, he would need to travel further afield. Several times during the early years of his correspondence with Collinson, Bartram left his capable wife to tend the crops, cattle and the four children they now had, and went off plant-hunting. He explored Pennsylvania and New Jersey, and rode along the Delaware River in search of the seeds, acorns and cones the English collectors wanted. Peter Collinson accompanied him every step of the way—in spirit. He sent Bartram paper for the dried specimens, a pocket compass for "exact observations of the course of the river" and a long list of instructions. Being English, educated and wealthy, Collinson felt superior to Bartram in every respect and, despite never having visited the colonies himself, insisted that he knew exactly how Bartram ought to travel. He would need a spare horse, he advised, with baskets on each side of the saddle to hold the paper and seeds, as well as for linen, provisions, insect nets and boxes. "Be sure [to] have some good covering of skins over the baskets, to keep out the rain," he added in patronising tone.

For his part, Bartram provided Collinson with detailed accounts of his travels. His regular reports and maps brought the strange continent into the comfort of Collinson's Peckham house, and allowed the Englishman to experience by proxy the thrill of the untamed landscape. Collinson was also fascinated by wild animals, and asked "is there any account of the panthers." And when Bartram included a map of the Schuylkill River, Collinson expressed his delight at being able to "Read & Travel at the same Time." Yet despite carefully noting the distances between the colonial cities and writing to Bartram, "[I] wish to bear thee Company," Collinson was very much an armchair traveller, conceding that the thought of rattlesnakes was a greater deterrent than the lure of curiosities.

Within a few years, however, Bartram had scoured much of Pennsylvania and the English collectors demanded new and different species. Comparing himself to the proverb "Like the parson's barn—refuses nothing," Collinson decided that Bartram ought to go to Virginia. And so, in December 1737, he once again used his colonial contacts and began to send letters of introduction to the learned gentlemen in and around Williamsburg. He wrote, for example, to John Custis, a wealthy plantation owner who had exchanged some seeds with Collinson in the past. Bartram was a plain country man, Collinson said, and warned Custis "not Look att the Man but his Mind" because "[h]is Conversation I dare say you'l find compensate for his appearance." Collinson also drew attention to the fact that Bartram's garden now contained many European plants which, he hinted, might come Custis's way should he help the farmer. Collinson hoped that this would tempt Custis, who regularly complained that "the seeds in generall wee have from England very often never come up."

The strategy worked and Bartram made preparations for his journey, aided as usual by Collinson's advice. "Pray go very Clean, neat & handsomely Dressed to Virginia" and don't "Disgrace thyself or Mee," Collinson wrote, sending some cloth and detailed instructions as to what should be worn during the expedition. Being the inhabitants of the oldest of the British colonies in North America and the descendants of many generations of settlers, the wealthy Virginians had a reputation for snobbish affectation. Many of the plantation owners had been on a Grand Tour, and had brought back European tastes and fashions. Some of them, Collinson, explained, "look phaps More at a Man's Outside than his Inside." Bartram was no doubt better pleased by receiving from Collinson a set of Mark Catesby's engravings to help him identify the plants he was being sent to fetch.

Finally, on 25 September, five years after he had gathered his first seeds for Collinson, Bartram saddled his horse, packed food, paper, boxes and ox bladders in which to keep the specimens fresh, as well as the clothes and compass that Collinson had equipped him with, and set off in the pouring rain. The journey to Williamsburg took fifteen days through rain that rarely let up, but eventually Bartram reached John Custis's house just as the sun was setting. His fortitude was rewarded. Not only did Custis own one of the finest houses in town, he also had a four-acre garden which was reputed to be one of the best stocked in Virginia. There Bartram saw a horse chestnut, a tree which he had been desperate to grow ever since Collinson had so vividly described the candle-like white blossom. The horse chestnut was not native to Britain, having been introduced from Albania and Macedonia only in the early seventeenth century, but it was soon one of the most popular trees. Cultivating it in Pennsylvania proved more difficult. Although the sapling in Custis's garden was only a foot high, Bartram, who had never succeeded in raising even a seedling, was deeply impressed. Interesting too were the twenty-year-old English hollies and yews, which were clipped into balls and pyramids—Custis adored in particular the species with variegated leaves, although he admitted that "I am told those things are out of fashion." But the combination of the severe winter of 1737 and the drought over the previous months had left the low box hedges that bordered the flowerbeds looking brown and parched. Even though three slaves had watered the plants every evening and built arbours to create shade, many European plants had been damaged, leaving the garden stripped, "[O]ur poor country grows more unseasonable yearly," Custis complained.

As Custis became acquainted with Bartram he was charmed: "he is the most takeing faceitious man I have ever met with and never was [I] so much delighted with A stranger in all my life," he wrote to Collinson. Bartram reciprocated the enthusiasm: "my dear friend Col Custice who entertained mee ... with extraordinary Civility & respect beyond what ever I met

Eastern Hemlock

with in all my travels before from a stranger." Custis was just one of the many people who became devoted to Bartram. Easygoing and well-mannered, Bartram had a talent for making friends with people of all classes, as he felt comfortable with almost everyone, especially if they loved botany. After two nights and one day in Williamsburg, Bartram left on 11 October and arrived the same evening at Westover, William Byrd's plantation near Richmond, Virginia.

Having spent much of his life in England—mainly in search of a rich wife—Byrd was an anglophile who loved both English plants and English garden designs. When Catesby had visited America for the first time, more than twenty years previously, he had advised Byrd on the latest garden fashions, and Collinson occasionally sent cuttings and seeds of European plants. Even though Byrd was wealthy, employing gardeners and owning slaves, he loved the practical aspects of horticulture and never went into his orchard without pruning shears. By 1738, the Westover garden had the reputation of being one of the best in Virginia. Bartram saw elegant wrought-iron gates with Byrd's monogram set amid the intricate swirls, and even a little greenhouse with fruiting orange trees. The two men spent a whole day in the garden "observing curious plants" and talking about their cultivation.

The following day Bartram left Westover to ride along the James River towards the Appalachian mountains, the western frontier of Virginia and the beginning of the uncharted territories. It was not long before his rambling bore fruit. Late one afternoon, he crossed the river and began searching the forest along the southern embankment. Here he discovered *Thuja occidentalis*—a tree that he had never seen but which had been one of the first North American coniferous trees to arrive in Europe almost exactly 200 years previously. On the same day, while riding up a mountain, he also found the tall Eastern hemlock, an elegant conifer whose gently arching branches covered with a profusion of soft needles and tipped with small dangling cones look like feather boas carelessly slung around a woman's neck. Bartram had already sent Collinson this tree two years previously but was delighted to find more cones because he knew that it was "much esteemed in ye english gardens." The shallow roots of the Eastern hemlock make it the perfect mountain tree, as they cling to rocks, ravines and the stony embankments of waterfalls. However, Collinson, who was the first to cultivate it in England, and his plant-collecting friends were successfully growing it in their gardens. Six years after Bartram sent the first batch of cones in 1736, seven specimens sold for £2 2s. each at a plant auction at an estate in Essex—a hefty price tag when £2 would also buy 1,000 two-foot saplings of native ash.

When plant-hunting, Bartram always rose with first light, "for I cant aford to loos any time," he told Collinson. One morning, having camped in the foothills of the Appalachians, he woke before dawn and climbed a peak with only moonlight illuminating his path. As he reached the crest of the mountain the glowing curve of the sun was emerging from the horizon and Virginia lay before him, stretching as far as he could see. "I had," he wrote to Collinson, "ye fines prospect of ye largest Landskip that ever my eyes beheld." The hills were blanketed in trees—each crown adding to the patchwork of autumnal red and rusty orange. Once the sunbeams began caressing the leaves, the whole landscape was set alight. The leaves of maples, scarlet oaks and dogwoods were like drops of amber clinging to the branches, against which were set the dark needles of the conifers. Beneath the trees grew thickets of evergreen kalmia and rhododendron. As Bartram rode along the Shenandoah mountain range there was such a great variety of plants that he could hardly contain his excitement, endlessly crossing ridges and streams in search of new treasures. For days he rode alone, accompanied only by his horse and the ravens that circled the summits. At such a height Bartram could see clouds weaving through the tree trunks and the distant chains of hills towards the west, unreal, like layers of bluish cardboard cutouts.

By now most trees had shed their seeds and Bartram decided to make his way towards Philadelphia. He left the Appalachians and arrived home on 26 October after riding 1,100 miles in only five weeks. But when he opened his bags, he found that all the seeds which he had wrapped so carefully had been mixed up by the movement of the trotting horse. There was nothing he could do, and in particular not immediately, because after his long absence there was urgent work to be done on the farm. It was not until December that Bartram finally found time to pack his seeds, including a note with the suggestion that they be sown together in one flowerbed to make the most of the jumbled mess.

In England Collinson and the consortium of collectors eagerly awaited the new seeds and expedition report. But even after Bartram dispatched them, there were delays. To Collinson's great frustration, one captain refused to carry the box on his ship at all. "I Long the arrival of the Ship with the Seeds," he wrote in February. Three and a half weeks later, he returned from the Pennsylvania Coffee House near his office, where he had been enquiring about the arrival of another ship, and wrote yet another letter to Bartram. "In pain for the Cargo of Seeds this year," he moaned. Despite these protestations, he had to wait until March for the ship to bring Bartram's journal, letter and plant boxes.

Even this letter left Collinson unfulfilled. He was desperate for information about his old acquaintances in Virginia and disappointed that Bartram had not sent reports about their families, estates and "tasts in Life." Bartram felt unfairly crit- icised. His descriptions had been purposefully short, he countered, since Collinson had too often complained about his lengthy accounts—"as thee often expresseth hath not ye leisure to read them." Over the next few months, Collinson and Bartram continued to bicker like a squabbling old couple. In a few lines they managed to blame, appease, tease, thank and insult each other. Accusations were peppered with declarations of friendship and gratitude, and these petty arguments could last a year or longer depending on the postal delays.

When Bartram mentioned, for example, that Collinson had sent him a "rotten mouldy" hat, Collinson was offended. "My Cap it's True, had a small Hole or Two," he replied, but "Instead of giving it away I wish thee had sent it Mee back again. It would have served Mee Two or Three Years to have worn in the Country in Rainy weather." When Bartram received the letter four months later, he replied that he hadn't believed Collinson could have put the disintegrating headgear into the box: "I thought some sory fellow had thrown it [in]."

On another occasion it was Bartram's attempt at generosity that caused friction. He sent an especially large load for

Collinson in gratitude for dealing with the seed distribution, but because it had been too large to place under the captain's bed, the rats on deck had made their nests in the box, and eaten and trampled the plants. To Collinson's complaint, Bartram responded that the dimensions were not larger than Collinson himself had advised, and he could not be blamed. But despite these accusations both men continued to address and sign the same letters as "my kind & generous friend" or "Thine very Truly."

In the first year after the expedition to Virginia, Bartram sowed the seeds he had not sent to England in his own flowerbeds so as to be able to harvest cuttings, cones and seeds for future dispatches without having to travel to Virginia, and so that he could provide Collinson with advice on their cultivation. He raised them in the five acres of descending terraces which he had set aside on his farm, a layout popular in the colonies. The upper terrace was just below his house and contained a flower and kitchen garden from which two sets of stairs led to a lower softly undulating terrace. This part of the garden sloped down towards the Schuylkill River, the gently rippling land forming little glades and scoops. Cut through this lower area were two 150-yard-long tree-lined walks which led from the riverfront to the house. Next to these avenues Bartram planted the trees that he had found on his travels.

Like many other American gardeners, who favoured productive trees over ornamental ones, Bartram combined the practical with the beautiful. Cherry trees lined avenues while crops were integrated into the landscape. His garden was cluttered with tubs of saplings, which were sent to Collinson because gardeners in England failed to raise some American species from seed. In this way kalmias and rhododendrons were nurtured for two or three years in order to be strong enough to survive the passage to England as living plants. Just a few months after his Virginia expedition, Bartram once again left his wife, Ann—this time heavily pregnant with twins—in charge of the farm and the children. He paddled up the Schuylkill River in a canoe to find rhododendrons, which he uprooted with sods of earth and then planted in boxes in his garden. Although Collinson had been the first to grow them in England, he continued to ask Bartram for more as "they seem to Die Dayly with Mee."

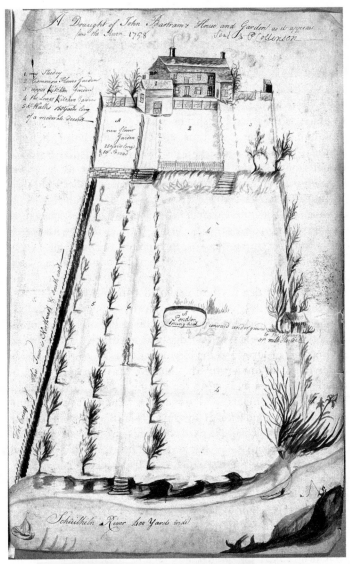

Bartram's house and garden as drawn by his son William in 1758.
John Bartram is probably the figure on the path with the walking
stick. The little shed (1) was Bartram's study before he extended the
house. It was located on the upper terrace between the "New
Flower Garden" (A) and the "Common Flower Garden" (2).

In his garden Bartram tried to imitate the natural habitat in which he had found his plants. His land was uniquely suited for his purposes because the dividing line between the sandy soil of the coastal plains to the east and the rocky outcrop to the west cut right through it. Thus plants that needed dry acidic soil, such as lady's slipper orchids or the pines that Bartram collected from the Pine Barrens in New Jersey, could thrive just as much as the trees and shrubs that he had brought from the mountains. At the same time, the undulating lie of the land provided protective pockets for the more tender plants—a micro-climate that allowed Bartram to grow trees such as *Magnolia grandiflora*, which usually needed a more southern climate. Visitors to the garden were not always favourable in their comments about this arrangement. To some it looked as if he had let nature run riot; rare shrubs were covered with weeds while valuable trees "were lost in common thicket." Fifty years later, on a visit to buy trees for Mount Vernon, George Washington admired the "curious Trees, Shrups, & flowers," but thought the garden was not designed with much "taste."

Others were less censorious, understanding that Bartram's garden was "a perfect portraiture of himself": just as Bartram's exceptional botanical knowledge was hidden by his daily life and farmer's routine, so the elegant orchids he cultivated were covered by the undergrowth. Bartram was not interested in creating a garden pleasing to the eye; he wanted the best environment for his plants, and sometimes that meant using land outside the boundaries of his farm. For Bartram, the whole world was his garden. He used the rocks, swampy patches and stony outcrops that lay between his farm and the town of Philadelphia to plant his treasures. "Every den is an Arbour, Every run of water, a Canal, & every small level spot a Parterre," one visitor observed.

Below the windows of his study, on the upper terrace, Bartram kept the flowers he received from Collinson, Miller and his other customers in England. As well as bulbs such as hyacinths, tulips, anemones, narcissi and lilies, there were many annuals, which he had raised from seed. Alongside typical English flowers like the foxglove grew more exotic species such as *Callistephus chinensis*, a white and purple flower from the daisy family which had been sent from China to France by

French missionaries and then on to Miller. Another Chinese plant was the seven-foot-tall persicaria, whose pendulous clusters of pink flowers would in time become a popular Victorian annual.

Whenever Bartram received large deliveries of seeds from Collinson, he bragged, "I shall show my friends that they never saw before." Just as Miller and Collinson craved plants from America, so colonial gardeners coveted European species—the seeds were, as Custis said, worth more than twenty times their weight in gold. And like Miller and Collinson, Bartram faced a huge challenge in keeping alive plants that were accustomed to a different environment. The colonial climate was volatile—some winter days being very warm and then, as Bartram explained to Collinson, "in A few hours most cold"; and the summers were hotter and drier than in England. To help, Collinson sent advice on how to protect the precious plants: "the beds should be Cover'd with a Good Heap of pea Straw" or "plant them in a South Shelter'd aspect." When a vine was once again killed by the Pennsylvanian frost, despite many pages of advice, Collinson reminded the frustrated Bartram that "patience & perseverance overcomes the Hardest things." Bartram struggled also with the majestic cedar of Lebanon. Collinson had sent the first cones in 1735, having procured them from Chelsea, where Miller was growing four of them. But despite trying every method in Miller's *Dictionary* or suggested by Collinson, Bartram repeatedly failed to bring them to maturity. Some seedlings he placed in the sun for a couple of hours in the morning, while others he kept in the shade, exposing them to dew at night. Some got neither sun, dew nor rain. To imitate their natural habitat he even planted a few in the shade of a great rock by a waterfall a few miles from his house. Over the next two decades he tried again and again, all to no avail, as they "perished & disolved to slime," he wrote to Collinson.

The letters between the two men were filled with a mixture of encouragement, horticultural knowledge and competitive zeal, but as Bartram grew more confident he began to challenge Collinson's dominance in their relationship. When Collinson sent common lilacs, for example, Bartram complained and demanded rarer species. In this instance, Collinson was defensive, writing, "[Custis] Desired some & then I thought possibly you might want them ... however, this shall be a Caution to send nothing but what you write for." Nevertheless, sometimes the moaning forced Collinson to pack a mollifying gift, sending seeds he had gathered in his garden and in the countryside in the hope of demonstrating to Bartram he was "no slothful forgetful fellow." At other times, having failed to deliver what Bartram wanted, he promised "to be better this year."

There was one thing about which Bartram complained repeatedly, and that was the lack of sufficient payment for his labour. The subscription only paid for five or six weeks of his time, and he argued that Collinson underestimated how long it took him to collect acorns, cones and seeds. When Collinson ordered "a Horse Load of cones," he must have imagined that America was a kind of abundant Eden from which Bartram could easily pluck as much as he could carry. The reality, though, was that Bartram had to work hard to comply, especially as many trees shed all their seeds within just a few days, forcing him to keep daily watch, an onerous task when the trees were some distance from the farm. Sometimes Bartram had problems in fulfilling the orders at all.

Reaching the cones of the "spruce firs," Bartram wrote, was "next to impossible" because they only grew on the top branches of trees which were between 100 to 150 feet high. The only solution was to fell the tree, which "for half A dozen seeds" was hardly worth the effort. Similarly, in 1737, the year before his Virginian expedition, Bartram could not supply enough of the popular American willow oak because only one in twenty trees had borne any ripe seeds that autumn, and those few acorns there were fell together with the foliage so that Bartram had to sift through huge heaps of leaves to find them. Three years later he again could not keep pace with Collinson's requests for the conifer *Juniperus virginiana** when blight destroyed all the berries. But as much as Bartram fretted and

---

* The dense conical shape of *Juniperus virginiana* made it the perfect tree with which to mirror the columns of the fashionable Palladian architecture in the landscape. But in the colonies they were used as agricultural rather than ornamental trees—Bartram, for example, planted a hedge in 1737 to keep out the cattle.

grumbled, he was not the only one involved in a time-consuming enterprise. Every time his boxes arrived at the docks in London, Collinson had to go through the same rigmarole at Custom House as well as distributing the contents to Bartram's various subscribers. "[T]hee art not sensible the Time & trouble it takes up, to gett these things," he complained to Bartram.

As Bartram's and Collinson's relationship matured, a rhythm developed: Bartram would make his expeditions in the autumn; the boxes would arrive in London throughout the winter and early spring. Bartram usually enjoyed his annual trips but occasionally they seemed cursed. When he arrived in New Jersey, in late August 1739, he noticed that most pine cones had already shed their seeds. Previously, on occasions when he had arrived too late, he had managed to gather some from the forest floors, crunchy with needles and cones, but these particular pines had clawed their roots deeply into the riverbank and their branches hung out over the swirling water of a rapid which had washed away all the seeds. Ever resourceful, Bartram scrambled up into the trees to shake the upper branches while holding his hat underneath the cones. But even this failed to collect sufficient quantities. To make matters worse, his horse was stolen and after a fruitless three-day search he had to return home empty-handed and on a rented mount. Two weeks later, when he set off on another expedition, he cut his foot so badly that he was confined to bed for a month—a great annoyance, as any illness was "a great affliction to mee who cant rest long without action." The moment he was able to walk again, even if only on crutches, he travelled by canoe to New Jersey to collect more pine cones.

During his lonely journeys Bartram encountered some of America's most spectacular scenery. Riding towards the west, he saw the rising peaks of the Appalachians like signposts to the wilderness that lay beyond. He quenched his thirst with the crystalline spring water that gushed from nature's rocky taps and bathed in the cold lakes fed by cascading rivers. When he followed the course of the Delaware, he found a pass through the mountains—the Delaware Water Gap—an entrance to the territory beyond. He saw sheer rock faces to which clung a few shrubs, swam with his horse across rivers as wide as 150 yards and slept on beds of quilted moss. It was a landscape few white men had beheld before. "[N]o mankind to be seen all ye way," he wrote. Not many settlers used these thick-forested Indian mountain trails. It was dangerous terrain, and often "dismall traveling" Bartram said, fearing not only nature's forces but also the Native Americans, whom he despised for murdering his father. One grabbed Bartram's hat and "chewed it," he reported to Collinson, "I suppose to shew me that thay would eat me."

At other times, when riding across the plains of New Jersey, Bartram became melancholic at the sight of the barren land-scape, devoid of trees and shrubs. Then a grove of pines loaded with cones would appear in the distance, and he would gallop towards it impatient to see the hoard. He waded through almost impenetrable swamps which spread across several hundreds of acres—once nearly getting stuck with his horse on a search for the white cedar. This narrow conical tree had caused much excitement when Collinson had received the first batch from Bartram in 1736. Nobody had grown it in England before, and it quickly became one of the most expensive trees. Only five years later, one of Bartram's young trees was sold for £2 2s.

The discovery of a new plant was always a sublime moment for Bartram. He would occasionally become so engrossed that he failed to notice sudden changes in weather, finding himself stranded in storms and darkness. Surprised by thunder and lightning, he once sought protection in a cave only to fall on to a narrow ledge of rock where he had to spend the night buffeted by thunder and rain. With no coat—he had forgotten it in his rush to explore the new terrain—he pulled off his trousers and put each arm in a leg to keep warm. But even stuck on this ledge 150 feet above a river gorge, Bartram enjoyed the expansive view and thought that if any Indians saw him they would have taken him as "one of ye Silvan gods."

As the years progressed and his plant export business ran more and more smoothly, Bartram increasingly resented working on his farm. At home he felt restless and unchallenged, and he longed for the moments when he could depart on an expedition. So great was his wanderlust, he explained later, that he "delighted most to dream of flying from ye top of one mountain to another." Even the most torturous journey brought him pleasure and he was happiest when he could follow "the path of wild beasts."

In 1741, for example, when he was the first plant-hunter to explore the Catskill Mountains, he was thrilled by the danger. Today a popular weekend retreat for New Yorkers, to Bartram they were the highest mountains he had ever seen. It was a vast bulging landscape—mountains as far as one could see. Like the Appalachians to the west, the Catskills were patterned with deciduous and coniferous trees, but here the ascent was steeper. Enormous boulders covered the forest floors, large ledges hung over cliffs, and Bartram was impressed by the "perpendicular banks of rocks."

The climb to one of the peaks took almost a whole day, and Bartram only descended as the sun was setting, with darkness enveloping him midway down. He continued "by runing & sliping & tumbling & yet it was so dark . . . that I could hardly see ye brambles or rocks before I run against them." When Bartram arrived at the foot of the mountain it was the middle of the night. Although bashed, scratched and bruised, he felt elated because the mountains were clothed in the greatest variety of trees and shrubs he had ever seen together in a single location. Bartram found white pine, scarlet oak and Eastern hemlock, as well as—at last—the balsam fir. Earlier that year Collinson had written, "thee must Look sharp after the Balm of Gilead Firr"—because the tall evergreen conifer was in great demand in England for its rarity.

The plant business was so profitable that Bartram bought some fertile land next to his farm, tripling his property to almost exactly 300 acres—half of which he leased out. In order to pay the interest on his mortgage, he asked Collinson to send the balance owed by his English clients not in goods but in cash. The money arrived in early summer 1739, "just in the very nick of time," Bartram reported. And he was right, for international politics was about to threaten the smooth running of his little enterprise.

In 1739 Britain declared war on Spain in order to break the Spanish trade monopoly in South America. As a consequence Collinson feared that the "Rascally Spaniards" would get hold of the boxes. An additional complication was that ships changed their routes and departure times because of the war, and vessels could be captured or delayed. "[T]his untoward Warr putts all things in Confusion and out of their usual Chanells," Collinson explained, when Bartram complained about not having received letters from his friend. But despite these problems, some boxes still arrived in London's Custom House and Bartram did at first not sustain any major financial losses.

His was an ordered life. After the annual plant-hunting trips in autumn, Bartram spent many of the long dark evenings in his study, an annexe at the back of the house, overlooking the garden. Secluded from the activities and noise of his large family, he turned to his "winters amusements," investigating his mounted duplicate specimens with a magnifying glass to compare them to others in his growing library of plant prints. Despite his humble education, Bartram was a voracious reader. He had soon worked his way through all the botanical books belonging to the Library Company in Philadelphia—some of which he was allowed to borrow for a week. But he felt starved of knowledge, because the Americans, Bartram bemoaned, had no interest in natural history and few books were available in the colonies.

At first, his requests for books fell on deaf ears. "[I]n Reading of Books there is no End," wrote Collinson, quoting Solomon's advice. But Bartram fired back that "if solomon had loved women less & books more he would have been a wiser & happier man than he was," and Collinson relented. Soon the other subscribers were contributing to his library. The Dutch botanist Gronovius repaid Bartram's parcels with a copy of *Flora Virginica* (a catalogue of all known North American flora), which had been published in 1739. Hans Sloane, who received regular batches of dried American specimens from Bartram via Collinson, sent his *Natural History of Jamaica*, while Mark Catesby sent parts of the *Natural History of Carolina, Florida, and the Bahama Islands*. Soon Bartram owned copies of the traditional botanical treatises by Nicholas Culpeper, William Turner and Matthias de l'Obel as well as scientific treatises by Carl Linnaeus, Nehemiah Grew and Richard Bradley. When one English collector sent three books as payment, Bartram replied that he would have preferred some others, as two of the books he had already read and he did not think much of the third one.

With his growing knowledge and expanding network of contacts, Bartram became bolder and more discerning. When he discovered, for example, that Miller's *Dictionary* had failed to credit him for introducing a plant to England, he did not hesitate to write to Collinson, "I think he has neither done me nor my province Justice." He was beginning to realise that he was as important to the European garden owners, plant-collectors and botanists who corresponded with him or bought his seeds as they were to him.

Part II

ROWTH

5

"All gardening is landscape-painting"

There is not a citizen who does not take more pains to torture his acre and half into irregularities, than he formerly would have employed to make it as formal as his cravat.

HORACE WALPOLE, *The World,* 1753

It was Peter Collinson's habit to spend his summers touring the country estates of his friends, to assist with the design, improvement and planting of their gardens. They all adored him, because Collinson was charming and entertaining, and nobody ever said or wrote a bad word about him. Each year he would visit such places as the Duke of Bedford's Woburn Abbey in Bedfordshire, the Earl of Jersey's Middelton Park in Oxfordshire and the Duke of Richmond's Goodwood in Sussex. But his favourite by far was Thorndon in Essex, the seat of his "most Valuable & Intimate Friend," Robert James Petre, the eighth Baron Petre and "a universal Lover of plants."

Although almost twenty years younger than Collinson, Petre had taken to the merchant from the moment they met. He looked up to the older man for pastoral advice and horticultural expertise, while Collinson enjoyed his paternal role, encouraging Petre's love of botany and gardening. They wrote long letters to each other which were filled with horticultural gossip and declarations of their mutual adoration, and met whenever Petre came to London. Whether they were dining together, attending meetings at the Royal Society, or simply having a "chat by ye fire," Petre listened attentively to Collinson's advice. And when, in 1732, he inherited Thorndon, at the age of nineteen, he had followed Collinson's tenet that "young people of fortune ... should as early as possible be initiated into some rational pursuit, and especially into a taste for all kinds of rural improvements," and made his garden his life's work.

In the decade that followed, Petre issued a stream of invitations for Collinson to visit Thorndon. "[H]ow much I wish it could be oftener & for a longer time" and "ye oftener you come ye better," he wrote, reluctant to make decisions about his garden without his mentor. Collinson's expertise was so great, one of his other friends reported, that he "often prevented young planters from committing capital mistakes." Few horticultural subjects were beyond him, from suitable types of soil to the design of large plantations. But Collinson also organised other aspects of Petre's life: the bill for the fishmonger, the purchase of marble basins, silks and cloths for his wife and children, a complaint to a fellow merchant for sending the wrong wine-glasses, and procuring some old fishing nets to cover the fruit trees. Most important, he also suggested that Petre become the first subscriber to Bartram's boxes.

When Collinson made a visit to Thorndon in August 1741, some of the trees and shrubs that had grown from Bartram's seeds were already seven years old because Collinson had shared the early boxes with Petre even before the subscription system had been set up in 1736. As soon as Collinson arrived, the two men immediately went for a walk through the nurseries, gardens and parkland to admire how Bartram's seeds had "grown to great maturity." Though still small, these American plants were beginning to shape a landscape and Collinson kept Bartram informed of the progress at Thorndon. Sometimes there was good news, such as "Lord Petre has raised abundance from thy seed," "a great Many of thy seeds are come up" or "the pine seeds & Okes came up as thick as grass," but at other times Collinson was less ebullient, writing "none is yett Come up," "I am afraid the Acorns will also Fail" or "Bad Luck attends the growing plants."

For Collinson, there was nothing more pleasurable than to see the treasures he had distributed thriving in such magnificent parkland. He wrote that he adored these plants as if they were the "Children of my own procureing and Raiseing up." The emerging American forest at Thorndon offered a particular pleasure because it was on such a grand scale. In Peckham, Collinson only had space to grow a couple of trees per species, creating an exquisite botanical collection rather than a landscape. Petre, by contrast, had ordered tens of thousands of seeds from Bartram to furnish the one thousand acres of his parkland. His requirements vastly exceeded those of other subscribers: besides the normal five-guinea boxes he wanted "a Bushel of Red Cedar Berries [*Juniperus virginiana*]," several thousand of the black walnut and liquidambar, all species of oak that Bartram could procure, and "Horse Loads" of pine cones.

Petre had also planted 900 tulip poplars, one of the most sought after species in England for its large blossom and leaves that looked as if the gardener had snipped them at the tip. Although John Tradescant the Younger had already brought the tulip poplar from Virginia 100 years previously, it had remained rare and was still one of the most expensive plants available in England. One nurseryman in London sold a twenty-five foot specimen for £21, an exorbitant amount when compared to the 15s. charged for a similarly sized American sycamore. One reason for the expense was that, since only a few tulip poplars had flowered in England, and none had produced seeds, the colonies were still the only source. By 1741, Bartram had sent thousands of tulip poplar seed cones to England, and it was at Thorndon that they could be admired in all their glory.

Petre raised the contents of Bartram's boxes in his nurseries, a series of rectangular walled plots hidden in a large grove of trees to the west of the house. They were the largest in the country. Here hotbeds, glass bells, heaps of manure and Miller's tanner's bark all aided the propagation of Bartram's plants. Countless saplings grew in orderly rows until they were mature enough to be planted out on the estate. In addition the gardeners spent their time sowing, layering, budding and grafting in order to multiply the stock—on such a scale that "20 Thousand Trees are hardly to be missed out of his Nurseries," Collinson boasted to Bartram.

With Bartram's plants, and those procured from elsewhere, there were more than 200,000 exotics at Thorndon—by far the most extensive and finest collection in the country. Even Miller, who had brought so many rare species to Chelsea, was often awed by Petre's plants. When he saw a rare palm from Java, he could not move away from it for half an hour as it was "ye finest Palm he ever saw" (Petre guarded his plants jealously, however, writing to Collinson that "I would not part with it for half ye plants at Chelsea if Mr. Miller would change with me"). Petre was also the first to procure *Camellia japonica* from Asia. When his two plants—one with white, the other with crimson double flowers—had first unfurled their origamied blossom in Thorndon, Collinson, in his excitement, had dashed off letters to his correspondents. Indeed, Thorndon became for Collinson a kind of marketing showcase. Despite its many plants from across the world, it was the huge numbers of American trees and shrubs that it was famed for—and the fruitful American harvest had soon convinced several other gardeners to place orders with Bartram.

Until Collinson had introduced Petre to Bartram's boxes, nobody had tried to create a park populated with such quantities of American exotics. Only a few years previously, Thomas Fairchild had sold his rarities individually rather than in the vast numbers supplied by Bartram and Collinson. Indeed Petre's former guardian, John Caryll, feared that the expense of procuring and raising the young trees and shrubs, together with the cost of digging canals, vales and hills, would ruin his protégé. As recently as 1740, the Thorndon gardeners had moved 10,000 American trees from the nursery into the park, where they mixed them with 20,000 European varieties and a few from Asia. Petre had even succeeded in transplanting some mature trees, anticipating a fashion for instant gardening that has been more often attributed to Lancelot "Capability" Brown. A "Herculean Undertaking," Collinson noted, as twenty horses had been required to pull a cart on which a sixty-foot specimen had been fixed upright.*

Over the previous few years there had also been much moving about of earth on the estate as canals and lakes were dug. At the northern end of the park the labourers had raised two mounts, one of which was more than ninety feet high and

---

* Petre might have been inspired by André Le Nôtre's tree-moving technique at Versailles. Lancelot Brown improved the method when he invented a contraption with a pole that he attached to the trees, levering them out of the soil. The trees were placed horizontally on the cart, instead of transporting them in an upright position.

planted with the choicest American exotics, re-creating the forest-clad mountains that Bartram described so vividly in his letters and journals. There were 230 scarlet oaks, 120 American sycamores and 69 tulip poplars intermixed with 1,100 three-year-old *Juniperus virginiana*, which Petre had raised from Bartram's berries, as well as many other evergreens. At the very top of each of the hillocks Petre placed a cedar of Lebanon which, although incongruous to the American theme, provided a suitably grand crown.

Collinson would write to Bartram after the summer visit of 1741 that all "his" plants were in "greatest pfection." This was remarkable given that Petre's ambitious project was almost halted by the cold winter of 1739/40, during which England experienced the worst frost on record. With the Thames frozen for eight weeks, London hosted for the first time in almost three decades a bustling fair on the river, and Londoners went shopping, drinking and gambling on ice. But the blizzards also brought devastation and destruction. To his great despair, Collinson saw hundreds of his choicest specimens in Peckham die. And other collectors lost many of their American rarities. Practically all callicarpas and *Magnolia grandifloras*, for example, were destroyed by the frost. In a desperate attempt to save plants, straw was wrapped around trees, baskets were placed upside down over saplings to shelter them and tubs were moved into hothouses. And because it was cold for so long, garden boys had to take night shifts to stoke the stoves, much to the annoyance of the frugal Collinson, who complained that he had "to keep Constant fires in My stove & Green house." Petre wrote to Collinson, "notwithstanding I have done every thing that can be done I am sensible I shall suffer in several places though I hope not considerably."

Yet most of Petre's tender exotics survived, largely because he had constructed such magnificent hothouses. Built with careful attention to Miller's instructions in the *Dictionary*, they were, like everything else at Thorndon, on a huge scale. "The Great Stove is the most Extraordinary Sight in the World," Collinson trilled. One hothouse was entirely given over to the production of large numbers of pineapples, which were still so rare that they were often not eaten but passed around to show off at parties. The sixty-foot-long "Tan Stove" or "Ananas Stove"—again exactly copied from Miller's pioneering instructions and detailed drawings in the *Dictionary*—was so sophisticated that not even the harsh frost interrupted the harvest: in January 1740, with temperatures well below zero degrees centigrade, Petre sent one pineapple by coach to London, so that Collinson would not miss the treat. One and a half years later, the gardens at Thorndon showed very little sign of damage.

What Petre achieved at Thorndon was truly remarkable and he would come to be hailed as the most innovative gardener of his age. Having come into his fortune young, he could have easily indulged in the life of gambling and drinking that was so typical for his peers. Instead, his obsession with gardening directed his life and he surrounded himself with men who loved horticulture—"my brother Gardeners," as he called them. At the core of this group were Collinson, Miller and the Duke of Richmond. The four men pooled their resources: Petre and Richmond provided wealth and land to experiment with; Collinson brought both knowledge and the trade connection to America; and Miller shared his hard-learned horticultural expertise.

Petre's relationship with Miller had begun when, as a young and inexperienced gardener, he had subcontracted Miller to manage his plants. And after Miller left his employ, Petre continued to send particularly difficult seeds to Chelsea "to the care of P Miller." Miller had also spent time at Thorndon, compiling a plant catalogue, and continued to organise the transport and transplantation of plants which Petre was swapping with Richmond. As Miller was the best gardener in the country, both Petre and Richmond eagerly paid him for advice, designs and seeds. At Richmond's estate, Goodwood, for example, Miller designed the flower borders (one of which was more than 400 feet long) as well as supplying cuttings, seeds and saplings.

This kind of work supplemented Miller's income from the Chelsea Physic Garden, but also appealed to his aspirations. He enjoyed the illustrious company to such an extent that he began to alienate old colleagues and friends. One complained to a friend after a visit to Thorndon where he had met Miller:

"as to Miller I deliverd yr compliments to him but I do asure you hel [he'll] not Look one [on] a poore man; nothing Lesse then a Ld or a duke is Company for him so if you take my advice never trouble ye Coxcombe with any thing more." Miller cared little about such criticism, as he now had the affection and support of aristocrats. Petre, for example, even sided with him in petty "botanic wars," by calling one of Miller's critics "an arrant ould woman."

So it was that, with the help of Miller and Collinson, Petre gradually dismantled the ideas that had underpinned the baroque garden. Petre eschewed the formal arrangements in which each specimen plant had stood alone against a backdrop of ebony earth, like the elegant marquetry work on a precious piece of furniture of the time. In the seventeenth-century garden, no branch or flower head had been allowed to grow unruly. These gardens had been like precious jewel boxes in which each gem was laid out side by side in order to be inspected and admired. At Thorndon, however, it was nature herself who created the shapes and patterns. Bartram's trees could offer columns, cones, pyramids or spheres without the need for pruning shears; branches feathered down or twisted towards the sun; some trees grew so bushy that they were green barriers, while others were as delicate as ornamental lattice-work against the sky. The bark could be smooth and almost waxen, or striped, grooved or peeling. It was like painting with trees. The different hues of leaves, Miller said, should be like "Lights and Shades in Pictures," while Collinson described the way in which Petre used a tree's foliage, its texture, bark, height and shape, as his "living pencils."

Thanks to Bartram, Petre had a wonderfully broad palette to play with. Once mature, the conical shape of Bartram's *Juniperus virginiana* or red cedar, for example, would provide vertical brushstrokes, while laurels and rhododendrons spread in looser, more horizontal lines. The smoothness of the silvery stems of the native birches was a contrast to the flaking bark of American sycamore or the scaly almost square plates of the shortleaf pine, a common pine in the northeast and southern states of America but one that Petre was the first to grow in England. Until Bartram began sending his boxes, autumn in England had been a fairly lacklustre affair. Now the falling of leaves was preceded by an extraordinary show. At Thorndon scarlet oak and white ash competed with the bloodied foliage of tupelo, and the glowing reds and oranges of the large fluttering maple leaves with the aubergine purple of liquidambar.

In front of them, providing endless interest and variety, grew shrubs that flowered in spring and produced berries in autumn. There was *Viburnum dentatum*, which Collinson had introduced only a few years previously and which was also called arrowwood because the Indians used its straight branches for their arrows; and *Rhus typhina*, one of the earliest American plants to have been brought to Britain. The latter remained colourful even after it had shed its bright red leaves in autumn, parading upright cones of clustered tiny red fruits at the tip of its naked antler-like branches (the shape of the branches gave the species its common name—staghorn sumac).

The creation of year-round beauty on such a scale was an entirely new art. Britain had, for example, only four native evergreens (Scots pine, holly, box and yew). Thorndon, however, now featured almost thirty different species of coniferous trees, of which more than ten were pines—Collinson's orders always included "all sorts of pines thee can gett." Petre admired in particular the white pine, for its unusual feathery plumes of long needles. It was also often called Lord Weymouth pine, after the man who had been the first to grow it in large numbers in England. Earlier this year Collinson had written to Bartram that "our people are insatiable after them" and Petre thought that Bartram "cannot send too much of ye Ld Weymouth Pine."

But it was not only the evergreens that provided winter interest. Against the backdrop of dark conifers or variegated hollies, the blossom and berries and even some branches of the American shrubs glowed red and orange on crisp winter mornings. Many of them had never been seen before in England, such as witch hazel, whose spidery yellow blossoms that clung to its naked branches surprised garden visitors in winter in the same way as the pale pink flowers of Bartram's *Rhododendron periclymenoides* did in spring. This American rhododendron— also called pinxterbloom azalea—is the parent of many modern hybrids. In Collinson's Peckham garden they tended to perish, but they thrived in Thorndon because, Collinson explained, "they seem to like Lord Petres Soil Better." Collinson was so enthralled by Thorndon's plantations of trees and shrubs that he wrote to Bartram, "when I walk among them, One cannot well help thinking He is in North American thickets—there are such Quantities."

In many ways, Thorndon was the first garden in which Enlightenment thinking found its visual expression. It was, as Collinson wrote, a garden based on "Hints Borrowed from Nature." For centuries, high walls had excluded untamed nature from the garden, providing protection from the landscape beyond. Now, with the Enlightenment and man's growing knowledge of the natural world, gardeners embraced the idea of letting nature reign. Petre had planted his trees and shrubs in such a way as to allow the visitor to think of them not as an object of singular delight but as an integral part of a larger landscape. He had followed Miller's advice that his plants should "appear accidental, as in a natural Wood," and grow like "Spontaneous acts of Nature." Clumps in the lawn like those at Thorndon created drifts of dense growth on the smooth grass, and unlike the complex patterns of baroque gardens, which were best enjoyed from the windows of the first or second floor, Thorndon could only be appreciated by walking along its meandering gravel paths.

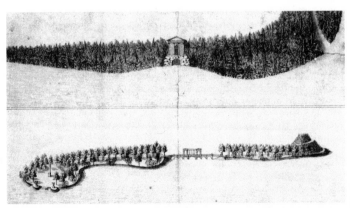

Details of Lord Petre's designs for sinuously shaped tree groves.

Petre had carefully tiered the trees and shrubs that lined these paths so that they gradually rose in height, almost like seats in a theatre, the smallest flowers at the front, the largest trees at the rear.* And whereas, in the past, gardens had been experiences for just the eyes, now they were for all senses. Some of the flowers came so close to the path that the hooped petticoats of the women visitors brushed against them, releasing the fragrant scents of the blossom. And instead of motionless rigid hedges that used to enclose parts of the garden like "Prison Walls," here at Thorndon the leaves danced to the maraca rhythm of the wind. Every corner turned revealed a new vista or arboreal scene. "The whole," Collinson wrote to Bartram in September 1741, "is planted in thickets & Clumps, and with these Mixtures are perfectly picturesque, and have a Delightfull Effect."

No other garden in England combined the new ideas of the informal landscape garden with such horticultural genius and arboreal diversity. For the first time encyclopaedic horticultural knowledge was given priority over the ability to use the protractors, dividers and levels which had long been the tools of the trade.

It was a reflection of the political ideas of the time. English thinkers praised their country as the seat of liberty and, if absolutism was concomitant with the geometrical patterns and clipped shapes of Louis XIV's formal gardens at Versailles, then English gardeners should invite nature into the garden as if liberty was to be allowed back in.†

At Thorndon each tree or shrub in a clump was like an

---

* Some of Thorndon's planting schemes are deduced from Petre's instructions for the Duke of Norfolk's estate at Worksop, where he designed sinuous clumps of trees that punctuated the lawn like islands. Twenty to 50 feet in diameter and up to 150 feet apart, they were planted in irregular density to make them appear more natural. They predated the style of the landscape park for which Capability Brown would become famous.

† This rejection of the formal French baroque designs was also inspired by writers such as Joseph Addison and the poet Alexander Pope. Already in 1712 Addison had condemned the "Marks of scissors upon every Plant and Bush," and a year later, in 1713, Pope had ridiculed the "Monstrous attempts" to cut trees and bushes into the "most awkward Figures of Men and Animals."

instrument in an orchestra which would play a solo at some seasons and disappear into the background at others—with Petre as their conductor. Petre was able to provide these time-based planting schemes because he had a knowledge of horticulture that was as good as any nurseryman's. As his friends said, Petre "understood the colours of every tree, and always considered how he placed them accordingly." He would not think of allocating them a particular location in the land-scape until he had studied the habitat, shape, growth and seasons of each plant species as well as numbering, labelling it and writing it up in his inventory. And unlike previous wealthy garden-owners, Petre was a hands-on gardener himself who knew every spot on his land and liked to walk or ride across his park, checking the progress and giving instructions. When he saw ripe seeds, cones or berries of some of the English native species in his woods, he stopped and collected them for Bartram's garden in Pennsylvania—"with his own hands," as Collinson reported. The foundation of his knowledge came from Miller's *Dictionary*, which had pride of place in his library—he had two copies, one for himself and one for his head gardener.

It was a time when this passion for practical horticulture was spreading among the aristocracy. The Duke of Richmond, for example, always turned melancholic in January and February because the "cruel weather" prevented him from doing any planting; instead he entertained himself by sending precise instructions to his steward at Goodwood about how to apply manure and straw as frost protection. Even Frederick, Prince of Wales, was known for making his family, guests and courtiers work in the shrubberies. So obsessed were both men that they eventually died from the consequences of working outside during bad weather: the prince when, in October 1750, he was surprised by a hail-storm while directing the planting of trees and caught a severe chill from which he never recovered, and Richmond in the same winter when he inspected progress at Goodwood during a particularly cold day.

As word spread about Petre's success and as plants became more available through Bartram's schemes, more and more people developed a passion for gardening. Miller's *Dictionary* communicated the new ideas to a wider audience and included long lists of the various plant types that could be used to create tiered shrubberies in a "natural" garden: "Hardy-trees and Shrubs . . . ranged according to their several Growths," "Evergreen Trees and Shrubs, with variegated Leaves" or "flowering Plants as will thrive under large trees." In subsequent editions, Miller added yet more detail to prevent novice gardeners making mistakes, explaining more fully "some of those Rules, which have been by several Persons misapplied."

As the new garden fashion spread there was an increasing demand for professional landscape designers. The most sought-after was William Kent, a self-professed italophile who called himself "Guglielmo." In the mid 1730s he had designed Prince Frederick's garden at Carlton House, where American trees and shrubs loosened the straitjacket of formality. Unlike Petre, though, the prince had bought all his plants in commercial nurseries—some 14,000 for the nine acres. This was far fewer than at Thorndon, because nurserymen could not provide the same quantities as Bartram, though some sold large American trees—occasionally more than twenty feet high—to create the instant impression of a mature garden.

Kent was later hailed as the man who "leaped the fence, and saw that all nature was a garden," and it was through him that the phrase "Nature abhors a straight line" came into being. But although Kent did much to disseminate such ideas through the many gardens he created, it was really Petre and his network who developed and innovated planting, fostering a passion for horticulture among the English aristocracy. Petre "carried it farther," a fellow garden-maker remarked, because Kent did little more than "mixing lighter and darker greens in a pleasing manner."

By 1741, when Collinson visited Petre at Thorndon, horti-culture had become so fashionable that the Duchess of Queensberry brought her garden with her to court. Dressed in a white satin gown that was embroidered all over with trees, hills and flowers, the Duchess was said to have looked like a walking landscape garden. The next day letters criss-crossed the country describing the most unusual dress of the season in every detail: it had "brown hills covered with all sorts of weeds," "an old stump of a tree" and "all sorts of twining flowers." And just as the wind swayed the branches of Petre's American trees, so the Duchess of Queensberry's every step swung the hoops of her petticoat, bringing alive the nasturtiums, honeysuckle and ivy on her skirt.

"Send no Seeds for him ... all is att an End"

The first Drudgery of Settling new Colonies, which confines the Attention of People to mere Necessaries, is now pretty well over; and there are many in every Province in Circumstances that set them at Ease, and afford Leisure to cultivate the finer Arts, and improve the common Stock of Knowledge.

BENJAMIN FRANKLIN, 1743

As the American saplings were maturing in Thorndon's landscape garden, Benjamin Franklin reminded his friend John Bartram when they met at the Philadelphia Library Company that his botanical knowledge could also be of use to the colonies themselves and encouraged him to find American subscribers for the expeditions. The purpose would not so much be to supply plants and seeds for gardens but to publish a book on America's flora. Who knew, said Franklin, what species lay undiscovered with medicinal and culinary properties that might be of economic and agricultural value to the colonies. Bartram immediately liked the idea. It appealed not only to his patriotic instincts, but also to his pocket: he was in need of extra income as Collinson was having difficulty extracting Bartram's payments from the English subscribers. "[I]t is very hard getting Money of great people," he wrote, "They are glad of the Cargo, but are apt to forget all the Rest."

Franklin knew many of the English subscribers personally. He had spent time in Britain during the 1720s and was well connected to the vibrant world of botany and science in Europe. He was a printer by trade but also a man of the Enlightenment, known for his writing and insatiable curiosity for the natural world. He was, and would continue to be, Bartram's greatest encourager in the colonies. Even when he later became the agent of the Pennsylvania Assembly in London and was involved in difficult political negotiations, Franklin always

supported Bartram's endeavour, believing that the indige-
nous flora would bring prosperity to America—a notion that
would become even more important after the colonies became
independent.

But already in 1742, as the proprietor of the *Pennsylvania
Gazette* and a co-founder of the Library Company, the thirty-
six-year-old Franklin was the ideal promoter of an
American-funded plant expedition. Their target was to raise
an annual £50, which would allow Bartram to "spend most
of my times for several years in searching & observing natural
productions." By early spring 1742, when Franklin had printed
the subscription proposal in the *Pennsylvania Gazette,* £20 had
already been raised.

But just as Bartram was getting excited about turning his
passion into a full-time profession, he received devastating news
from England. "[T]he man I Loved & was dearer to Mee than
all men—is no More," Collinson wrote to Bartram on 3 July
1742. On the previous day, Lord Petre had died suddenly of
smallpox, aged thirty. Only a few days before his death, Petre
had written a letter in which he excused his failure to meet
Collinson in London because of "a great cold which . . . has
put my head out of order." What Petre didn't know was that
this was the first symptom of the disease which would kill him
within ten days. "All our Schemes are broke," wrote Collinson
to Bartram. "Send no Seeds for him . . . all is att an End."

The loss of a man like Petre was a severe blow to Bartram.
Not only had he purchased so many of the seed boxes, he had
encouraged others to do so. Without him, Collinson feared,
the whole enterprise might be "Lost in Embrio." The garden
at Thorndon was not finished, the trees immature, and Bartram
was left with only the Duke of Richmond, Philip Miller and
the Duke of Norfolk as regular clients. Collinson immediately
proposed to help Bartram by offering to try to sell any seeds
Bartram had collected already, and writing to friends and
acquaintances to ask them to become subscribers. Catesby's
illustrations in his *Natural History* might function as something akin to an order catalogue, he suggested. But five months
later, in December, Collinson still had not managed to find
more English clients, nor was Franklin's scheme for American
subscribers further advanced.

But it wasn't just Bartram that Collinson worried about. He was intent on rescuing the trees and shrubs he had helped to assemble at Thorndon. If Petre had lived, Collinson lamented, "all round Him would have been America in England." As it was, Petre's son and heir was only nine months old at his father's death, and Lady Petre felt unable to look after the gardens. It was decided that most of the American species and exotics should be sold. Collinson was dispirited, "Oh what will become of his Collections?," he wrote to a fellow gardener, "I am affraid all Stove Plants will go down," as no other private garden had such advanced hothouses. Collinson feared that all his efforts had been in vain, for "Young Trees are like young Children, they require frequent Looking after."

He needn't have been troubled—at least about the hardy American species—as three garden owners were readying themselves to battle for Petre's collection. The Dukes of Bedford, Norfolk and Richmond were all laying out their estates according to the new fashion for landscape gardens and were keen to take on Petre's plants. It was simply a question of which ones to take and how much to pay.

Since Philip Miller acted as consultant to the Duke of Bedford about his estate at Woburn in Bedfordshire, it was inevitable that Bedford should turn to him for advice about the trees and shrubs at Thorndon.* Five months after Petre's death, Richmond's uncle, another keen gardener, reported meeting Miller at Woburn and wrote to Collinson that they "had a great deal of discourse concerning ye nurseries of Thorndon." Collinson, too, was drawn in: the Duke of Bedford insisted that he join him on a day trip to Thorndon to help him to choose which plants to purchase.

The Duke of Norfolk was similarly eager. Because Petre himself had designed Norfolk's garden at Worksop in Nottinghamshire, the Thorndon trees would exactly match the planting scheme. Norfolk had already subscribed to the five-guinea boxes in the previous year, but the advantage of the Thorndon

---

* Miller had moonlighted since March 1741 for the Duke of Bedford, subsidising his income with an annual twenty guineas for going to Woburn at least twice a year.

collection was that many of the American plants were already seven or eight years old and had acclimatised to England.

Unlike the newcomer Norfolk, the Duke of Richmond was, along with Miller, Bartram's most longstanding customer. Having increased his Goodwood estate from 200 to 11,000 acres, Richmond was in urgent need of trees and was therefore very excited by Lady Petre's offer of "some of the Curious plants." Knowing that Collinson had an encyclopaedic knowledge of every specimen, and was in close contact with Lady Petre, Richmond bombarded him with a series of letters, each filled with enquiries, orders and invitations to meet. Less than three weeks after Petre's death, Richmond asked Collinson to stay with him at Goodwood, enticing him with the offer of a free ride in "a vacant corner" of the Royal Society president's coach. A few weeks later, he summoned Collinson to his London residence, in the morning "between the hours of nine & ten, butt if you come sooner I shall be in bed, if you come later, I shall have business."

The jockeying for position among the aristocrats reached its height when five months later, in December 1742, Collinson announced that he was planning to visit Thorndon. He received an avalanche of letters. Richmond asked for 100 five-foot *Thuja occidentalis*—the coniferous tree which Bartram had sent regularly—to match the ones he already had. Then, two weeks later, Richmond asked if Collinson could procure forty to fifty tulip poplars, writing that he would pay any price for them because "they are not to be gott any where else." He also requested other evergreens for his grove and 200 cedars of Lebanon for a bleak hill north of his park which he wanted to transform into "a mount Lebanon." When Collinson replied a few weeks later that Lady Petre was willing to send all the trees he had ordered, Richmond became even more greedy. "I fear I am still craving more," he wrote, and asked for eighty tulip poplars instead of forty. His greatest fear was that "the Dukes of Bedford & Norfolk will sweep them [the trees] all away."

Perhaps it was the competition over the Thorndon trees (or the fact that, when the three men eventually shared the plants among them, they did not have enough trees to furnish their parkland) that enabled Collinson to convince Richmond,

Norfolk and "an other Gentleman" to order some more American species from Bartram. Whatever the reason, a year after Petre's death, and a few months after the Thorndon auction, Bartram received a one-off order of thirty-five guineas from them with the news that Collinson had also been able to sell the box of *Juniperus virginiana* berries which had been intended for Petre. To Bartram's great relief, the new order meant that he could continue his business. In his excitement and gratitude he dashed off a letter to Collinson, thanking him profusely and writing, "[i]f ever I come to pay you A visit I would bring abundance of trees & shrubs with me"— all those difficult ones which perished unless someone "that understands them ... takes particular care of them in their pasage." And though Bartram knew that this would probably never happen, when he left his farm two weeks later, in July 1743, he was happy and optimistic about life and the future.

His summer expedition this year was not only a plant-collecting one. It is indicative of Bartram's growing status among the colonists that he joined Conrad Weiser and Lewis Evans on a diplomatic mission to talk with Native American leaders of the Iroquois Confederacy and, according to Bartram, "to introduce A peaceable understanding between ye Virginians & ye five nations" (the "five nations" being in fact by then six tribes that had formed the Iroquois Confederacy). Relations between the colonists and the Native Americans had turned sour the previous summer when white settlers attacked an Iroquois hunting party.

Bartram went as the naturalist; Weiser, who had lived as a boy among the Native Americans, as the interpreter and negotiator; and Evans as the cartographer. The three men travelled north to meet the council of the Six Nations at Onondaga (near Syracuse). The talks laid the foundation for a treaty that was signed the following year in Lancaster, Pennsylvania. This turned out to be one of the most influential agreements of colonial America, as the colonists exploited it to claim the Ohio Valley, but it also paved the way for an alliance between the British and the Six Nations against the French. Bartram and Evans used the two weeks of diplomatic talks to go on day excursions and on a trip to Fort Oswego at Lake Ontario. Evans used his observations for his famous series of maps of

British America* and Bartram—though not returning with many specimens—wrote a journal that was eventually published in 1751 describing the agricultural potential of the territory.

But the following year, Bartram's business was threatened once again. As Collinson needed to attend to his shipping office, he could not spend all his time advertising Bartram's boxes, and so only Richmond and Miller had subscribed for that year. With so much effort going into finding seeds and the unpredictability of the English customers, Bartram sometimes felt like giving up altogether. Collinson, however, refused to let him: he adored the new English landscape peppered with American trees and shrubs, and assured Bartram that he was "looking for New Subscribers."

Collinson was certain that, if other gardeners saw the American plants, they would be eager to grow them, but they couldn't because they had not heard of Bartram's service. The solution, Collinson believed, was to sell directly to nurserymen who would be able to raise the seeds and sell them on to other gardeners. He was confident that he could sell probably three or four of the five-guinea boxes, but only if Bartram took the time and trouble to enclose in each box an exact list of names and numbers of all species included, as well as paying for the freight himself.

Collinson also reminded various of his international correspondents, such as the Swedish botanist Carl Linnaeus and the Dutch Johan Frederik Gronovius, "not to forget the pains & Travel of Indefatigable John Bartram," hoping that they would reimburse the farmer—at least partly—for the many seeds and specimens he had sent. It was hard work for little reward, and subscribers remained scant.

Instead of planning great expeditions to supply the gardens abroad, Bartram now concentrated on the advancement of science. He was particularly concerned to demonstrate that science in the colonies could be as accomplished as any that

---

* Evans published in 1749 *A Map of Pennsylvania, New Jersey, New York and the Three Delaware Counties* (revised in 1752) and in 1755 *A General Map of the Middle British Colonies in America*. His maps were used as guidelines for boundary disputes and were of great military importance in the Seven Years War.

took place in Europe. Collinson had already been "Surprised att the progress of Botany" in America, and Bartram for his part was becoming increasingly confident as a botanist. Connected to the world through a group of international correspondents such as Sloane and Gronovius, Bartram now felt so self-assured that he also enjoyed his "secret pleasure of modestly informing them of some of their mistakes." But Bartram was at his most sanguine when he suggested that the equivalent to the Royal Society be established in America.

Collinson had initially dismissed the idea on the grounds of the "Infancy of your colony." But Bartram, together with Benjamin Franklin, persevered, because they believed it to be important to promote "Useful Knowledge." Consequently they founded the American Philosophical Society, which by March 1744 had held three meetings attended by several "Curious persons." Such a society was only possible, Franklin insisted, because the "first Drudgery of Settling new Colonies . . . is now pretty well over" and it was time to bring together all like-minded men. Some plantation owners and tradesmen were now sufficiently wealthy to dedicate time to cultivating their interests in the arts and sciences. Moreover, improved roads and postal services made scientific collaborations and communication a feasible reality for the first time. Franklin even applied to become the Postmaster General, a position that allowed him to grant free postage to the members of the society.*

Almost immediately, though, Bartram and Franklin became frustrated by the "poor progress" of their society. Bartram believed that the problem lay in the make-up of colonial society. In true Linnaean spirit he classified the colonists into three categories to explain the reasons: the ones who were only interested in laying out their large estates; the ones who indulged in luxury (often the children of plantation owners) and, finally, those who, although often "ye most curious" about science, couldn't afford to pursue their interests because they had to work hard to support their families. Similarly, Franklin also complained about the "very idle Gentlemen" who failed to send

---

* Postage was expensive and calculated by sheet and distance; one sheet from New York to Boston was charged at one shilling, while from Charleston to Boston it was two shillings a sheet.

letters to the "headquarters" in Philadelphia. The timing of the foundation had been unfortunate, Bartram feared. Britain had become drawn into the War of the Austrian Succession in 1740—allied to Austria against the Bourbon coalition because George II feared for his beloved Hanover. Because France was involved, it was inevitable that the war would impact on the French and British colonies in America.* "Most our thoughts," wrote Bartram, are occupied by "ye tumultuous reports of wars Invasions & reprisals."

Thus, Bartram's two schemes—the seed boxes and the scientific society—were limping along without great progress. But it got even worse when the French began to capture British boats. Bartram worried for his cargo and how to feed his growing family of nine children. The ship on which he had sent the handwritten journal of his expedition to Onondaga had been seized and he feared the same would happen again. As a precaution, Collinson suggested that the boxes be addressed to the brothers de Jussieu, keepers of the Jardin des Plantes in Paris, in the hope that the French would refrain from throwing overboard trunks destined for their compatriots. "War keeps the Muses silent," Bernard de Jussieu bemoaned, and he offered to send any captured boxes to England.

The plan worked—at least sometimes. When French privateers captured a ship with Bartram's cargo and saw the label "de Jussieu" attached to the boxes, they forwarded the boxes to the botanic garden in Paris, from where they eventually reached England. But more often than not boxes were captured and lost. Collinson now advised splitting the cargo between two ships (which was difficult as so few vessels left for Britain) and to insure it, which though expensive, was better than losing everything, he explained, as "half a Loaf is better than no Bread."

The danger of loss was particularly irritating because, finally, Collinson was managing to drum up subscribers. Less than five years after Petre's death Collinson found customers for

---

* In Britain this war was deeply resented as politicians and citizens felt that the government squandered public funds, and that British soldiers lost their lives for a war that was not theirs.

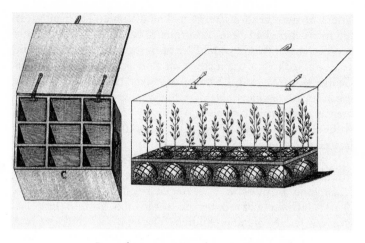

Boxes for transporting plants overseas.

eleven boxes—the largest order since Bartram had begun his enterprise. In addition, the harvest of forest seeds in 1747 was so exceptional that it would have been easy for Bartram to fulfil the order, had there not been a war. As the conflict shifted from the European theatres to the colonies, British colonists now feared that a French invasion was inevitable, in reciprocation for the British capture of the island fortress at Louisbourg off the coast of Nova Scotia. Bartram wrote to Collinson, "our people is daily exercising & learning ye Martial discipline," but to his annoyance most of his fellow Quakers remained pacifists—"fools" as he called them, refusing to protect country, liberties and families with the sword.

While all this was going on, Collinson kept in close touch with Lord Petre's widow, and even took her daughter to a London doctor to get a tooth pulled. At the same time Miller had tried to look after those plants that remained in the Thorndon hothouses by supplying tanner's bark. But, Lady Petre wrote to Collinson, "many plants droop there [*sic*] heads, from your long absence." She felt it impossible to stop the decline of the collection and as a consequence she decided— or maybe Collinson persuaded her—to sell the remaining healthy plants from the Thorndon nurseries. As many of the hardy trees and shrubs were now well matured, they would perfectly augment the park of one of the larger estates.

In October 1747, almost exactly five years after the first auction, Collinson asked the Duke of Richmond if he would be interested in buying the remaining Thorndon plants for Goodwood, emphasising that "the like opportunity will not perhaps be again—and these you know are not to be mett with in any Nursery." Over the past years Collinson had become Richmond's chief supplier of plants and seeds, handling plant lists, orders and carriage instructions—for example, he sent "2 Basketts & 5 Bundles of Trees as by Inclosed Acct.," wishing Richmond a "pleasant Time of planting."

When Richmond received the sales list he complained about the prices, but the prospect of a whole collection of maturing trees and shrubs was too tempting for him to be parsimonious.* At £2 2s. a piece, *Magnolia virginiana* was one of the most expensive trees in England, Richmond wrote to Collinson, but "I must have them, tho I beleive [*sic*] nobody else will be fool enough to buye any at that price." Thousands of Bartram's trees went to Goodwood: *Magnolia grandiflora* (only introduced to England in 1734); white cedar, the conifer that Bartram had found in the swamps of New Jersey and which most English gardeners had failed to raise; and shrubs such as flowering dogwood and witch hazel. By spring 1748 the plants had been delivered and Richmond wrote to Collinson, "I hope you don't forgett that you are to go down with me to see them planted."

Slowly the landscape gardens of England were becoming home to Bartram's maturing trees. Similarly, Collinson's garden at Peckham was overflowing with foreign flowers and shrubs. Richmond might have had great numbers of plants, but Collinson had a greater variety. For more than a decade now, Bartram had used every opportunity to repay him for his help. "Curious species for thee [are] packed about here & there to fill up vacancies," Bartram wrote. And Collinson bragged to one of his correspondents in America: "If I may boast my garden

---

* Lady Petre and her gardener had compiled a price list for Collinson's information. When some nurserymen got hold of it Lady Petre was upset because the prices had not been intended for the commercial trade.

can show more of your Vegetables then perhaps any in this Island." It was a sentiment confirmed by one of Linnaeus's pupils who visited Collinson in the summer 1748. There was "scarcely a garden in England in which there were so many kinds of trees and plants," he observed.

The small garden in Peckham consisted of a square plot, bisected by a path. At one end was a greenhouse in which Collinson kept tender exotics. The layout was arranged so that most of the flowerbeds caught the morning sun, which was, according to Collinson, important especially for the American plants. The borders were edged with thousands of bones from horse and ox legs—sunk vertically into the soil so that only the soft curves of the white knuckles were visible. Though it seemed an eccentric method to his contemporaries, the lines of bones prevented the earth from slipping on to the paths while, at the same time, releasing phosphorus and calcium—an effect similar to many of today's organic fertilisers that contain animal bones.

At one end of the garden Collinson had clipped the branches of an elm to form the roof of a little pavilion, and opposite it, at the other end of the garden, he had placed a bench and trained the branches of a horse chestnut into a leafy canopy. It was here in the shade—as he hated the heat—that Collinson sat admiring his American plants at the end of the day, thinking, "I am no Stranger to America."

The rigorous order of his garden was also something that Collinson cherished. Unlike the cultivated wilderness at Thorndon, in Peckham plants were grown in long flowerbeds between three and four feet wide and all were labelled with numbers painted on pales or stakes placed in the soil, or drawn directly on the wall in the case of climbers and espalier fruit. These numbers were cross-referenced to a catalogue of names.

Many of our most popular garden flowers grew in Peckham for the first time. The clump-forming perennial *Monarda didyma*—the parent of most monardas today—began its English life in Collinson's garden. One of the common names is Oswego tea because Bartram had found this bright scarlet flower near Oswego on his expedition to the Iroquois Confederacy in 1743. There was also the graceful *Dodecatheon meadia*—the shooting star—with its dangly dart-shaped magenta-pink petals. For the two years before it flowered Collinson had thought it a vegetable, because the "Leaves are so much Like the first Leaves of Coss lettice."* Bartram had collected its seeds in the Appalachian mountains of Virginia and they flowered in Collinson's garden for the first time in England in the summer of 1744.

Another of today's favourite border plants, the pink, purple and light-blue species of phlox with their sphere-like clusters of blossom, brightened up the Peckham beds in late summer. Collinson later wrote that all species of phlox in England "came from my Garden where they first Flower'd."† Several sorts of asters, lilies, rudbeckias and sunflowers added yet more colour, along with *Yucca filamentosa* with its sturdy lance-like leaves and large spikes of white flowers. John Tradescant the Younger had originally introduced the yucca from Virginia a century previously, but Collinson claimed that his was the first to flower in England. Various non-American species also made their first show in Peckham. There was stately *Delphinium grandiflorum* from Siberia, and Collinson insisted, "I received the first double Spanish broom that was in England."

"Every particle of Earth goes through my Fingers," Collinson wrote to Bartram, and he became so passionately attached to these plants that when moles dug through his garden he declared war. Each morning and evening he waited for them to stir, and as the earth moved the short, slight Collinson, who had hands as small as a woman's, rammed in a spade in order to drive the mole above ground. Once out, Collinson boasted, they were "Easily kill'd. Many a One has Mett with its fate this way." He was also infuriated when other little intruders munched through his treasures, writing to Bartram that "most unfortunately a Black Snail Eat of the flower bud before it was discover'd." Similarly, he smoked out beetles or squirted them with a "hand Engine" filled with tobacco water (water in which tobacco leaves had been soaked), and he may even have had seagulls with clipped wings to eat the worms and snails, like some of his

---

* Horace Walpole had a similar problem with his exotic flowers, writing to Henry Seymour Conway on 29 August 1748, "[I] talk very learnedly with the nurserymen, except that now and then a lettuce run to seed overturns all my botany, as I have more than once taken it for a curios West Indian flowering shrub."

† Collinson was certainly the first to grow *Phlox divaricata* and *P. maculata*, but Miller had *P. glaberrima* already in 1725 at Chelsea.

acquaintances. Gardening was, Collinson explained to a friend, "how I Employe all my Leisure Hours, I may say Minutes, from Business."

When friends and visitors admired his garden, Collinson praised Bartram's boxes. When young aristocrats showed an interest in gardening, he applauded them and recommended Bartram's services. When nurserymen needed American species he offered to get them from Bartram. And so, in between the time he spent on his business, gardening and country visits, the tireless Collinson worked diligently on Bartram's behalf.

In spring 1749 he had recruited so many new customers that he sent an order of thirteen boxes. And to Bartram's relief, it was possible, for the first time in years, to fill the boxes and put them on a ship without worrying about losing his money, because the war had come to an end with the signing of the Treaty of Aix-la-Chapelle in the previous autumn. By spring 1750 Bartram's new customers included aristocrats such as Lord Lincoln, who—encouraged by Collinson—was "quite mad after planting," and Sir Hugh Smithson (later the first Duke of Northumberland). Both had shared with Richmond the remaining plants of Thorndon's second plant auction, in 1747. Collinson had even kept his promise and procured four London nurserymen as new subscribers: John Williamson, who would become Bartram's best client; Nathaniel Powell, who ran his business from Fetter Lane; Lord Petre's former head gardener, James Gordon, who had a nursery in Mile End; and Christopher Gray, a friend of Miller and Collinson, who was based on the King's Road at Chelsea. Collinson wrote in delight that "the Laudable Spirit of planting prevails Here."

The most prestigious order came from the Prince of Wales for his gardens at Kew and Carlton House. Keen to raise some of Bartram's seeds, acorns and cones himself, the prince requested a five-guinea box. His adviser had, as a fellow gardener heard, "seatled a correspondance in Asia, Africa, America, Europe" and was planning a 300-foot-long hothouse. But before the prince could fulfil these dreams, he died in early 1751. Collinson lamented the loss of "our Late Excellent Prince," but whereas after Petre's death, a decade earlier, Collinson had feared for the continuance of the infant landscape gardening, now he was optimistic.

And though the death of Prince Frederick had "cast a great damp over all the Nation," Collinson was sure that "the good thing [gardening] will not Die with Him for there is Such a Spirit & Love of It among the Nobility & Gentry." The love for gardening would be "a lasting Delight," Collinson trumpeted; nothing could stop the growing obsession of England's gardeners.

7

"Commonwealth of Botany"

Every Wednesday and Saturday during the summer months, the inhabitants of the small Swedish town of Uppsala, some forty miles north of Stockholm, were subject to a minor invasion. First they would hear the rhythmic sound of French horns and kettle drums; then they would glimpse the flash of colourful banners through the trees as the noise grew louder. Past the dilapidated medieval town walls and undulating burial mounds of their Viking ancestors, 300 men—and a few women—came marching, led by a small, lean man, his face lit up in a bright smile, his dark brown eyes sparkling. When they eventually reached the town, they swarmed into nearby gardens and on to the turfed roofs of Uppsala's wooden houses. Their uniform, made of light linen, was a short coat, loose breeches and a wide-brimmed hat. Their weapons were insect nets, pins, magnifying glasses and knifes; and their trophies were butterflies and flower garlands which they attached to their clothes and hats. They were Carl Linnaeus's "army of botanists."

General Linnaeus led his troop with military discipline. They would set off at 7 a.m. exactly; lunch was at two o'clock, followed by a rest at four—any latecomers were punished. Experienced participants were honoured with official ranks such as "Annotator" (responsible for recording the day's observations), "Prefect" (in charge of discipline) or "Sharp-Shooter" (the official bird-hunter). There were also captains and lieutenants, while the rest were divided into groups and ordered to scour the meadows, forests and hedgerows for botanical treasures, bringing them back to their master for judgement and classification. At the end of the day, they would return to their headquarters, the botanic garden. "Vivat Linnaeus!" they shouted as their leader wandered alone along the gravel path through the flowerbeds towards the house where he lived with his wife and two children.

The opprobrium that Linnaeus had faced on his return to Sweden in 1738 had turned out to be short-lived. By 1741 he was Professor of Botany at Uppsala University, and by the late 1740s he had become the most famous botanist in Sweden. Students came from across Europe to learn the Linnaean method of classification from the master himself, because his lectures and garden brought the revolutionary sexual system expounded in his *Systema Naturae* alive. To peruse his rectangular flowerbeds was like reading a living book. Starting at the orangery, where tender plants from the first class, with one stamen and one pistil, such as *Canna indica*, were kept, they could see each class laid out in meticulous order, arriving finally at the last bed outside, which contained non-flowering horsetails and ferns in the twenty-fourth class. In one corner of the garden was Linnaeus's house and sometimes visitors saw the face of the master himself through the windows on the first floor. Here, from his study, Linnaeus could view at a glance the order he had brought to the natural world.

Since 1741, he had studiously transformed Uppsala's ramshackle botanic garden into one of the finest collections in Europe. Starting with a mere 300 species, he had increased it to 3,000, with much begging for seeds from his acquaintances in Britain, Holland and Germany. "I brought the natural sciences to their highest peak," wrote Linnaeus, with his usual lack of modesty. There was a new orangery and hothouse, both filled with exotics from across the globe; and a large collection of American plants supplied by John Bartram, at first via Peter Collinson, but later by direct correspondence. Unlike Collinson, who particularly adored "all fine, showey specious plants,"

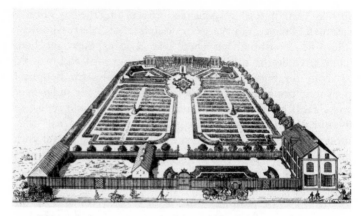

Linnaeus's Botanic Garden at Uppsala. His house is to the right
and the orangery with the hothouses is at the far end.
The rectangular beds to the left and right are for perennials
and annuals planted according to the sexual system.

Linnaeus wanted every species available. No matter how small,
ugly or insignificant it might appear to other gardeners, Linnaeus
needed it in pursuit of knowledge and completion.

Almost everybody he had met during his travels in Europe
had been happy to part with some seeds and plants, and though
many botanists still had not accepted Linnaeus's system of classification by stamens and pistils, they supported his endeavour
to record and classify the natural world in its entirety. To look
after his collection, Linnaeus had headhunted the gardener of
his former Dutch patron, George Clifford—an ungrateful repay-
ment to Clifford for financing Linnaeus's studies in Europe and
supplying Linnaeus with duplicate specimens from his own
herbarium. Unsurprisingly Clifford had been upset, writing to
his former protégé, "I did not deserve such treatment."

This was not the only complaint to arrive in Uppsala, as
Linnaeus often refused to reciprocate his fellow botanists' gifts.
The professor of botany in Oxford, Johann Jacob Dillenius,
lamented in October 1746, "What have I received in return
for all the seeds sent you, but those of one plant." And even
the ever-patient Collinson wrote in spring 1748, "The seeds &
specimens I have sent you from year to year, but not the least
Returns," adding that as a friend he felt obliged to inform him

that it was "a General Complaint that Dr. Linnaeus Receives all & Returns nothing." Linnaeus quickly gained the reputation of being a botanical miser who safeguarded his own seed cupboard with cold determination: nobody had a key to it but he. His pupils—answering requests from other botanic gardens—knew that the cupboard was opened only at sowing time, but, even then, one wrote, it was not certain that Linnaeus would part with any.

Although Linnaeus often failed to reciprocate, he was part of a network that he called the Commonwealth of Botany. They were connected by ships that crossed the oceans laden with bags full of letters, with boxes packed with hundreds of books, with lenses for microscopes, with descriptions of experiments, diagrams and grand theories. From London to Leiden, from Birmingham to Vienna, from Oxford to St. Petersburg and from Uppsala to Philadelphia, these ships shuttled ideas between centres of learning and the new commercial hubs of the Western world. Bartram wrote to Dillenius, Collinson to Franklin, de Jussieu to Linnaeus, Miller to Bartram, Gronovius to Collinson, and so on. Several European scientific societies exchanged their publications with one another, and most learned men prided themselves on their international correspondence. Bartram's observations about America's natural world, for example, were presented by Collinson at the Royal Society in London, as were Franklin's experiments with electricity.*

For Linnaeus, this network was a crucial part of his greatest project, the *Species Plantarum:* a survey of all plants known to man. He asked every European botanist he knew "to send me plants complete with flowers . . . of the comparatively rare plants that I have not described." In addition, he dispatched his pupils—or "apostles" as he called them—to America, Egypt, Spain and China on plant-collecting missions. Their returns were the most exhilarating moments in Linnaeus's life. When

---

* Collinson triggered Franklin's interest in electricity when he sent equipment for the experiments in 1746. Later he read Franklin's letters to the fellows at the Royal Society and published *Experiments and Observations on Electricity, Made at Philadelphia, by Mr. Benjamin Franklin, and Communicated in several Letters to Mr. P. Collinson, of London, F.R.S*, cementing Franklin's reputation as a scientist.

Pehr Kalm, for example, arrived in Uppsala after two and a half years in Canada, New York and Pennsylvania, where he had met Bartram and Franklin, Linnaeus was so impatient to see the spoils of the expedition that he wrote to a friend, "I long for him like a bride for one o'clock at night."

When the *Species Plantarum* was published in 1753 it became the most important botanical publication of the century, more so even than Linnaeus's *Systema Naturae*. This was not only because it provided such a comprehensive list of plants, but because it offered a solution to that fraught subject—the naming of plants. Until its publication, the naming of new species had provoked constant debate. A "new" American plant, for instance, could receive different names in Paris, London and Leiden; and the botanists who compiled catalogues struggled to work out, from their different names, which plants were actually the same. Several attempts had been made to make names more "rational": already in the previous century, for example, botanists had replaced vernacular names such as "welcome-home-husband-though-never-so-drunk," "priest ballocks," or "mares fart" with a Latin nomenclature, while Miller's and Fairchild's short-lived Society of Gardeners had attempted to introduce standardised names thirty years previously. More recently, Thomas Fairchild's neighbour, the nurseryman John Cowell, had suggested a rather whimsical method in which the plant name revealed the flower's colour by sharing the same initial letter: for example a white flower marked with crimson streaks could be called William the Conqueror—the "W" in William standing for white and the "C" for crimson. But none of these proposed systems managed to win international consent.

The problem was particularly evident to readers of Miller's *Dictionary*, which printed all the names that various authors had given each plant. For *Magnolia grandiflora*, for example, Miller listed his own preference, *Magnolia foliis lanceolatis persistentibus, caule erecto arboreo,* the name *Magnolia altissima, flore ingenti candido,* used in Mark Catesby's *Natural History*, as well as common names such as "Greater Magnolia" and "Larger Laurel leave'd Tulip Tree."

According to the old systems, such as John Ray's, each name included lengthy descriptions of the species' habitat, leaf shapes and calyx. Today's *Kalmia angustifolia*, for example, was called *Chamaedaphne sempervirens, foliis oblongis angustis, foliorum fasciculis opposites*—meaning "evergreen dwarf laurel, with oblong narrow leaves growing in bunches, which are placed opposite." The trouble was that when a new species was similar to an old one, additional traits had to be added to the old name to make it distinct, making the names even longer. And as, at a time of expanding empires, the number of discovered plants rapidly increased, some names ran to half a page.

Linnaeus "abhorred" names that were "1 foot long" because with each newly discovered plant, communication between botanists became more difficult. He proposed to impose order on the natural world by bringing together "[t]hese widely scattered NAMES ... reducing all to one system." The rationale for this was not only practical but also philosophical, because he knew that without permanent names there was no permanent knowledge.

Linnaeus's solution was easy and straightforward: he gave every plant a two-word name, like a first name and surname. The "surname" was the genus, such as *Magnolia* or *Collinsonia*, which often commemorated a friend or the genus's discoverer. To this he added a second word (like a Christian name) such as *grandiflora* or *canadensis* to signify individual species. Thus the white pine, which had previously gone under the unwieldy name *Pinus Americana quinis ex uno folliculo setis longis tenuibus triquetris ad unum angulum per totam longitudinem minutissimis crenis asperatis* was reduced to *Pinus strobus*. Linnaeus also applied his system to animals, and was the first to name humans *Homo sapiens* and to classify them as primates.

In his introduction of this so-called binomial nomenclature, Linnaeus was not so much an innovator as a consolidator: his brilliance was to bring together in a single system disparate ideas that had been circulating for centuries. In the sixteenth century botanists such as Otto Brunfels and Leonard Fuchs had already used two-name tags, but they had done so arbitrarily, without a system. Linnaeus's genius was to understand that a name did not have to include details of the plant's characteristics, such as colour, leaf shape or habitat: this information could be looked up in botanical encyclopaedias. Rather the binominal name was a label or a point of reference. Standardised and universal, it was easy to use and at the same time gave access to the other information available, like a name on a door that could be opened if further insights were necessary.

Linnaeus enforced his order with a rigid set of rules, which read like commandments: "every plant-name must consist of a generic name and a specific one," and, above all, "only genuine botanists have the ability to apply names to plants"— making sure that not too many people could interfere with his business. By the time Linnaeus published the *Species Plantarum* he had named and classified 7,700 plants, the culmination of two decades of work which included new descriptions for each species written in a plain and simplified botanical Latin. With these standardised definitions Linnaeus inspired a new generation of thinkers, such as Antoine-Laurent Lavoisier, who, almost forty years later, erased the vestiges of alchemy from chemistry when he introduced a nomenclature for chemical elements which, like the Linnaean system, was simple and logical.

Every time we garden or talk about plants today we are in the company of Linnaeus. The *Species Plantarum* is the universally acknowledged starting point of modern plant names—those with the abbreviation "L" after their name indicate that Linnaeus invented or validated them, while those without the "L" were given after his death (but according to his method). Today's botanical names are regulated by the *International Code of Botanical Nomenclature* (ICBN), which takes the publication of Linnaeus's *Species Plantarum* in 1753 as its starting point. Only the International Botanical Congress, held every six years, has a right to change the rules governing the scientific names of plants. Because of Linnaeus's efforts botanists across the world can now communicate unambiguously about plants. They can monitor biodiversity and endangered species, and gardeners can buy flowers by mail order and exchange information about their cultivation across languages and continents—the names are the same in Cornwall, Tokyo, Atlanta or Sydney.

The *Species Plantarum* was a monumental work, enormous

in scale, and a methodological feat that distinguished Linnaeus as the most ingenious classifier of his time. Inevitably, however, it was not easy for the Swedish botanist to persuade his international colleagues to accept a method that would entail a complete renaming of the natural world, visionary as it was. Once again, the English were particularly difficult to convince. Just as they had refused to adopt the sexual system of classification expounded in Linnaeus's *Systema Naturae*, now they proved resistant to the idea of a binomial nomenclature. Linnaeus must have had an inkling that this would be the case from their reaction to the publication, fifteen years earlier, of his *Hortus Cliffortianus*—the catalogue of Clifford's plant collection in which he had given many of the new exotic plants new generic names. This had provoked a howl of protest from botanists, who had accused him of considering himself a "second Adam."* Though Dillenius conceded that the old nomenclature was a complete mess—"an Augean stable"—he still thought that Linnaeus "overturn[ed] every thing" without reason. Similarly, another botanist feared that Linnaeus's predilection for making new names would "lead to worse than the confusion of Babel," while Miller ranted that the changing of names was "unpardonable." It cannot have helped that Linnaeus chose some names which were explicitly linked to sexual organs, such as *Phallus impudicus* for the stinkhorn and *Clitoria* for the butterfly pea.

The reactions to the *Species Plantarum* were even more angry. Within a year Miller's old friend Charles Alston, the professor for botany in Edinburgh, had published *A Dissertation on Botany*, a long tirade against Linnaeus's methods which, according to him, were for the "most part useless, and frequently deceitful." Even the future Prime Minister, the third Earl of Bute, one of Britain's most influential amateur botanists, was so irritated that he felt compelled to write to Collinson,

---

* Botanists were upset that Linnaeus had given some of the new exotics names that appeared in classical texts. He had taken the name *Thuja*, for example, from Theophrastus and used it to replace the old *Arbor vitae*. This, botanists thought, might cause great confusion because, so Dillenius wrote to Linnaeus on 18 August 1737, "the day may possibly come when the plants of Theophrastus and Diocorides may be ascertained; and, till this happens, we had better leave their names as we find them."

complaining that Linnaeus was "so Vain as to imagine he can prescribe to all the World."

Collinson also found the new names more confusing than helpful. "Butt if you will be for Ever Makeing New Names & altering Old & Good Names," he wrote to Linnaeus, "it will be Impossible to attain to a Perfect Knowledge." Instead of appreciating the simplicity of the new nomenclature, Collinson, who had turned sixty years a few months after the *Species Plantarum* had been published, resented the effort and time it would take to learn all the new names. "Thus Botany, which was a pleasant Study, and attainable by Most Men," he complained to Linnaeus in April 1754, "is now become by alterations & New Names the Study of a Mans Life." Not only might all noblemen and gentlemen who had shown an interest in botany give up in confusion, but the parsimonious Collinson also resented the fact that botanical publications had to be reprinted, writing to a fellow collector, "the Azaleas he [Linnaeus] has turn'd into Kalmias So that Every book he prints will require a New Edition." Four years later, Linnaeus reached the very pinnacle of disgrace when the Vatican included his publications in its catalogue of forbidden books.

But in a sense, none of the criticisms would have mattered had Miller approved of the system, because through the best-selling *Dictionary* he had the greatest influence on the gardening community. Linnaeus's pupil Pehr Kalm had asked many garden owners what book they used daily, and "all answered with one mouth, Miller's *Gardeners Dictionary*." The influence of the *Dictionary* had by now spread beyond Britain to the Continent because it was available in Dutch and German. If Miller used the new binominal names, everybody else would follow. Unfortunately for Linnaeus, Miller did not approve, railing instead that Linnaeus had "the Vanity of being the Law Giver."

Linnaeus, though, pretended not to care about the rejection, as he had learned to endure what he called "the most bitter abuse." Instead, he drew up an imaginary army of botanists in which he, of course, held the position of commanding general— while Miller was allocated one of the lowest ranks in the hierarchy, with only Linnaeus's old arch-enemy Johann Georg Siegesbeck below him. Linnaeus's overblown self-confidence also helped him bear the slights, because he continued to believe he had a deeper insight into the three kingdoms of nature (plants, animals and minerals) than anybody else. Even before publication he had predicted that the *Species Plantarum* would "be terrific," and without scruple he wrote reviews in the Swedish and international press in which he described his own work as "masterpieces" and "jewels." Later, when he dictated his autobiography, he called the *Species Plantarum* "the greatest work in science"—a somewhat conceited comment for a man who insisted that "self-praise smells bad, in fact stinks."

As with the sexual system, Linnaeus's names received a far warmer welcome in North America. "Such neatness! such regularity!" one admirer wrote to Linnaeus, for it was "more easy for beginners." Cadwallader Colden, colonial politician and friend of Bartram and Collinson, who had, according to Bartram "attained to ye greatest knowledge in botany of any I have discoursed with," was an enthusiastic advocate. His daughter Jane Colden became so familiar with Linnaeus's ideas that she produced more than 300 drawings of the local flora of New York, all accompanied by the new binomial names and classified according to the sexual system—the first woman to do so in America. Here in the colonies Linnaeus was called "the favoured priest of Nature."

Although it was known in Britain that Linnaeus had made several proselytes in the colonies, this advocacy did not further his cause among the English, who for the most part regarded American thinkers and natural philosophers as mere suppliers of observations—recorders of the flora and fauna in the New World—rather than scientists in their own right. Their work and opinions had often only marginal impact on the wider scientific community in Europe. When Colden, for instance, sent his thesis on gravitation to Collinson in order to circulate it in Europe, one fellow of the Royal Society thought it must have been plagiarised. No American, he believed, was learned enough to write something so sophisticated—rather it must have been the "papers of Some Ingenious European" that had fallen into Colden's hands.*

---

* Collinson was ashamed about this accusation and urged Franklin to keep this information a secret so as not to insult Colden.

Bartram, who with his growing enterprise was becoming increasingly confident, tried to broker a peace and explained to a furious Miller that "there is no general rules without some exceptions, so it is with Lineus sistem." But when Bartram criticised Miller's classification of American birch trees, for example, Collinson warned him not to offend "our Great Botanist" and wrote: "[P]ray run not the risque of harming thy Little finger for a Paper Birch." Miller, now in his mid-sixties, was buoyed by his success, but was also becoming increasingly obstinate and belligerent. It was observed that "[w]hen the greatest lords drove out to their estates, he [Miller] often drove out with them in the same carriage," but as much as the aristocracy cherished Miller's company and advice, many fellow gardeners were annoyed about his arrogance. Even old friends and colleagues were turning against him.

Indeed, the anger that many felt for Miller became fuel for the acceptance of Linnaeus's system, because Linnaeus eventually found some dedicated allies in Britain—even if they only joined his ranks to spite Miller. The most outspoken advocate was John Ellis, a merchant acquaintance of Collinson and an amateur naturalist, who believed that it was time for Miller to leave centre stage. In 1757, four years after the publication of *Species Plantarum*, Ellis was delighted to inform Linnaeus that Miller had been "obliged to change most of his obsolete names and descriptions of plants, to your more intelligible and accurate ones." In Ellis's mind Miller was a good gardener but not a great botanist.

The long-lasting dispute between Ellis and Miller escalated in a row over the classification of American and Asian poison ivy. The argument soon enveloped the whole of botanical London. Miller, who saw himself as a botanical oracle, was summoned to the Royal Society but feigned illness. But Ellis was confident and trumpeted, "I shall be able to triumph over Miller's blunders" (which he did when Miller corrected his mistake in the eighth edition of the *Dictionary*). Ellis predicted Miller's waning fame, and concluded that soon Linnaeus's nomenclature would be fully accepted in Britain. Since "those people that used to laugh at it as chimerical" were now slowly conceding, it was time to publish the Linnaean system in English,

Ellis suggested, so that even beginners and amateurs with no knowledge of Latin could use it.*

Slowly, over the next few years, even without Miller's validation, more and more botanists began to see the value of an international naming standard, the free flow of information thus taking priority over vanity and competition. And so, as one after another accepted the new system, turning to Linnaeus for the final decisions on what plants should be called, the Swede really did become the "second Adam" of botany. Collinson was one of the first to acquiesce when, in the summer of 1755, he called upon Linnaeus to resolve a dispute over the "true name" of a plant. "Now, my dear friend," Collinson wrote, "it is left to your profound judgement, what it is, and what name it shall bear, and to what class it belongs." Six months later, Collinson asked Bartram's son William to produce some botanical drawings with dissected flowers "after Linnaeus Method which seems to be the prevailing Tast[e]."

From his new position of power Linnaeus was able to confer immortality on fellow botanists by naming plants in their honour, which he believed was "the only prize available to botanists." The more beautiful, important or large a genus was, the greater the homage. To celebrate Pehr Kalm's triumphant expedition to North America, for example, Linnaeus changed the name of one of Collinson's favourite shrubs, which had been previously called either chamaedaphne, chamaerhododendron or azalea, to *Kalmia*.† The French botanist Pierre Magnol, who had invented the concept of plant families, was honoured by being associated with the genus *Magnolia*.

For Miller, however, Linnaeus chose a small-flowering and weedy member of the daisy family—*Milleria quinquefolia*, a genus whose only species had been sent to Miller from Central America in 1731 by the plant-hunter William Houston. It was important, Linnaeus explained, that there was a link between the botanist and the plant. In this case, the short sepals of the *Milleria*, when closed, reputedly alluded to Miller's plump

---

* All of Linnaeus's publications were written in Latin, the international language of botany.
† Linnaeus split chamaedaphne, chamaerhododendron or azalea into the genera *Kalmia* and *Rhododendron*.

figure, while the fact that the calyx entirely enclosed the seeds recalled Miller's "labour over acquiring rare American seeds [and] preserving them" (and maybe it also reminded Linnaeus of Miller's reluctance to part with any of them). Linnaeus enjoyed this marrying of characteristics when it came to people and plants. His pupil Pehr Forsskål, for instance, was commemorated with the *Forsskaolea tenacissima*, a plant that attached itself to wood and fingers, which, according to Linnaeus, echoed Forsskål's character as he was "truly tenacious of his opinions."

For his old teacher Olof Rudbeck, Linnaeus chose the popular *Rudbeckia*. The tall flower reflected Rudbeck's stature, Linnaeus explained, and the ray-like petals bore "witness that you shone among savants like the sun among stars." The *Linnaea*, a diminutive forest flower from Lapland,* Linnaeus described as "lowly, insignificant, disregarded"— just like himself "who resembles it," because like the *Linnaea* he felt forgotten and ignored.

His worst insult he saved for the hated Siegesbeck, who, as punishment for having so relentlessly criticised Linnaeus's sexual system, will be forever remembered in *Siegesbeckia*, a stinking weed that thrives in wasteland. He had chosen a plant that expressed what he believed to be Siegesbeck's "natural abilityes," Linnaeus claimed: the plant had so many tiny sticky hooks that "all wool and hair remains attached to it." Linnaeus's anger over Siegesbeck's criticism of the sexual system had lingered for years. When Linnaeus found a packet of *Siegesbeckia* seeds at the botanic garden at Uppsala he relabelled them with the made-up name of "*Ingratus cuculus*"— ungrateful cuckoo. Although intended as a private joke, the mockery spun out of control when the seeds were dispatched to Siegesbeck by mistake. Unsurprisingly Siegesbeck was offended. Fellow botanists tried to mediate but Linnaeus refused to retract. Not even a hoard of Siberian plants would tempt him, he announced: "If someone said 'if you apologise to Siegesbeck, I will give you a large or rare collection,' I must admit that it would be impossible." Never would he be

---

* *Linnaea borealis* was later also found in North America. Linnaeus was almost always pictured holding the *Linnaea* in his hands.

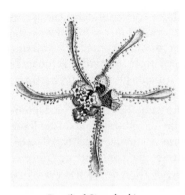

Detail of *Siegesbeckia.*

able to pardon Siegesbeck, Linnaeus wrote. Instead, he would "laugh at the idiot and fool who wants to be a *botanicus,* something he will never be."

In the succeeding years the naming of plants became sometimes a troublesome business for Linnaeus. Though his early revenges had been fun, the process of honouring members of the botanical community was complex and soon involved tantrums, emotions and competitive zeal. Letters arrived in Uppsala filled with complaints and requests to change plant names. When Ellis heard that Linnaeus had selected a tiny American flower to be the *Ellisia,* he pretended that he did not care that most people regarded "a little mean-looking plant as reflecting no honour" but nonetheless suggested it be exchanged for another plant that grew in Britain. His reasons, he explained to Linnaeus, were that a native plant would be more readily available for his friends "to put them in mind of me." Coincidentally the one he wanted was a beautiful evergreen shrub with fragrant flowers!

Collinson canvassed continuously for Bartram to receive a plant name, suggesting the beautiful *Dodecatheon meadia,* for example. In the end, however, Linnaeus made *Bartramia* a genus of a tropical burr weed (*Bartramia rhomboida,* which Linnaeus later joined with the genus *Triumfetta,* renaming it *Triumfetta rhomboida*). Today Bartram is commemorated by a genus of moss. Collinson was grateful to get *Collinsonia,** and thanked Linnaeus for "giving Mee a species of Eternity

---

* According to the botanical painter Georg Dionysius Ehret, the *Collinsonia* had been named by the French botanist Bernard de Jussieu, who had seen it in Collinson's garden. Linnaeus, though, was the first to publish it in the *Hortus Cliffortianus* in 1738—without acknowledging de Jussieu. "When he [Linnaeus] was a beginner," Ehret wrote on 27 October 1758, "he appropriated everything for himself . . . to make himself famous."

(Botanically speaking)," though he was confident enough to add that "Some thing I think was Due to Mee from the Common Wealth of Botany for the great number of plants & Seeds I have annually procur'd from Abroad." In fairness it should have been called *Bartramia,* he wrote to Bartram, "for I had it In the very first seeds thee sent me." It was a small woodland plant from the mint family with a lemon-scented bloom which gave Kalm "a pretty violent head-ach whenever I passed a place where it stood in plenty." Nevertheless Collinson requested more seeds from Bartram to send to his friends. He still spent much time in his garden when "retreated from the Hurries of the Town," working the soil, nurturing his seedlings, and in the evenings he compiled a list of his plants.

Being a plant-collector rather than a botanist, Collinson frequently asked eminent botanists to help him classify his plants, providing duplicates in recompense. In 1737 he had longed to have Linnaeus help him classify four books of dried plants that Bartram had sent from Philadelphia. So it was that, when Linnaeus suggested that Collinson publish a catalogue of all his plants—using the binominal names of course—Collinson replied that the task would be beyond him "unless I had one of your Ingenious Pupils to Digest and Methodise it for Mee." It made sense to Linnaeus to send one of his "apostles" to England at this point, to cement the binominal nomenclature and encourage the complete acceptance of the sexual system. And so he settled upon Daniel Solander.

Solander was Linnaeus's favourite pupil. He had been only seventeen years old when he had arrived in Uppsala in 1750, but had quickly become the best student. He had made "greater progress in Natural History . . . than anyone else I know of," Linnaeus insisted, and he was soon one of the family in the Linnaeus household. He spent the Christmas of 1756 with them, and though he did not actually live in Linnaeus's house, he visited daily and said that "I almost spend more time there than at home." Unlike his mentor, Solander was so popular and sociable that his fellow students did not begrudge his preferential position—one even congratulated him "that Herr Linnaeus has counted you as his best swain in the subject."

COLLINSONIA. *Hort. Cliff.* 14. *Sp.* 1.
a. *Flos magnitudine naturali.* b. *Idem a tergo visus.*
c. *Calyx sub florescentia constitutus.* d. *Idem fructu prægnans.*
e. *Germen.*

G. D. EHRET del.   J. WANDELAAR fecit·

Collinsonia canadensis

Solander delighted everybody with his natural charm and kindness. Everything about him was gentle and friendly. He was chatty and modest, and his round face was always animated by a smile. He was insouciant and charismatic, but also a mine of information—be it about matters trivial or serious—in short, "a philosophical gossip," as an acquaintance later remarked. Unlike the hasty and impatient Linnaeus, Solander exuded an aura of calmness and reassurance. But while his stocky physique and chubby face could have let him appear quite placid, his pointy nose and lively dark eyes, continuously in search of something new, revealed an agile man full of curiosity, energy and humour.

The news that Solander was coming to Britain was greeted with enthusiasm by Linnaeus's admirers. "Pray desire him to study English immediately," Ellis wrote to Linnaeus, and offered to help introduce the young man to all the botanists and gardeners in the country.

But it took longer for Solander to arrive than anyone had expected. He began his journey in the spring of 1759, bidding farewell to his adored teacher and to sixteen-year-old Lisa Stina, Linnaeus's eldest daughter, who had been promised to him in marriage. In his pockets he had letters of introduction in which Linnaeus recommended Solander "as I would my own son." Yet seven weeks later, he had failed to appear. On 30 May, Linnaeus wrote to Ellis, assuming that Solander had arrived in London and "found tranquil asylum in your friendship," only to be disabused. Two months later Collinson wrote to Uppsala that Solander was still missing. "Is no certain Advice come of his Fate?" Collinson asked, but nobody knew why and where he had disappeared to. Letters crossed the Baltic but there was no word from Solander, who seemed to have vanished.

Then, more than six months after Solander had left Uppsala, Linnaeus received a letter from Skåne, a province in south Sweden. It turned out that Solander had not yet even left the country. The reason for his delay, Solander explained in his letter, had been an inflammatory fever—a type of malaria which recurred every few days. For fear that his parents—who had been against the journey to Britain in the first place—would order him back to Uppsala, Solander had not told anybody of his whereabouts. Now he was making Linnaeus complicit, urging him "not to let them hear of it." With winter approaching and no ships leaving for London, he was concerned that, if his parents knew he was still in Sweden, he would never leave its shores.

In April 1760 Linnaeus wrote to Ellis that he regretted that "my dear Solander" had been kept for "a whole year from the benefit of your society," but assured him that his pupil was now on the way. Finally, on 29 June, almost exactly four years after Collinson first asked for assistance, the young Swede arrived in London. And immediately found himself embroiled in a botanical conflict.

The battlefield was the garden of Miller's friend Richard Warner, who lived in Woodford, Essex, ten miles from the city of London; the opposing forces were—as so often—Ellis and Miller; and the reason was the naming of the most sought-after plant of the time: a showy evergreen shrub that had been introduced to Britain from South Africa in 1754 by Warner. When it first displayed its perfumed ivory blossom in Warner's garden in 1758, a pilgrimage of botanists and horticulturists, including Collinson, Miller and Ellis, had rolled up at Woodford. Not since the introduction of *Magnolia grandiflora*, two decades previously, had there been such excitement about a new plant. So great was the furore that London's most respected nurseryman, James Gordon, made the small fortune of £500 from the 100 plants he had raised from four cuttings. "Every body is in love with it," Ellis wrote to Alexander Garden, an American physician whose name was at the centre of the battle.

When Solander arrived in London, the row had already been running for two years. Miller had classified the new plant as *Jasminum* but Ellis was sure it was a new genus. To settle the dispute Ellis had sent a dried specimen to Linnaeus, who had confirmed Ellis's taxonomy. But Miller, who hated to be contra-dicted, insisted that his own classification was correct. The fight came to a climax when Ellis asked Linnaeus to name the new plant *Warneria* to celebrate its first owner. Poor Warner was torn, as he felt that he ought to side with his old friend Miller, who insisted the shrub was a species from the jasmine genus. In the end he chose friendship over immortality, and asked Linnaeus not to name it after him. Ellis had dithered over several other names and then eventually, two weeks before Solander arrived, suggested *Gardenia* after Alexander Garden, hinting to Linnaeus that the American "may be a most useful correspondent."

Within two days of his arrival Solander knew all the gossip, as Ellis made sure to indoctrinate him properly. And before Solander had even visited Chelsea, he parroted Ellis's opinions when he wrote to Linnaeus that even the nurseryman James Gordon had "far more insight than Miller." It was obvious that Ellis despised Miller, Solander explained to Linnaeus, because he thought him to be "capricious and vain." In Uppsala, detached from the minutiae of the dispute, Linnaeus grew impatient with the bickering English botanists. There was no reason, he felt, to call the new plant *Gardenia*, as Garden had neither discovered nor cultivated the plant. Linnaeus declined Ellis's request and wrote to Solander, "Let Mr. Ellis . . . bravely understand this so that he does not get angry; he who is already somewhat angry in this matter." Meanwhile, Solander was trying to placate Linnaeus on behalf of the English botanists and gardeners, sending one of the rare living specimens over to Uppsala and underlining that it was worth six guineas.

Ellis, who had always boasted of his influence on Linnaeus, now faced the public humiliation of having to admit defeat— at least on the choice of names. To save face he begged, threatened and flattered, promising Solander that he would henceforth dedicate himself to the Linnaean cause more than any other man in England, if only Linnaeus would do him this one favour. For months Linnaeus refused to change his mind, explaining to Ellis, "I wish to guard against the ill-natured objections, often made against me, that I name plants after my friends, who have not publicly contributed to the advancement of science." Then in autumn of the same year, Linnaeus complied, probably to repay Ellis's kindness towards Solander and to keep sweet his greatest English advocate. Ellis was relieved, because now his "botanical reputation" was redeemed, Solander wrote to Linnaeus, otherwise "his enemies would have rejoiced at his ignorance, and discovered how little he could achieve." Miller, though, stubborn to the last, refused to be appeased. Eight years later, in 1768, he listed the *Gardenia* under *Jasminum* in his eighth edition of the *Dictionary*, explaining that Linnaeus had been wrongly influenced by a description "taken by some hasty people"—namely Ellis.

Linnaeus, though, remained confident. He declared himself the messiah of botany. His science, he claimed, was "the light that will lead the people who wander in darkness."

"The English are all, more or less, gardeners"

By these, the prickly-leaved oak you see,
And, with frontated leaves the tulip-tree;
Here, yellow blows the thorny barberry-bush,
And velvet roses spread their bright'ning blush;
And here the damask, there the provence rose,
And cerasus's, double blooms disclose;
With rip'ning fruit domestic raspberries glow,
And sweet americans their scents bestow;

JAMES WOODHOUSE, "The Leasowes," 1763

The London in which Daniel Solander arrived in the summer of 1760 was a city of gardens. There were private gardens, nursery gardens, market gardens and pleasure gardens as well as residential garden squares and parks where Londoners walked along meandering paths, found shade in little pavilions and enjoyed the soothing trickle of fountains. These were the sanctuaries from the chaos of the traffic and crowds, retreats from the coal-smoke which clouded the air. Gentlemen greeted each other from horseback, aristocratic ladies observed one another from their carriages, foreigners strolled along the gravel walkways and, on Sundays, maids promenaded in their best frocks while children played skittles on the lawns. In St. James's Park one could buy fresh milk produced from the grazing cows and rest on seats in the avenues to admire the colourful shrubberies.

Even the bustling streets of London were not devoid of plants: on every corner old women and girls sold small bunches of flowers to passers-by, and almost every house had a little yard either at the front or the back which was often crowded with flowerpots and beds. These urban gardens brought, one visitor thought, "the pleasant enjoyments of a country life in the midst of the hubbub of the town." By now gardening and horticulture had become a defining part of the English way of life, much more so than on the Continent. The nation's mild climate and fertile soil, along with the scientific and technological advances being made, the lively exchange of observations and specimens among its botanists, the many horticultural publications and the fierce competition between gardeners all combined to make it so.

The superiority of British horticulture was one of the reasons why Linnaeus had sent Solander abroad. Not only could his pupil spread Linnaeus's ideas to the English, but Solander could learn as much as possible about English horticulture and recruit English gardeners to join the Uppsala army, as "I am in quite great need of them," Linnaeus instructed. Despite his conceited belief in his own botanical genius, Linnaeus was self-knowing enough to admit that English gardeners were much more advanced in practical horticulture than the Swedes. When, for example, a few months after Solander's arrival in Britain, some bulbs from South Africa failed to thrive in Uppsala, Linnaeus asked Solander to "[m]ake yourself informed with Mr. Myller how he keeps his *bulbos capenses* [Cape bulbs] so that they flower." Linnaeus's opinion of English gardening skills was so high that when seeds failed to germinate or plants refused to flower in England, he believed, "it will never be done."

English gardeners, both professional and amateur, kept more plants alive in England than anywhere else in Europe. Peter Collinson could now, for example, order 1,000 cedar of Lebanon saplings from as mundane a source as a butcher in Barnes (who ran a little nursery business on the side), whereas only twenty years previously the Duke of Richmond had needed to beg Lady Petre for this tree because not even the best nurseries sold it. Moreover, the number of foreign plants growing in England had risen dramatically since Linnaeus had visited London twenty-four years previously. In his latest edition of the *Dictionary* Philip Miller had written tantalisingly of "this extraordinary Multiplication of the Varieties in Plants." "Your country seems ... the kernel of the whole globe," Linnaeus wrote to John Ellis, feeling buried at the "fagg End of the World" and dependent on the goodwill and generosity of others. He envied Collinson Bartram's regular shipments, Miller his collection at Chelsea and every Londoner who could obtain the rarities he so coveted by a quick visit to a nursery. Solander, he hoped, would be able to get hold of the species that were missing from his garden at Uppsala.

Solander was initially a faithful pupil, providing Linnaeus with regular reports on what he saw and learned. Nor did he waste any time. On 30 June, the day after he stepped off the boat, he went to James Gordon's garden in Mile End, to the east of the city, one of the best nurseries in the country. Here, Solander was told, he would find many of the plants his master had requested, since, for decades, Gordon had specialised in American flowering trees and shrubs, first as Lord Petre's head gardener at Thorndon and then in his own nursery. He was also a disciple of the Linnaean method and Solander was gratified to find that, when he arrived at the nursery, Gordon was examining some plants with the help of Linnaeus's *Systema Naturae*—probably in an attempt to impress his visitor. But it was the vast number of thriving American plants that won Solander's instant admiration.

Never before had he seen so many of them in flower. There was fragrant Carolina allspice—*Calycanthus floridus*—which had first been brought from the colonies more than three decades previously but had remained rare in England until Bartram sent it from Carolina. "Imagine if one could get *Calycanthus*," Linnaeus wrote longingly, as he had never seen it in blossom. There were two different sorts of kalmias, as well as rhododendrons. But, to Solander, the magnolia was the most striking of all—though he wasn't sure what struck him most: its beauty or its price, which Gordon said was as much as £15. This was Solander's first encounter with England's obsession with American plants. "[W]here in Sweden would we find anyone willing to pay 10 plates [of silver] for a rarity?" he asked Linnaeus. What surprised Solander most was that Gordon had become "rich through his garden."

Solander believed Gordon to be the man from whom to "gain the most . . . in the art of planting." In fact Gordon's cultivation methods were so effective that most of the saplings were growing "uncovered," their only protection a raised ring of soil filled with dung in order to provide nourishment, warmth and retain moisture. When Gordon offered lodgings at his nursery and unrestricted access to the flowerbeds, Solander accepted with excitement. Here he would be able to procure all the rare plants Linnaeus wanted—and there were even plants Miller did not have at Chelsea.

Living at Gordon's, Solander could witness the transformation of the horticultural world and the English garden from the inside. Gordon was part of the burgeoning nursery trade that was feeding the new-found English obsession with gardening. His talent for raising "the dusty Seeds of Calmias, Rhododendrons, Azaleas" into flourishing plants meant that he had the largest selection of American exotics in the country, and he distributed them both to private garden-owners and other nurserymen. He had been the first nurseryman to order Bartram's boxes, and by now bought and produced so many seeds that he even had a seed shop in Fenchurch Street, near Collinson's office in the city.

When Thomas Fairchild had founded his Hoxton business, in 1690, there had been only fifteen nurseries in the whole country. Now there were thirty in London alone, in addition to a dozen seedsmen. Most of them were Bartram's regular clients. In the past decade he had delivered boxes to almost 100 customers, many of whom were commercial plant businesses. Without Bartram's seed boxes these nurseries would have had few foreign plants to sell. As it was, at the time Solander was inspecting the nurseries around London, one of Bartram's new clients offered for the first time a selection of almost 600 seeds of flowering annuals and perennials—a spectacular increase from the 200 he had advertised only a few years previously.

Spread by nurserymen, Bartram's plants now grew in every landscape garden and shrubbery in England. One of James Gordon's clients, for example, was the Duchess of Portland, who ordered many American exotics. Her garden at Bulstrode made proud display of such plants as collinsonia, phlox, *Echinacea purpurea* and "Bartram's Aromatick Vine a new species of Aristolochia from Canada" *(Aristolochia macrophylla)*. The Earl of Northumberland at Syon House received many American trees and shrubs from Bartram's best client, the nurseryman John Williamson, and planted an array of pines and maples as well as balsam fir (which Bartram collected in the Catskill Mountains). The actor David Garrick had a shrubbery of American shrubs at his villa, Hampton House, and Horace Walpole adored his black locusts and his white pines at Strawberry Hill.

The network of nurseries was making England, for the first time, the leading garden nation in the western world. Where once English gardeners had visited France, Italy or Holland to improve their skills, copy layouts or buy plants, now the horticultural trade had shifted across the Channel and continental collectors and garden-owners were ordering their shrubs and trees directly from England's nurseries.* There was great demand, particularly from Germany, as ten years previously the translation of Miller's *Dictionary* had been so successful that Germans now believed that "the English are all, more or less, gardeners." German gardeners used Miller's plant lists almost like a mail order catalogue when requesting plants from English nurseries, and the trees and flowering shrubs that created the thickets and clumps of the English landscape gardens soon became known in Germany as "der Englische Wald" (the English forest), albeit that they consisted of Bartram's American trees.

Miller attempted to profit from the interest created by his *Dictionary* by establishing a commercial nursery, some two miles away from Chelsea, from which he could send Bartram's harvest to the Continent. He had, so Miller boasted to a German client, "a large Nursery of Trees for sale . . . in which are the greatest variety of trees and Shrubs to be found in any Garden for sale in Europe." The only trouble was that Miller, who was by now almost seventy, ran his nursery in dilettante fashion, and therefore always struggled to earn enough money. He dealt with his commercial clients as he had done with his network of plant-collectors over the previous four decades. For him plant-collecting was still a gentlemanly pursuit which involved politeness and gossipy letters but also forbearance because he was such a busy man. "I have been so much in the Country upon business . . . as not to have leisure to look over my seeds," Miller wrote when he once again failed to dispatch an order. His old correspondents understood that he was often too

* When Miller published the sixth edition of his *Dictionary* in 1752 he changed the entry for "nurseries": instead of referring to the area of an estate where plants were raised, he now described them as commercial enterprises.

occupied with the Physic Garden, his *Dictionary* or some gardening consultancy for a great country estate, but Miller's commercial customers were not as forgiving—in particular the German garden owners. They thought Miller unprofessional and unreliable. His nonchalant attitude to his commercial enterprise was interpreted as slack, inefficient and unpunctual. "Mr. Miller at Chelsea near London is cheaper [than other nurserymen] but very tardy in correspondence and dispatch," one German client complained. Miller was, they speculated "maybe too old and too rich" to be a reliable dealer.

Unlike Miller, Bartram was very active in supplying the new market he had created. Over the past decade he had streamlined his business by developing a standardised packing method in order to supply more customers. Whereas in the early years Collinson had sorted and distributed the contents of the boxes to his friends, now each box was identical, which saved a great deal of time, allowing Collinson to send the boxes to the recipients without even lifting the lid. Each box was priced at five guineas and contained seeds of 105 species, which were kept in separate cloth bags labelled with a number that referred to the enclosed list.*

As a result of the standardisation, some nurserymen became also seed merchants, acting as wholesalers. This rapidly increased the influx of American plants—not just to England but also to the Continent. Johann Bush, for example, a German gardener who had founded a nursery in Hackney, exploited the orderly packing method to attract many of Miller's disgruntled German clients. Simply by adding a plant catalogue in English and German, Bush made a three-guinea profit per box. With such profit margins Bush became one of Bartram's best customers, dispatching to Germany, as one of his clients said, "whole forests from England."

Collinson remained the driving force behind the enterprise. In pursuit of glory, beautiful gardens and Bartram's profit, he had recently persuaded the editors of the *Gentleman's Magazine* to publish Bartram's complete plant list. Thus the business

---

* There were some initial problems in the 1750s when a number of nurserymen complained about mouldy paper bags, too few seeds and "Such wreched Cargo," but by 1760 the business ran smoothly. Only when customers ordered specific seeds in large quantities did they receive special boxes.

was advertised from Dundee to Canterbury. And, as the numbers of customers grew, so did profits—to such an extent that Bartram sometimes failed to withdraw his money quickly enough from Collinson. The merchant in Collinson reprimanded Bartram for being slack with the accounts, "for I Love to Ballance accounts & keep them Even," he wrote to Philadelphia.

It wasn't always easy, though. With the greater number of clients came an increased number of complaints, and Bartram and Collinson had their usual squabbles when deliveries disappointed customers in Britain. Collinson, for example, accused Bartram of being a "Careless person" when a box arrived that was almost empty (in fact it was plundered because the Spanish had captured the ship), but he received a stern reply from Philadelphia. "[T]his was not ye 1st or 2d or third time that I have been rashly censored & reproached," Bartram complained, "when I have hazarded life & limb both my own & children." But despite their occasional bickering over business matters, their friendship remained strong—"the fire of friendship is blazeing," Collinson wrote.

Collinson's greatest sign of friendship was the way in which he protected Bartram's business interests in England—without financially profiting himself. When, for example, he discovered that the white cedar which Bartram had introduced to England was easily propagated by cuttings, he promised that "Gordon & I keep [it] a great Secret," because if gardeners and nurserymen in England found out they would not need to order seeds from Bartram—"it would Spoil Trade." And when an invoice from another colonial nurseryman fell into his hands by accident, he forwarded it to Bartram "for thy Amusement," adding smugly, "He makes but a poor figure when Compared with thine." At the same time Collinson used his garden to promote Bartram's seed boxes, encouraging enthusiastic gardeners to come and see the American species in maturity. "[M]y Garden ... is Enriched with many Rare Plants & Trees not to be seen Elcewhere," Collinson had written to Linnaeus, who had urged Solander to visit the merchant as soon as possible for a thorough inspection.

The garden that Solander visited two days after his arrival in London, on 1 July 1760, was no longer in Peckham. Ten years earlier, Collinson had moved to Mill Hill, some ten miles to the north-west of the city, where he had painstakingly re-created his Peckham garden. It had taken him two years to transplant every flower, shrub and tree—most of which were so precious that he had planted them out himself, enjoying the peaceful world of his garden away from the buzz of London. "[T]he Robin Red Breasts serenaded Mee as I was planting," Collinson had written to Bartram. To reach this rural idyll, Solander and Ellis boarded a coach at noon—probably the daily service from the Bull Inn in Holborn near Collinson's office—and drove out into the countryside. From Collinson's garden, which sloped down from the house towards the west, they could see the woods that belted London, and could just make out the royal palace of Windsor Castle in the distance.

Though the garden was full of Bartram's American species, Collinson also had some curiosities from elsewhere. He had been given seeds of Tree of Heaven by a Jesuit priest from China, and he was the proud owner of "a most beautiful" specimen of the evergreen *Arbutus andrachne* from the Levant—also called Greek strawberry tree because of its red strawberry-like fruits. It remained such a rarity that one mature tree sold twenty years later for £53 11s. Collinson enjoyed showing off his treasures, like a proud father parading his impeccably behaved children. And Solander's visit fell at the moment of the best blooms. Only a few days previously Collinson had dashed off a letter to Bartram about the simultaneous flowering of some of his most precious plants: "I am Charm'd, nay in Extasie."

Solander caught a waft of the Carolina allspice's scent before he saw it, because, Collinson reported to Bartram, "its fragrance is smelt at a great Distance." Another perfumed glory in Collinson's garden was the fringe tree, which for decades had flowered every year—its blooms were creamy white feathery clusters that clad every branch and almost entirely hid the emerging leaves. Also in the garden were Collinson's beloved lady's slipper orchids as well as magnolias, kalmias and a vast array of colourful annuals. Solander could admire the beautiful contrast between the bright green foliage and the dangling rose pink petals of *Robinia hispida,* one of the rarest sights in England and "a glorious show," as Collinson boasted. The vigorous climber *Campsis radicans*—trumpet vine—hugged the greenhouse, covering it with its orangey red tubular flowers. Although it had been introduced to England from America more than 100 years previously, it was only now being grown more widely and would soon become one of the most popular shrubbery plants. Solander saw sorrel tree, also called the lily-of-the-valley-tree for its delicate tiny urn-shaped flowers. Showing off all his plants and their unique traits, Collinson probably told Solander that the glossy foliage turned deep red in autumn and that the branches were decorated during the winter months with seeds in the shape of thousands of upright capsules. The ground below the trees was covered with the delicate white blossom of the shade-loving *Saxifraga x geum*.

One plant Solander had never seen in blossom was the strange pitcher plant, which Collinson kept in a perforated pot filled with moss and placed in a tub of water—an artificial bog that imitated its natural habitat in America. This carnivorous insect-devouring flower, which in Miller's words was "so different from all the known plants," had originally been plucked from the swamps around Jamestown in Virginia by John Tradescant the Younger in 1637. But ever since then, English gardeners had struggled to keep them alive. Even Miller had little success, but Collinson brought them to blossom every year. The crowning glory of Collinson's garden tour was *Rhododendron maximum*—currently his favourite shrub. He adored its exuberant display of tufty white blossom, which contrasted elegantly with its glossy, dark-green leaves, and envied Bartram "the Ravishing Sight" of vast landscapes "alive with this rich embroiderie." In the Appalachian and Allegheny Mountains it covered the slopes in impenetrable thickets often growing to well over twenty feet.

Every one of the plants in the Mill Hill garden told a story, and called to mind Collinson's "Absent Friends." And so, as Solander strolled among the flowerbeds and trees, he passed exotics he had never seen and heard the stories of men he had never met. The cedar of Lebanon, for example, recalled the late Duke of Richmond; balsam fir was a reminder of Lord Petre because Collinson had received the first sapling from Thorndon; while another of Bartram's clients, the Duke of Argyll, was immortalised by the much adored longleaf pine.

But it was Bartram who was most represented: magnolias, rhododendrons, kalmias and azaleas. All of these, Collinson promised Solander, would be provided as seeds or cuttings for Linnaeus.

Collinson took an immediate liking to Solander. Unlike the ageing and self-congratulatory Linnaeus, who had offended the English with his brashness and arrogance, Solander charmed everybody. Collinson and Ellis intended to pass him around the gardening circle—but first they sent Solander to the country to learn more English, because he found that everybody spoke too "fast and indistinctly." A quick learner, Solander's language improved within a few weeks and by August Ellis was able to invite him to visit a gardening friend in Somerset. Soon everybody wanted to meet the young Swedish botanist, who seemed to offer an indirect and therefore more pleasant contact to his curmudgeonly master. And so Solander flitted around the country like a bee from flower to flower—being a careless dresser, though, he was reminded by Ellis to bring an extra pair of trousers and shirts as well as "a pair of shoes extraordinary" when visiting the great country estates.

D.ʳ SOLANDER, F.R.S.

By October Solander had run out of money. "[I]t is terribly expensive to live here," he complained to a friend, and wrote a desperate letter to his old teacher asking him to advance some of the royal stipend that Linnaeus had organised to fund the trip, out of "fondness for me." "[O]ne must live like other people," wrote Solander, to explain his expenditure of £40 for three suits and a coat and his travel expenses, since in England "[o]ne cannot look at anything without money." To sweeten the request, Solander added a long list of the plants he had already procured—free of charge from Gordon, Collinson, Miller and their acquaintances—which included tulip poplars, magnolias, rhododendrons, kalmias and hydrangeas. In addition he had "received promises of a great number of shrubs and trees in pots," which would be ready for transport by spring. While awaiting funds from Linnaeus, Solander followed Collinson's advice and toured the private gardens of England in order to charm the owners into donating plants for the botanic garden at Uppsala.

Over the next year Solander visited gardens filled with Bartram's American plants. No one, Solander insisted, could imagine "the multitude of foreign trees and shrubs" that populated the English landscape gardens. At Goodwood, the Duke of Richmond's estate, he saw 400 different American species all planted as wilderness. Some of the tulip poplars, he told Linnaeus, were higher than forty feet and a *Magnolia grandiflora* was thirty feet. "I have a good idea of the new English taste in planting, which doubtless is the best," Solander reported, impressed by the latest fashions. He would have agreed with another garden tourist, who noted that landscape gardens had "become the absolute Necessities of Life, without which a gentleman of the smallest fortune thinks he makes no Figure in the country." Solander would have also noticed that a landscape garden filled with Bartram's trees and shrubs had become *the* way to show one's wealth and taste.

While Solander walked in awe under the ruffled canopy of American trees, others began to ridicule the new English obsession. Horace Walpole, mocking himself as much as his fellow gardeners, was so amused by the preoccupation with meandering rivers and "shrubberies planted of all kind of exotics" that he was waiting for someone to propose the alteration of Jerusalem "in the modern style." Others were even more specific and lampooned garden owners such as Collinson's old acquaintance the Duke of Argyll, who adored his garden at Whitton near Twickenham so much that he had planted it before he built his house.* A newspaper depicted him as an obsessive gardener who dragged his grudging and hungry guests, still wearing their slippers, through his groves. Solander, who was given a tour by the Duke himself, was not so reluctant; quite the opposite, because nowhere in the country had he been able

---

* The Duke of Argyll explained to Pehr Kalm in May 1748 that he had decided to prepare soil and plant trees first because "Nature must have its time." The house could wait because "I can always build the most handsome Castle in one year."

to see the American species at such towering heights. *Juniperus virginiana*, which Bartram had collected for all his early customers, was now more than thirty feet tall, as was scarlet oak, while some of the white pines were almost fifty feet high. There were so many different sorts of exotics that it was described as "another world," and Argyll promised to give some of his plants to Linnaeus.

Solander was impressed not only by the sheer number of mature foreign trees and shrubs in the English garden but also by the "peculiar manner" of planting them. The garden which displayed this to perfection was Painshill in Surrey, a pleasure ground that Solander continued to visit regularly. Painshill house was set in a valley surrounded by gently rising hills, the tops of which were dressed in dark "hanging" woods. Lower down the trees were plants in clumps, tapering out into single trees on the sunlit lawns of the valley floor. Beginning his garden in the 1730s, Charles Hamilton—one of Bartram's customers as well as a subscriber to the *Dictionary* and Catesby's *Natural History*—had transformed a barren plot of land into this botanical Eden.

The garden was famed for Hamilton's painterly use of American exotics to create different "moods." Shaded pathways gave way to perfumed shrubberies, while, in autumn, the red and orange brushstrokes of Bartram's deciduous trees contrasted with the thousands of American evergreens that Hamilton had raised from seed. These made a mottled tapestry of greens, ranging from sombre dark shades to almost yellow: a picture so perfect that Hamilton enjoyed it twice, seeing it reflected in the silver surface of the lake that cut through Painshill. Colourful flowering shrubs and scented perennials were woven into these plantations. Hamilton had also created dense thickets of conifers from across the world. Underneath their soaring branches snaked narrow paths, cushioned in a thick layer of crunchy needles, so that one visitor compared it to walking on "the finest Persian carpet."

These American groves had become part of the landscape, Solander observed, and Horace Walpole would note a few years later that they "contributed essentially to the richness of colouring so peculiar to our modern landscape." This was in large part because of the general hardiness of the American species. Previously exotics had been grown in individual pots and only brought out of their gilded glass cages when the mild summer air warmed them, but the American species had clawed their roots into English soil and were thoroughly naturalised, growing side by side with native trees.

The southern catalpa, for example, which Catesby had introduced four decades previously, in 1722, had become a popular flowering tree "to adorn Pleasure-gardens" and was available in most nurseries. Its large translucent light green leaves and white summer blossom enlivened many gardens, including Painshill, Whitton and Goodwood, and today it grows everywhere in Britain, most famously at Parliament Square in London. Similarly, the fringe tree had become "more common," Miller wrote, as had the black locust. Honey locust, with its strangely twisting long seed pods, had been "raised in great Plenty," as had sugar maple. And to accelerate propagation because of the "great Demand," *Fraxinus americana*, the white ash, was often grafted on to the rootstock of common ash. One of the most striking additions was the flowering dogwood, a shrub-like tree that was completely wrapped in white flowers in spring before the leaves emerged. In autumn its foliage glowed bright red and purple. Bartram had sent it every year and, by now, many nurserymen stocked it. Another popular plant in the shrubbery was the American wisteria, the seeds of which had been so rare and coveted when they had arrived in 1724 that only a few "curious Persons in London" had got hold of them. Now it was readily available in the nurseries around London.

Even Capability Brown, the man who would be remembered for creating the archetypal English landscape garden, incorporated many of Bartram's American trees and shrubs into his schemes. At Petworth in Sussex, for example, Brown created "a heavy-timbered American forest," and at Tottenham Park in Wiltshire he planted evergreen American cedars, white pines and balsam fir, as well as spring-flowering tulip poplars, and sumacs that turned flamboyantly red and orange in autumn. Later, Brown would also tinge Burton Constable, in Yorkshire, with brilliant autumn colours from American ash trees, sugar maples and scarlet oaks. For the more botanically minded collectors, Brown's attempts were regarded as amateurish, as he tended to plant the more common species rather than the expensive rhododendrons, kalmias and magnolias. Ellis haughtily told Solander, "I met ... the famous Mr. Brown that lays out gardens for the Nobility; his Talents lie in that, not in botany or the Culture of Plenty."

In these first few months in England, Solander thrived on being surrounded by people who adored plants. How different this was to Sweden, he wrote to a friend in Stockholm, because here in England "people have taste in such things." He made many friends and learned to adjust to English idiosyncrasies, such as people constantly changing their minds, an attitude that was, Solander believed, influenced by the daily newspapers, which displayed the "same fickleness of opinion." His chameleon skills eased his way—"Throw him where you will, he swims," James Boswell later wrote. And Solander quickly discovered that flattery was the way to navigate among the English: "[T]ell them that everything you have seen in England is better than anything you have seen before." Although he had planned to stay for only a few months, he now decided to stay at least a year longer, though he did not know how to tell Linnaeus, who was impatiently awaiting his return. Solander hated conflict and always deferred confrontation: "[O]nce one begins to postpone one does it so easily," he had admitted to Linnaeus at a previous occasion—and so it was unsurprising that he just stopped writing to Uppsala.

Solander's extended sojourn in England caused consternation in Uppsala. Linnaeus had sent his favourite pupil to gather plants and to promote his system but had not intended to lose him entirely: "I wish I had had you at home," he wrote in September 1761, complaining that it was now six months since he had received a letter from Solander. At the same time Linnaeus also tried to contact Solander via Ellis. "Pray let me know if my friend Solander is well," he asked. His entreaties were to no avail. Three months later, there was still no letter. "What sirens or furies kept [you] there the whole year I do not know," Linnaeus raged, reminding Solander that it was not prudent to live in such an expensive city for so long. In just a few paragraphs Linnaeus scaled a whole spectrum of emotions: he accused Solander of neglect—"Do not let place and time change [your] love"—reminded him of the duties of a son, as his old mother was "tormented by uterine haemorrhages," and begged to remember the once important teacher "who [was] getting old." The hypochondriac Linnaeus, at fifty-four clearly feeling his age, worried that his anointed successor was wasting time in England.

Solander could not understand his teacher's fury. Had he not done everything asked of him? He had sent all the requested plants to Uppsala and the sexual system was now fully accepted by the reluctant English. Indeed, during the previous twelve months, a summarised English translation of the Linnaean classification had been published, and even Miller had finally acquiesced and endorsed the sexual system by including an introduction to the method in his twelfth edition of the *Gardeners Kalendar* in 1760—albeit implying in the preface that he had been pressured into doing so: he was, he wrote, "advised to subjoin a short Introduction" and "prevailed on to undertake it."* Linnaeus, however, believed Solander should "look at the future." Employment opportunities in Sweden having slipped, Linnaeus was now lobbying for Solander to become the professor of botany in St. Petersburg. "[I]t would make all the rest so much easier," Linnaeus wrote to Solander, because after a couple of years in St. Petersburg, the young botanist could return to Uppsala with sufficient qualifications to succeed Linnaeus.

Linnaeus still did not receive an answer. "I have been longing daily like a bird for daylight to get [your] letter," Linnaeus wrote in March 1762, almost two years after Solander had arrived in London. Solander, meanwhile, made the most of his independence. Instead of living in the shadow of the controlling Linnaeus, with his army of uniformed pupils, Solander, with his penchant for brightly coloured waistcoats, was now his own master. "[T]he more he is known, the more he is liked," said Ellis, who also appreciated Solander for procuring some seeds and cuttings from the Chelsea Physic Garden "without letting it known to whom it is"—well aware that Miller would never part with seeds destined for a man with whom he was still embroiled in botanical disputes.

---

* It would be another eight years before Miller finally adopted Linnaeus's binominal names in the eighth edition of the *Dictionary*. But even then he refused to use some of them. For example, instead of Linnaeus's *Dodecatheon meadia*, Miller continued to list *Meadia dodecatheon*, while the loblolly bay, Linnaeus's *Gordonia*, remained *Hibiscus* in the *Dictionary*.

Callicarpa americana
This was one of Mark Catesby's drawings that made English
collectors and gardeners excited about colourful American plants.

Lady's slipper orchid (*Cypripedium calceolus var. parviflorum*)
This was one of Peter Collinson's favourite flowers. He noted in his copy of Catesby's *Natural History* that this lady's slipper orchid had been sent by Bartram and that it flowered in Peckham in April 1738.

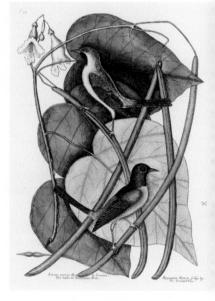

Catalpa bignonioides
Mark Catesby introduced the southern catalpa to England, and Bartram collected them in North and South Carolina.

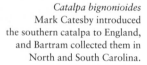

Kalmia angustifolia
Catesby wrote that Collinson was "excited by the view of its dryed Specimens, and Descriptions of it [and] procured some Plants of it from Pennsylvania." Bartram dispatched them regularly, but it took almost three decades until Collinson succeeded in making it flower in his garden.

White pine, also called Lord
Weymouth pine (*Pinus strobus*)
This was one of the most popular
American conifers in the English
landscape garden, adored for its
long feathery tufts of needles and
straight growth.

Tulip poplar
(*Liriodendron tulipifera*)
Although John Tradescant
the Younger had already
introduced the flowering
tree with its strangely
shaped leaves in 1638,
it was Bartram who was
responsible for the large
numbers of specimens in
England's parkland.

S.Edwards del   Pub. by W.Curtis St Geo: Crescent. Jan. 1. 1802.   F.Sansom sculp.

Monarda didyma
Bartram found this flower during his expedition to the League of the Iroquois
near Oswego in 1743 (hence its common name Oswego tea). It flowered
first in England in Collinson's garden three years later.

Liquidambar styraciflua
Bartram always included the seeds of sweet gum in his five guinea boxes because the English gardeners adored the magnificent autumn foliage of this deciduous tree.

Platanus occidentalis
The American sycamore was introduced by John Tradescant the Younger in 1638 and is one of the parents of the London plane (*Platanus X hispanica*). Bartram collected the seeds in Pennsylvania and dispatched them regularly.

Carl Linnaeus in the Lapland dress that he wore while courting his wife-to-be, Sara Elizabeth. He also wore this costume in the Netherlands in order to impress the Dutch botanists.

Linnaea borealis
This delicate pink forest flower was named after Linnaeus who was almost always painted holding it (see his portrait above).

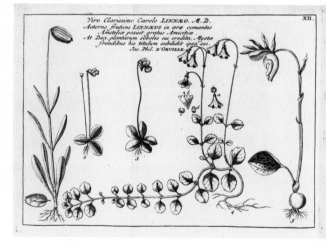

The frontispiece of Philip Miller's *Gardeners Dictionary* (1764) depicts a brick
greenhouse, which is flanked by a stove with a sloping glass roof—
almost certainly the one at the Chelsea Physic Garden.

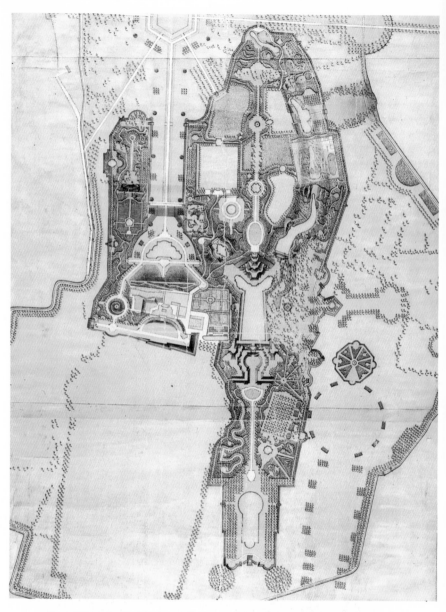

Plan of Thorndon Park in 1733. There are still formal elements left in the garden and park, such as the straight canals and avenues, but many of the groves were laid out in an irregular manner with serpentine walks cutting through them. This is a proposed plan. Lord Petre died before he could implement all elements of this design.

Unfortunately, the fact that Solander was well liked didn't prevent him from being broke. Assisting his English friends with their intrigues and collections did not pay for his expenses. So with no employment opportunities in England, Solander was forced to agree to Linnaeus's suggestions. In May he finally wrote a letter accepting the St. Petersburg offer and in late summer Linnaeus was able to inform Solander that he had been appointed professor of botany. Yet, by the end of November, Linnaeus had still not received a reply from Solander—which even considering transport delays or storms was an inexplicably long time. Once again Linnaeus tried to contact his wayward pupil through Ellis: "[B]y your regard for me . . . give me some account of him by return of post," he implored. However, Linnaeus was unaware that Collinson and Ellis had been plotting to keep the charming Solander in England and that a letter to this effect was already on a ship on its way to Uppsala.

When Linnaeus read the first sentence of Collinson's letter, he knew that this was unlike anything he had received from the merchant before. Instead of polite phrases, botanical chit-chat or gossip about acquaintances, Collinson's words were precise and to the point: in St. Petersburg Solander's accomplishments would be "Lost & Sunk into Supiness," Collinson argued, because with no Russian trade to the East or West Indies the opportunities to receive rare plants from across the globe were slim. Solander would be "Buried & Lost in Obscurity," he continued, and his talent would be wasted on the ignorant "Russian Bears." Rather Collinson and some friends had proposed "a Scheme" for Solander to work in London. Collinson revealed no details, only that its success would be known within four or five weeks. Linnaeus was furious. He had trained Solander to become his successor, an honour which he had gilded by promising his oldest daughter to Solander. He had not expected his investment to be picked off by the English.

Collinson's scheme was for Solander to sort and classify Hans Sloane's vast collection of dried plants, insects and minerals according to the Linnaean system. Sloane's curiosities had been bequeathed to the nation after his death in 1753. Now held in Montague House in Bloomsbury, they formed the basis of the British Museum, a project with which Collinson had been closely involved since its foundation. The collection was useless to students of natural history, Collinson thought, because it was uncatalogued and therefore "the Philosphick World is deprived of an Inconceivable fund of Knowledge." In October—six weeks before he informed Linnaeus about his plans—Collinson had written to one of the trustees of the Museum, suggesting Solander's unique assistance. "An opportunity now offers that may never happen Again," Collinson urged, reminding the trustee that Solander was a pupil of Linnaeus, "the Greatest of Professors."

Collinson felt no guilt in keeping Solander away from Linnaeus; after all, it would help to make England Linnaean. And after years of sending floral duplicates to Uppsala without receiving much in return, Collinson must also have thought he was owed a few favours. Even that September, a few weeks before he wrote to the trustee of the British Museum, Collinson had complained once again at Linnaeus's failure to send the latest edition of *Systema Naturae*. "That is not Kind," he had written to Linnaeus, "must your Oldest Friend be served last?" (He would write again on 17 September 1765 asking, "Must I not have the pleasure of Seeing your Noble Work the Systema Natura before I Die," but a year later he still had not received the book.) In any case, Collinson enjoyed the regular visits of "our beloved Solander" to Mill Hill too much to feel regret. Surrounded by piles of papers mounted with Bartram's dried specimens, they drank "a Chearful Glass of Wine," toasted botanists they admired and classified Collinson's collection— it was as enlightening as it was amusing. Collinson thought that Solander was the "most knowing naturalist of this age."

In December, Ellis wrote to Uppsala on behalf of Solander— who once again avoided direct confrontation—confirming what Linnaeus had feared: "[A]fter consulting many friends here, he [Solander] is determined not to accept of that professorship [in St. Petersburg]." Three months later, in March 1763, Solander took up his new position as Assistant Keeper at the British Museum, severing his ties to Sweden, his teacher, Lisa Stina— Linnaeus's daughter—and his family.

Again and again Linnaeus wrote to Solander, but he had to wait more than six years before he heard from his pupil.

"See what a complete empire we have now got within ourselves"

Look around ... Observe the magnificence of our metropolis—
the extent of our empire—the immensity of our commerce and
the opulence of our people.

CHARLES JAMES FOX, 1771

In 1756, four years before Solander arrived in London, Bartram found himself unable to make his annual seed-collecting expedition into the wilderness. "Ye barbarous inhuman ungratefull natives [are] weekly murdering our back inhabitants," Bartram wrote to Collinson, furious at the fact that he was bound to the environs of Philadelphia and the more populated areas of the colonies because of violent clashes between settlers and the Native Americans, who were allied to the French in the war against Britain. The conflict between the British and the French colonies had been simmering for a long time, and was now beginning to hamper trade. Miller, for example, could deliver only a few American seeds to his German clients because of, so he explained, "the War which is carried on." The animosities had escalated when the French had begun to build a string of forts along the Ohio Valley, a tract of land which the Virginians claimed to be theirs. British regiments had been deployed to protect the forts and trading posts, but some of the first battles had ended disastrously. When General Braddock lost two-thirds of his soldiers in July 1755 because they fought in a rigid linear formation unsuitable for the densely forested areas at the western border of the Appalachians, even Bartram eschewed horticultural for political news in his letters to Collinson. "Oh stupid obstinate Briton: that would not be moved By his soldiers falling round him in garments stained with blood," he wrote after the battle.

These skirmishes were the prelude to what would become

the Seven Years War, a conflict that would impede Bartram's ability to travel and Collinson's trade with the colonies. As with the previous war, a decade before, ships were again captured, the cargo lost to the French. But whereas, before, Collinson had remained a largely resigned spectator, this time he did everything he could to persuade the Lord Treasurer of the importance of the Ohio territories and the danger of the French encroachment. By providing the Duke of Newcastle, who would later become Prime Minister, with up-to-date information from America, via his colonial contacts, Collinson hoped the British government would be swayed. "The Preservation and Enlargement of our Northern Colonies is of highest importance as they are sources of the greatest part of our wealth and trades," Collinson urged Newcastle. If the English did not defeat the French, "they will eat us up by piecemeal," Collinson warned, knowing that a loss of the colonies would be the end of his import and export business as well as Bartram's plant trade.

On 18 May 1756, the threat of French power in America but also in Europe was regarded to be so great that Britain declared war. The same day British regiments embarked across the Atlantic. The colonists who had long asked for more assistance were frustrated that help had taken so long to come but, as Collinson explained, "This is the old English Way." As well as to America, English soldiers were dispatched to many other parts of the world to fight the French. This was a global war in which the British battled for dominance over India, the sugar production in the West Indies and the slave trade in West Africa.

As the war progressed, its effects were increasingly felt within the horticultural community. "[T]his Cruel Warr . . . Interrupts all Sorts of Commerce," Collinson complained, and Miller was now prevented from sending even a few American seeds to his customers in Germany because, during the winter months of 1756, only one out of eight vessels from the colonies arrived in Britain. The situation did not improve when, in January 1757, nineteen ships out of the twenty-one that had sailed from Carolina to England were taken. To circumvent these problems Bartram resumed the method he had used during the previous war: to address some of the boxes to French botanists in the hope that they would forward them on to Collinson—

if not, he wrote, at least his precious cargo would be in "learned [rather] than ignorant hands." Bartram also distributed the boxes over several ships to increase the chances that at least some would arrive in London.

It rapidly became apparent that the British were doing badly in the war. At the end of the second year, as more forts were conquered by the French, Bartram lamented, "Our affairs from year to year grows wors & worse." The British army, though outnumbering the French, failed again and again. Bartram was frustrated and wrote to Benjamin Franklin that despite all the money spent on the army it had "nothing done to ye advantage of ye king or countrey." Franklin was in London at the time, where he had come, as an envoy of the Pennsylvania Assembly, to argue against the right of the Penn family (the proprietors of Pennsylvania) to overrule legislation voted for by the Assembly.

But then, at the end of the third year of the war, Britain's fortunes turned when the strategically important Fort Duquesne was wrested from the French. In the beginning of 1759—the so-called *annus mirabilis*—Collinson dared to be optimistic again, reporting to Bartram, that "Wee have great Hopes as fort Duquesne is in our hands." More victories followed. In September, Quebec was captured, and a year later Montreal surrendered—Britain had ousted her enemies from Canada. By the end of 1759, the British controlled the sugar island Guadeloupe while the French position in India was weakening. At home the Royal Navy destroyed two French fleets which had been ready to cross the Channel, averting the dreaded invasion of Britain.

After years of military failure, the British felt buoyed by their success, even when their victories did not bring an end to the war. "Our bells are worn threadbare with ringing for victories," wrote Horace Walpole, while Daniel Solander was so amused by what he called the "unnatural boasting of the success of their arms" that he mocked the English as a "conceited Nation" who were deluded about their military powers; how else could they really believe that they "could subjugate the whole world if they wanted to"? Though he admitted that this inflated confidence was a kind of strength.

After years of being stymied by the battles, Bartram and

Collinson were excited by the thought of an approaching peace. "Thou will be able to sally forth again on New Discoveries," Collinson promised Bartram when the British captured Fort Duquesne. When it was rebuilt as Fort Pitt (the present-day Pittsburgh)—named after the Secretary of State William Pitt— Bartram decided to take advantage and planned an expedition to the fort in autumn 1761. He was "heated by ye Botanick fire," he wrote to the secretary of the Royal Society, a flame which could only be "quenched" if he could explore all floral treasures in North America—though he still feared the "skulking Jealous Indians."

Bartram didn't like the Native Americans and described them as lazy, drunken and violent savages: the "most barbarous crea- tures," who "skip from tree to tree like monkies." Anxious that Bartram's fear of the Indians might dissuade him from going, Collinson flattered—"thy Penetrating Eye will bring Hidden things to Light"—and cajoled, emphasizing how the English nurserymen were ordering huge amounts of seeds. Many wanted several double boxes in order to fill up their empty stores, but, Collinson warned, they wanted new species because "they say they are Tired of old ones." Many of the American plants had become so common in the English landscape that gardeners needed new species to parade as rarities in their shrubberies— the flipside of the abundance Solander had admired on his first tour through English nurseries during the previous summer. Bartram was furious. Having spent much of the last three decades scouring thick forests and climbing mountains to find new plants for the English garden, sending "seeds of allmost every tree & shrub from Nova scotia to Carolina," Bartram wrote to Collinson, "do they think I can make new ones." But yes, he would go to the hinterland of Fort Pitt, risking his life, even "if I die A martar to Botany Gods," he added.

Bartram's frustration was exacerbated because, at sixty-two, he was beginning to feel his age. The year before he had strained his hips so badly when moving the seed boxes that it took him several months to recover. In the same winter he also broke his arm when he fell from the top of a tree while collecting holly berries and still had problems dressing on his own. "[C]limbing trees is over with me in this world," Bartram said; in the next world he would choose "to fly like an angel" instead.

But despite these ailments and the danger of the forthcoming expedition, Bartram knew he had no choice if he wished to keep in business. Indeed, the market for American species had grown so much that other colonial gardeners were offering their services to nurserymen in England. Alexander Garden wrote from Charleston, South Carolina, to London on behalf of a young gardener who was wishing to collect plants, enquiring which species fetched the highest prices. Collinson repeatedly reminded Bartram of the competition, especially that a fellow gardener from Pennsylvania was also selling seed boxes and plants in London. "Dear John don't be outdone by this Fellow," Collinson had warned in May . . . and again in June . . . and in August.

And so in autumn 1761, despite the ongoing war and clashes between the Indians and settlers, Bartram embarked on his expedition. He rode west across the Appalachians for more than 300 miles, arriving eventually at Fort Pitt, the last outpost of European civilisation, the furthest point of the frontier. But in the whole of his journey, he did not discover a single tree or shrub that was not already growing in his garden. After travelling 800 miles he returned disappointed to his farm, where his mood declined even further when in the following spring a cold spell "disrobed" all his Carolina evergreens and shrubs except those he had kept in his greenhouse.

The death of his Carolina plants was particularly annoying because they were the "flowering Shrubs that grow quick & Show," which the greedy English gardeners ordered in large numbers. "Oh Carolina Carolina A ravishing place for A curious Botanist," Bartram said, as its warmer climate produced the most flamboyant blossoms in the colonies. From this southern colony Bartram had procured species such as the red-flowering *Robinia hispida*, the perfumed Carolina allspice and the elegant *Magnolia grandiflora* which Solander had admired in Collinson's garden. Although Bartram had obtained most of these by swapping his plants from Pennsylvania, Virginia and Europe with correspondents in the South, he had once made a trip to Carolina himself. Many people he had met there had been charmed by his "gaiety" and "happy Alacrity." The widow Martha Logan, for example, ran a nursery in Charleston and would, Bartram bragged, "pass thro fire or water to get any

curiosity for mee."* Until now he had dispatched what he propagated in his garden but after the destructive frost he had few
Carolina plants left.

At about the same time Bartram received a large order from
a London nurserymen for southern plants. "I would have all
these Chaps that wants quantities of these curious shrubs to
go to Carolina themselves to fetch them & then thay will know
how to get A proper value of them," he complained to
Collinson, explaining—as so often—that he could not just pluck
the plants from a flowerbed but was risking "life or limb."
But he knew that he had no choice but to undertake another
trip—although, this time, he would take his twenty-three-year-
old son William with him to climb the trees.

Over the previous years Bartram had worried much about
his son's future. William was a talented botanical painter but
Bartram wanted him to learn a "proper" profession, even
though William's attempts to set up as a merchant had failed.
Bartram planned that, after the trip, William would stay in
Bladen County in North Carolina to work with his uncle.

In autumn 1762 father and son set off. The plant Bartram
wanted more of was the wondrous Venus flytrap, or "tipitiwitchet sensitive" as he called it. It was the most exceptional
flower he had ever seen, because at the tip of each leaf was a
mechanism to catch insects that looked just like a bear trap
writ small. Sweet-smelling nectar lured flies or spiders into a
reddish mouth rimmed with sharp spikes, which snapped shut
over its prey to allow it to be slowly digested. The plant seemed
to bridge the world of flora and fauna. A French visitor who
came from Montreal, Bartram said, "burst with laughing" on
seeing the toothed trap closing. His English customers, Bartram
was sure, would pay him handsomely for such an unusual
plant. He had only a single root which had been sent from
South Carolina. All summer he carefully nurtured it, watering
it every night through holes in the soil so that not a "drop of
water" touched the leaves.

* Collinson teased Bartram about Martha Logan's dedication. On 1 August
1761 he wrote, "I plainly see thou knowest how to fascinate the Longing
Widow by so close a Correspondence ... now I shall not wonder if thy
Garden abounds with all the Rarities of Carolina."

Bartram rode to South Carolina and back in an enormous loop passing his familiar haunts in Maryland and Virginia, then travelling on from South Carolina to Georgia, before returning along the Appalachian Mountains. He bathed in warm springs, crawled through a seven-mile-long cave and admired a 200-foot waterfall. "I had ye most prosperous Journey that ever I was favoured with," he told his family when he returned in early November, adding that "my Gardien Angel seemed to direct my step to discover ye greatest curiosities." His harvest included coveted rare shrubs, a "glorious evergreen," as well as seeds and specimens of the Venus flytrap. Back at his farm he packed up his treasures carefully, pressed and mounted the Venus flytrap. Then he delivered them to the docks in Philadelphia and looked forward to hearing of Collinson's delight.

Meanwhile Bartram's previous letter, in which he had described the Venus flytrap, had arrived in London. Yet Collinson was not impressed: Bartram had sent neither a dried specimen nor even a little sketch. "While the French Man was ready to burst with Laughing I am ready to Burst with Desire," wrote Collinson, warning Bartram that if he did not send a specimen or at least a drawing in his next letter, then "never write Mee more for it is Cruel to tantalise Mee." Two and a half months later, in February 1763, Collinson had still not received the cargo from Carolina, the "Terrestrial Paradise." A

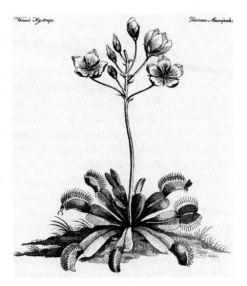

month later he heard news that the ship with Bartram's boxes had been captured by the Spanish. For once, though, they were lucky: the vessel was returned to the British, since it had been taken just as the long war had come to an end with the peace treaty of Paris. In May, six months after Bartram had sent his harvest, Collinson travelled to Custom House to retrieve it. Together with Solander, Collinson examined the much awaited Venus flytrap. It was a new genus, Solander assured him, which nobody in Europe had ever seen. And so Collinson parted from this only specimen, his "Jewel," and dispatched it to Sweden to have it named and described. "Linnaeus will be In raptures," he wrote to Bartram. Perhaps it was also a botanical peace offering in restitution for having kept Solander in London.

Solander identified nearly half of the collection to be new species, Collinson trumpeted, so that "All Botanists will Joyn with Mee in thanking my Dear John for his unwearied Pains." He gave the few seeds of the Venus flytrap to the nurseryman James Gordon, "for his skills Exceeds all others," and at the same time tried to get more from his long-time correspondent Arthur Dobbs, the governor of North Carolina. When the seventy-three-year-old Dobbs refused to help, Collinson believed the reason was the governor's marriage to a woman more than fifty years his junior, writing to Bartram, "It is now in vain to write to Him for seeds or plants of Tipitiwitchit now He has gott one of his Own to play with."

As Collinson had waited for Bartram's cargo, the political map of the globe was being redrawn. Having defeated France, Britain was now the largest colonial power in the world. The territories of the empire now stretched from India to Africa and from the West Indies to North America. "Now my dear John," Collinson wrote to Bartram, "look on the map and see by this glorious peace what an immense country is added to our long, narrow slip of Colonies . . . [w]hat a grand figure it will now make in the map of North America." Suddenly Bartram's world was larger than ever. Spain had lost Florida to Britain while Canada had been wrested from the French, as had the land east to the Mississippi. Instead of a thin strip along the east coast that was bordered by the Appalachian Mountains, the British colonies now occupied a vast territory that stretched from Newfoundland and the Hudson Bay in the north to the Gulf of Mexico in the south and across the Ohio valley all the way to the Mississippi.

"[D]oes not the ardour of curiosity burn in thy mind to explore the wonders of Louisiana?"* Collinson asked Bartram two months after the peace treaty. With their warmer climate, the newly acquired southern territories promised to hold horticultural treasures that had never been seen in English gardens. It was, Bartram agreed, the "very palace garden of ould Madam flora." Even before he had received Collinson's letter, Bartram had discussed joining his friend Colonel Bouquet's expedition to the Mississippi "when peace is proclaimed." Here he was certain to "find more curiosities" in six months than "ye English french & spaniards have done in 6 score years."

As Bartram prepared for new adventures, Collinson retired from his business, moved permanently to Mill Hill and enjoyed the fruits of their long friendship. After more than twenty years he even succeeded in bringing his *Magnolia grandiflora* to flower. Collinson had already seen, in his friends' gardens, how the tightly closed buds of the magnolia unfolded their flawless white petals among shiny evergreen leaves, but nothing equalled the pleasure of seeing "the most beautiful Tree in America" flower in his own shrubbery. An even greater achievement was the flowering of the kalmia. Placing a blossoming sprig in a vase on his desk, Collinson wrote to Bartram, "it stares mee in the face ... saying, or Seems to say, as you are so fond of Mee tell my Frd J Bartram who sent Mee to send some More to keep me Company." During the first winter after the war Collinson reread the thirty years of correspondence with Bartram, reliving their triumphs, their failures and their growing friendship. Bartram had sent Collinson a sketch of his house and garden done by his son William, and Collinson wrote, "My Dear John I have been Walking in *thy Garden.*" Sometimes, though, when Collinson leafed through the more recent letters, he seemed to forget that Bartram, like himself, was already in his sixties, complaining that his friend had encountered no beast wilder than a rattlesnake. Why had no "Panther ... sprung

---

* In fact the French would give Louisiana and New Orleans (and the vast territory west of the Mississippi) to the Spanish to compensate for the loss of Florida. Only in 1803 did Louisiana become part of the United States, when Thomas Jefferson bought it from Napoleon for $15,000,000.

out of a Thickett" or at least "a bear [been] wakened from his Den," he asked Bartram from the comfort of his soft armchair.

When Collinson read their letters he must have realised how much Bartram had changed during the past three decades. Where once he had been an isolated provincial farmer, Bartram now counted politicians, scientists and generals—the most influential men in the colonies—as his closest friends. His observations on the natural history of America were respected in Europe and he corresponded with leading thinkers and botanists from across the world. Bartram was confident and successful. His daughter had married well, his sons Isaac and Moses ran a successful apothecary business, he earned well over £100 annually just from his plant export and owned several properties in and around Philadelphia.*

His house at the Schuylkill River stood as a sign of his acquired wealth, taste and social standing. Over the previous decade, especially during the years in which the war curtailed his expeditions, Bartram had transformed his farmhouse into a large country mansion—a handsome grey stone building with three imposing two-storey-high columns at the entrance. He had finished the columns with scrolled capitals, carved the window surrounds with bold curving lines, and decorated lintels and stones with stylised flowers, fruits and leaves as well as simple arabesques. For the design of the house, Bartram had brought together all he had learned about fashionable architecture. It was inspired by the grandeur of the large country houses he had visited during his travels across the colonies, and informed by conversations with like-minded men such as his friend Edmund Woolley, who had built the finest public building in Philadelphia, the Pennsylvania State House (later called Independence Hall). Bartram had used the pattern books from the Library Company, quoting and mixing elements from baroque and Palladian architecture such as the arabesques and columns. In England a comparable house would have been built in smooth white Portland stone, but Bartram had used a coarse, grey local stone which gave the classical designs a rough,

---

* In 1751 Bartram earned £116 and in 1752 £133. The English orders remained similar throughout the 1750s and 1760s—he averaged about twenty five-guinea boxes a year, but sometimes sold more. In 1757, for example, he dispatched twenty-six boxes.

"SEE WHAT A COMPLETE EMPIRE WE HAVE NOW GOT"

almost rustic look. The result was a highly idiosyncratic version of Georgian architecture—but nonetheless grand by colonial standards, where columns and carved stone were still rare.

The interior of the house consisted of a spacious hall where the family and guests dined, with a large kitchen on one side and a comfortable parlour on the other. Bartram always delighted in offering his visitors a wholesome meal and inviting them afterwards to his study, or "Chapel" as he sometimes called it, where a bottle of Madeira and a collection of fossils, minerals and dried plants awaited them. Next to his natural history specimens, Bartram also displayed mementoes that showed his friendship with the finest scientists and thinkers in the world: there were, for example, botanical prints that Collinson had sent, portraits of Linnaeus and Franklin, as well as a silver cup that Hans Sloane had sent him twenty years previously. To Bartram's great pride, Sloane's name was engraved on the cup "so that when my friends drink out of it they may see who was my benefactor." On a warm summer evening Bartram would sit with guests on his porch, where *Bignonia capreolata*, a twining evergreen vine, hugged the stone columns that framed the view. From here they would look down towards the river. "My garden now makes A glorious appearance," Bartram told Collinson.

One of the shrubs of which Bartram was most proud was the evergreen Portugal laurel, which the Victorians later used—trimmed into neat balls—in imitation of orange trees. Bartram was the only one to grow it in the colonies and would not even "take ten guineas A piece for them." He also adored his horse chestnut,* having waited three decades to see the white, candle-like flowers that Collinson had once likened to a pyramid of hyacinth blossom—"one of the finest sights in the World."

Bartram was particularly fond of old-fashioned English flowers and was proud of his yellow, red and "iron colored" foxglove, which grew "lustyly." There were also fritillaries—an Elizabethan favourite for their chequered petals—as well as aconites (monkshood), tulips, anemones, irises and ranunculus. He was also pleased with the carnations, which flowered in

---

* These were possibly the horse chestnuts which Collinson had given Bartram's son Moses when he had visited him in London in 1751.

"ye brightest colors that ever eyes beheld," and double sweet Williams "of such A high blush as is dasling to look at." Some plant introductions, however, had been too successful. Thus when yellow toadflax (*Linaria vulgaris*)—a much loved flower in Britain—invaded the landscape like a weed (as it still does), Bartram was worried about the havoc. He was as troubled by these invasive species as today's conservationists are by the rapid spreading of *Rhododendron ponitcum*, Spanish bluebells and Japanese knotweed across Britain's countryside or by kudzu and Tree of Heaven that choke the landscape of large parts of the United States.*

Many of the Carolina plants were also thriving. There was *Halesia carolina*, also called silverbell because it was bejewelled in spring with drooping clusters of white bell-shaped blossoms, and the purple-berried callicarpa, which remained rare in England as it was not hardy enough. When the Carolina pitcher plant flowered, Bartram's wife took it to Philadelphia, where, so he told Collinson, it "was much admired."

The garden at the Schuylkill River also contained plants of Bartram's own creation: hybrids such as a larkspur which united upright stature and double blossom. "A gigantick monster of A flower," Bartram had boasted, obviously not tormented by any of the blasphemous thoughts which had troubled Thomas Fairchild almost half a century earlier. Next to the house was a greenhouse that was filled with geraniums, auriculas, bulbs from the Cape and Persian cyclamen—and heated with a "Pennsylvania fireplace," a cast-iron stove which Franklin had invented twenty years previously. The greatest horticultural success was Bartram's thriving pineapple plant, a feat that only the best gardeners in Europe achieved. "[T]hese I am proud of," Bartram immediately reported to Collinson, "as I think I am ye onely Proprietor of them in ye country."

"I can chalenge any garden in America for variety," Bartram boasted, and his stock of English and non-American plants was

* Similarly invasive were Norway maple, which Miller had sent Bartram in 1757, and paper mulberry. Collinson and Miller had received paper mulberry from China from a Jesuit priest and had forwarded it to Bartram. Its tolerance to pollution and dust makes it a particularly hardy species in American cities and roadsides, where it often forms thickets as it spreads from root suckers.

now so great that other colonial nursery owners ordered from him rather than from England. Bartram's collection and superior skills were well known in the colonies: fellow gardeners asked for horticultural tips and Bartram supplied customers from Fort Pitt to South Carolina. One from Charleston, for example, desired old English favourites such as sweet peas, wallflowers and carnations as well as seeds of *Callistephus chinensis*, which Bartram had received in one of the first dispatches from Collinson. As the American trade increased, Bartram grew reluctant to send living Carolina plants to England because he could sell them for double the price in Pennsylvania and Virginia and, he explained to Collinson, "have my mony directly without any risk or ensurance."

By turning towards the growing American market, Bartram was acknowledging a change in the colonies. As America shifted from being a wilderness frontier to an incipient nation, there was, for the first time, a class of wealthy colonists who had the leisure and money to spend time on their gardens. The colonies were, as Collinson said, "emerging out of their Infant State & begin to Walk alone." But there was also a change in Bartram. He, too, was becoming more independent. When the nurseryman Nathaniel Powell, for example, complained about the content of a seed box, Bartram dismissed the criticism, writing to Collinson, "I think powel need not make such A bluster." And when he felt cheated by the ungenerous horticultural gifts of his English correspondents, Bartram complained to Collinson "that you send us all ye worst & will not let us have ye rare ones." Indeed, he felt so frustrated that he even threatened to dig up all his flowers and build a pyre to make a "burnt sacrifice." Eventually Bartram's complaining paid off when Collinson dispatched boxes of "curious roots," to which Gordon added bulbs such as martagon lilies and hyacinths as well as some choice alpine plants.

With his growing success, Bartram was confident enough to demand to be treated as an equal. He engaged in intellectual debates with his correspondents, and, in tandem with the horticultural ripostes, was happy to contradict Collinson's opinions on colonial governance and politics. So while Collinson was optimistic about the stability of the peace, Bartram insisted that the Native Americans "will not keep to any treaty of peace."

The only method to keep the truce, Bartram explained, was to "bang them stoutly." And when Collinson said that "Wee are the Agressors" because the Indians were made drunk and cheated off their property, Bartram fired back: "we have more reason to destroy them & possess thair land then you have to keep Canada [from the French]." Bartram's staunch rebuttals revealed a man who no longer thought of himself as a subordinate.

Similarly other colonists were insisting on being treated as equal subjects. There was a growing unease at the lack of control the colonists exerted over the military and economic issues that affected their country. Already a decade previously Franklin had pointed out that they were being treated "as a conquer'd People, not as true British subjects." Now he made more vocal demands of the British government: "give us Members in your Legislature, and let us be one People," he insisted. But instead of accommodating these colonial concerns Britain chose to ignore them.

Weighed down by her own problems, Britain was struggling with the costs of managing a global empire. The initial euphoria at winning the war quickly waned as the country was burdened by war debts, unemployed demobilised soldiers and the effects of bad harvests. To alleviate the deficit, George Grenville, the new Lord Treasurer, proposed that colonists should be made to pay—the first time they would be taxed directly to raise money for Britain rather than to regulate their own trade and commerce. Grenville's idea was to impose a duty on the import of sugar,* and then, in March 1764, he notified the House of Commons of his intention to introduce a stamp tax—a duty on paper which, given its use for newspapers, legal documents, almanacs and even playing cards, affected almost everybody in America. In response the colonists mounted a campaign insisting that as inhabitants of a colony they had the same rights as the subjects in the mother country—to be taxed by their own representatives.

Shortly after Grenville's proposals Franklin wrote to Collinson, "We are in your Hands as Clay in the Hands of the Potter," but warned that "the Potter cannot waste or spoil his

---

* The Sugar Act was passed in April 1764. It imposed duties on French West Indies sugar, which was essential for the production of American rum—the most popular drink in the colonies.

Clay without injuring himself." A sentiment which John Adams, one of the future Founding Fathers of the United States, repeated when he wrote in the *Boston Gazette*: "Admitting we are children, have not children a right to complain when their parents are attempting to break their limbs, to administer poison, or to sell them to enemies for slaves?" Though these were apprehensive times, the colonists still believed themselves to be loyal subjects of King George III and were not talking about independence yet. Franklin, for example, still believed in what he called "the importance of a firm union between the two countries" and Bartram, in particular, thought the greatest honour was the title of "royal botanist."

In September 1764, just as protest at the stamp duty was beginning to grow, Bartram heard that his neighbour and fellow gardener William Young had been nominated Queen Charlotte's botanist. Bartram complained about the "sudden preferment" of a man who in his opinion had just "pick[ed] up A few common plants," compared to the risks and productivity of his own thirty-year career. Franklin was similarly outraged at Bartram's "Neglect." He suggested that Bartram pack up a box of seeds for King George, and promised that, if Collinson couldn't deliver the box to its royal addressee, he would use his own connections in England to do so. At the same time, Franklin dashed off a letter to Collinson saying, "I wish some Notice may be taken of John's Merit." A few days later, in October, Bartram dispatched some of his finest seeds to Collinson, to be passed to King George. "I hope thee will find some way to forward ye box," he wrote, asking if, in return, Collinson could send him a list of what Young had delivered to the Queen.

Having sent the box to England Bartram could do no more than wait. By the end of the year he had already become so impatient that he asked Franklin, who had once again travelled to London as the agent of the Pennsylvania Assembly, to question Collinson directly. "Our Friend Peter is not dead, as you apprehended; but . . . hearty, brisk, and active as a Youth," Franklin replied in February 1765, and reassured Bartram that the King had been pleased with the seeds. Two months later, Collinson was able to send news that Bartram had been made King's Botanist.

As Collinson's letter was loaded on to a ship to inform

Bartram of his royal appointment, so were other letters and newspapers for the colonists, all filled with reports that the Stamp Act had been given Royal Assent.* Horticulturally America had never been closer to England, with English shrubberies and parkland populated by plants from the pine forests in Canada to the Carolina swamps, but politically fissures were growing.

Although the colonists had always respected British rule and been fond of "Old-England," with the introduction of the stamp duty this was "very much altered," Franklin explained. A few months later, in early summer 1765, as Bartram prepared an expedition to Florida in his new position as the King's Botanist, many colonists rallied together to oppose the Stamp Act. They burned effigies of British officials, picketed royal offices, destroyed one tax collector's house and raided the homes of others. The "Sons of Liberties," as they called themselves, questioned Britain's right to tax the colonies without parliamentary representation but quickly began to discuss wider issues of liberties and rights. Alexander Garden, the man who had lent his name to the gardenia, was certain that the "die is thrown for the sovereignty of America!"

Since most of his customers were English, Bartram distanced himself from the conflict. Collinson for his part was concerned about the "riotous Mobs" in the colonies mainly because the boycott of British goods ruined his trade. He hoped that parliament would find "some happy Medium," an idea he discussed with Franklin at Mill Hill during the winter before Franklin was questioned for four hours in the Commons in January 1766. While Bartram roamed through Florida in search of new plants for his English clients, Franklin told British MPs that the colonists would never submit to this kind of taxation, and that the insistence on the Stamp Act would mean a complete loss of that respect for Britain, as well as the end of "all commerce that depends on that respect." It was this testimony, many believed, that swung parliament to repeal the Act. Thus, on 18 March 1766, almost exactly a year after it had received

---

* The House of Commons passed the Bill on 27 February, the House of Lords on 8 March and the Royal Assent was given on 22 March 1765—all without much opposition in Britain.

Royal Assent the Stamp Act was abolished, and Collinson hoped that "all Animosities will Cease & Trade & Business be restored." On the same day Bartram left the Spanish settlement St. Augustine in Florida, making his way up north to catch a ship in Charleston which would take him home to Philadelphia.

Back at his farm after his nine months' absence, Bartram picked up his correspondence with Collinson and once again horticulture took precedence over politics. "I have brought home with me A fine Collection of strange florida plants," Bartram wrote to Collinson, assuring him that he was about to dispatch a "little box directed to thee." On his way from Savannah to St. Augustine he had found "a rare tupelo with red acid fruite called limes which is used for punch"—*Nyssa ogeche*—and other trees that thrived in the wet conditions of the treacherous swamps. There were also different sorts of magnolia, "A shrub like the dogwood," and a "Cypress in deep water." He was very excited about "A very odd Catalpa with pods round as an acorn & short" which was probably *Pinckneya bracteata*, a tree with a showy pink blossom. There were many others: "A monstrous convulvolus," "a shrub aster," as well as a "very curious shrub" from near the Alatamaha River that he had never seen before. Bartram was sure that the latter was a new species if not a new genus, but he could not find any seeds to take home. Years later his son William would return to collect seeds of the plant, which would be named in honour of Benjamin Franklin: *Franklinia alatamaha*.*

As Bartram packed the Florida seeds, thieves emptied Collinson's flowerbeds at Mill Hill for the third time in as many years. The robbers must have been commissioned, Collinson told Bartram, because they only went for "my most Curious plants": lady's slipper orchids, magnolias, rhododendrons, the carnivorous pitcher plants and "thy kind present of Loblolly Bays [*Gordonia lasianthus*]." The loblolly bay, an evergreen with fragrant white flowers, was the greatest loss, as it was the rarest plant of all. Collinson thought it "next in Beauty to

---

* The beautiful *Franklinia alatamaha* was only ever found in Georgia, in the location where Bartram saw it in 1765, but within a few decades it was extinct in its natural habitat. Reputedly all of today's specimens across the globe derive from William Bartram's seeds and the seedlings sent from the Bartram nursery. John Bartram never saw it blossom.

the Magnolia." With the English craze for American trees and shrubs beginning to reach the same proportions as the "tulip fever" that had gripped gardeners and collectors in the previous century, such thefts were becoming increasingly common. To deter criminals, Collinson and his friends had a Parliamentary Act passed that same summer whereby plant thieves could be punished with transportation to one of the American penal colonies. But it did little to assuage Collinson's grief at his losses. "I am Ruined," Collinson lamented. "[O]nce I bore the Bell," he wrote to Bartram, but now he believed he was outdone by many other gardeners.

Though Collinson would make repairs to his garden, and continue to derive huge pleasure from it, things were not as they used to be. He was feeling increasingly old and tired, and, in December 1766, the winter after Bartram's Florida expedition, he signed his will.* He was seventy-two and Bartram was sixty-seven. Slowly reports of declining health and ailments began to enter the two men's letters. In Florida Bartram had battled with fevers and jaundice. He also seemed more prone to falls and, that winter, his legs were covered with infected sores. One leg was bruised, Bartram wrote to Collinson, and "I cut the other shin with an adze." Collinson sent "Boxes of pills" and told Bartram how he suffered more than ever under the heat of the summer, as well as having bought spectacles for his failing eyesight. Instead of discussing their ambitions, they now gossiped about their old acquaintances or described their gardens. "Our good frd B Franklin grows fat & Jolly," Collinson joked, while Bartram reported how the Queen's botanist, William Young, "struts along ye streets whisling with his sword & gold lace." Collinson, who in the past had often written hurried letters because of the demands of his business, now "scrawl[ed] on" to indulge Bartram.

Collinson rarely ventured into town anymore; instead he welcomed his friends at Mill Hill—"Think how happy I am," Collinson wrote to Bartram in July 1767, "to have the Two Doctors, Franklin & Solander [as] My Guests, for a few Days."

* Collinson's will was short (less than half a page) and concise: his sole heir was his son Michael. The only specific items Collinson listed were "my Six Cabinets of Natural Curiosities," "Glasses of Animals," "colour'd prints and drawings" and his books—nothing else was of any importance to him.

His garden gave much joy too. There was the *Robinia hispida*, which was loaded so heavily with its tassels of pink blossom that Collinson had "to prop up the branches," and in winter he loved walking through his perfume-laden greenhouse while the garden outside was powdered in snow. "I should be ashamed to write such insignificant stuff to anybody Else," he admitted to Bartram; "commit it to the flames."

With the ageing of the old friends, an era that had seen new discoveries, seismic shifts in the understanding of the natural world and the beginning of the English love affair with gardens seemed to be coming to a close. Miller, still irritable and bad-tempered, was at last accepting the Linnaean nomenclature in his eighth edition of his *Dictionary*, which was published in 1768 when he was seventy-seven. Fairchild's blasphemous Mule was now described as "extremely beautiful" and "proper to adorn court-yards." Solander was assisting many collectors to classify their herbaria, minerals and shells according to Linnaeus's method, as well as cataloguing and describing Hans Sloane's gargantuan natural history collection at the British Museum, where it was on show for tourists and accessible in the reading room for scholars.

Much had changed during the past five decades. Bartram's trees and shrubs were easily available for amateur gardeners in any of the numerous nurseries in the country. Gardeners now produced cucumbers at Christmas, peas and beans in February, and grapes in May—and for the outrageous sum of a guinea or two the spoilt customer could even buy a pound of cherries in March. "O, Botany Delightfullest of all Sciences," Collinson rejoiced.

Flowers had become an "infatuation" in England, Collinson believed, "so great is the Itch that a poor Raged Shoemaker a Weaver or a Baker will give half a Guinea or a whole for a New flower." And species travelled so fast across the globe that Collinson was astonished to hear that Bartram had already seen the blossoming of the South African gardenia in America— less than ten years after it had first flowered in England.

One of Collinson's greatest pleasures was to inspect the great parks and pleasure grounds that he had helped to create, meticulously noting the age and height of maturing trees in his commonplace book. And once again it was Thorndon that made the ageing Collinson the happiest. Although nothing was left of the great hothouses and the nurseries were overgrown, the young Lord Petre had become interested in his father's endeavour. When Collinson ordered a ten-guinea "Double Box" from Bartram on behalf of Petre's son, he was overjoyed: "Little did I think when I gave thee the first like Order for his Valuable Father . . . that I should Live to give the Like for his Son." In the summer of 1768, Collinson decided that he had recovered enough from a severe attack of gout to travel to Thorndon to see how it was progressing. He marvelled at the way the saplings that had not been sold in the 1740s had transformed the parkland into an American forest. But while surveying the extraordinary landscape he had helped to create, Collinson was taken ill. Weakened by his previous illness and the brisk walks across the park, he was—so his physician reported later— "seized with a total suppression of urine."* On 11 August 1768 Collinson died.

The devastating news took three months to reach Bartram. With the death of his great friend, the strong bond that had stretched across the Atlantic, binding him to Britain, was broken. He grieved for Collinson and fretted about what to do with his boxes. Not hearing from any of his English correspondents or from Collinson's heir, his son Michael, Bartram turned to Benjamin Franklin. "I have not any so intimate or capable as my dear Benjamin to take care of ye Box," he wrote, forwarding his annual seeds for the King. Months later, though, Bartram was still unsure how to manage his business in future. Franklin had succeeded in delivering the box to the King, but had no information as to who might take over Collinson's role. "I am at A loss to know whether I must send any more plants or seeds," Bartram wrote back on 10 April 1769, a few weeks short of his seventieth birthday. He was exhausted; his eyesight was so bad that he could not distinguish his plants in his garden "without stooping to view them near" and all winter he had battled with painfully sore ankles and ulcers. "[M]y travails as to any distant provinces is at an end," he wrote to his old friend.

---

* It was most probably the gout that triggered the kidney failure, as it is caused by high levels of uric acids which deposit around joints but can also form kidney stones and block the kidney tubules.

Bartram would never again truly regain control of the English market. Two years after Collinson's death, he complained that he was still "utterly in ye dark relateing to my affairs in England," and that Collinson's son Michael had still not authorised the transfer of his outstanding balance of several hundred pounds nor had seeds been ordered for the English clients. The old man felt that life was slipping by. William Young, his old adversary, had announced in the newspapers across the colonies that he was the botanist of the Queen *and* the King, and even his family troubled him, as his son William, after his failed experiment as a plantation owner, had "absconded," so Bartram told one of his English correspondents, but "I know not whither."

In spring 1771 Bartram finally retired, leaving his nursery in the capable hands of his son John Jr., who increasingly catered for the growing American market. Bartram just survived to see his wayward son William return in 1777 from his long travels in Florida—his bags full of specimens and drawings of plants that Bartram had never seen before. And after writing a long letter to his now grown-up children, reminding them to be moderate and virtuous, John Bartram died after a brief illness in September 1777, aged seventy-eight.

Part III

ARVEST

10

"Ye who o'er Southern Oceans wander"

The Pacific is the desert of waters—we seem to have sailed out of the inhabited world.

CAPTAIN SIR HENRY BYAM MARTIN, July 1846

> *Ye who o'er Southern Oceans wander*
> *With simpling B——ks or sly S——r;*
> *Who so familiarly describe*
> *The frolicks of the wanton Tribe*

ANONYMOUS, "Transmigration," 1778

While Peter Collinson lay dying that summer of 1768, Daniel Solander was involved in frantic preparations for a journey. At the end of June he had been granted leave by the trustees of the British Museum, where he had lived and worked for the past five years, giving him only a few weeks to pack up his personal belongings, brief his deputy at the museum and organise the necessary provisions. It must have seemed a short time, for, if Solander had uprooted himself to come to Britain, it was as nothing to the voyage he was about to undertake. In an age when most people never left the vicinity of their village, Solander was intending to travel to the other side of the globe, into territories where no white man had ventured before. He had joined the most daring expedition the British had ever planned, and in just a few days, would board the *Endeavour* and set sail into the unknown.

Along with Daniel Solander would go ninety-three other brave men. There was the usual crew of sailors, a carpenter, a boatswain, a surgeon and officers, all paid for by the Admiralty, but also a small party of civilians, of whom Solander was one, which included a botanical draughtsman, a landscape artist, four servants, a secretary with some knowledge of natural history, and Joseph Banks, a wealthy gentleman who had financed their passage.

At twenty-five, Banks was ten years Solander's junior, dashingly handsome, rich and confident. His face was beautiful, perfectly shaped with full lips, dark eyes and a strong chin adorned with a dimple. He was tall and energetic, a man who had always enjoyed outdoor activities such as swimming and fishing and had happily ignored his studies while at university in Oxford, instead wandering the country lanes in search of plants.

Banks had been fascinated by botany since childhood, when he had, according to a friend, taken to the subject "in a most violent degree." When, in 1761, his mother had moved to Paradise Row in Chelsea, the eighteen-year-old Banks had spent many hours in the Physic Garden, admiring Philip Miller's collection and dreaming himself into the distant countries which were conjured up by the alluring smells, exotic petals and strange fruits that grew there. By this time he had already inherited extensive estates in Lincolnshire which made him wealthy and independent. Aged twenty-three, he became a Fellow of the Royal Society, the beginning of a scientific career that would see him become the most influential man of the British Enlightenment. Unlike other gentlemen who were active in the Royal Society, Banks was willing to risk his life in the name of science. He was driven by a thirst for knowledge but also for adventure, and was ready to swap his luxurious and comfortable life for the dangers and deprivations of exploration.

For the past four years Banks had been a regular visitor of the British Museum, where, in the dark, damp reading room, he devoured books about taxonomy and the science of plants. It was here, over Sloane's herbaria, that Banks had met Solander. Like everybody else, Banks was charmed by Solander's easygoing manners and impressed by his botanical knowledge—he was "esteemed as a savant," one of Linnaeus's other pupils wrote after a visit in London, "and loved as a companion, full of spirit and life." Like other botanists and plant-collectors of his generation Banks was a "Linnaean." He was planning a visit to the master himself in Uppsala when he heard about a ship that was departing for a very different destination, and made a radical change to his plans.

The declared purpose of the *Endeavour*'s expedition was to

observe the rarely occurring transit of Venus in order to learn more about distances in the solar system. However, when, a few months before the _Endeavour_'s departure, another ship— the _Dolphin_—returned from a voyage to the South Sea and its captain reported sighting a landmass to the south, the Admiralty added a secret instruction to the _Endeavour_'s brief: to find this elusive continent of _Terra Australis Incognita_ and to return with information about its soil and plants, and the "temper" of the natives.*

Banks had already shown his adventurous side by taking part in an expedition to Newfoundland two years previously. If asked whether, like other English gentlemen, he would be touring the ancient treasures of Europe, he reputedly answered, "Every blockhead does that. My Grand Tour shall be one around the world." Nevertheless, trying to persuade the Royal Navy to take him and the learned group he wanted to travel with aboard the _Endeavour_ proved difficult. In the end, he had to sidestep the First Lord of the Admiralty—who had been against his participation—to get his way, gaining the approval from the First Secretary through family connections. It helped that Banks was willing to pick up the £10,000 bill for his botanical adventure himself—around 100 times the yearly wage of the _Endeavour_'s captain. Even with an annual income of around £6,000 it was a hefty price for Banks to pay but he believed it worth every penny. "I am a man of adventure," Banks declared later, and he knew that an opportunity such as this might never occur again.

Solander first heard about the _Endeavour_ at a dinner party, where Banks talked about the planned route and the "excellent opportunity . . . to improve science and achieve fame." Jumping up from his chair with an excited glimmer in his eyes and passionately gesticulating, Solander burst out, "I want to go with you"—an offer Banks was delighted to accept, for Solander would be not only a pleasant companion during the long months at sea but also invaluable for his unparalleled botanical knowledge.

Meanwhile at Deptford the _Endeavour_, a vessel that had been

---

* Ptolemy had suggested the existence of a southern continent in the second century A.D. The theory was that the continents of the northern hemisphere had to be counterbalanced by a landmass in the south. _Terra Australis Incognita_, it was believed, would mirror the climate and soils from the northern continents, promising wealth and commerce to the discovering country.

used for transporting coal along the British coast, was being refitted for her long voyage. At about 100 feet long, the three-masted ship was smaller than a naval frigate but because she was rounder and bulkier she offered a larger storage capacity. One of the most important advantages for an expedition into unchartered territory was that she only drew fourteen feet of water when fully laden, allowing her to be steered into shallow coastal areas. She was sturdy, and, as the crew would soon discover, the perfect boat to ride out a storm. To prepare the *Endeavour* for her adventures, she was re-rigged and covered in a mixture of

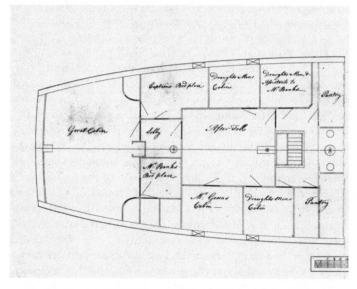

Details of the *Endeavour*'s cabin layout after the alterations.
Note Banks's cabin next to the Great Cabin.

pitch, tar and sulphur to protect her against tropical water worms. To armour her further she received an extra layer of thin planking at the hull below the waterline. Inside the ship, space was precious and Banks's party of nine pushed the limits.* The captain, James Cook, was probably not pleased when he saw that the lobby

* Banks's party consisted of his four servants, Peter Briscoe, James Roberts, George Dorlton and Thomas Richmond, the artists Sydney Parkinson and Alexander Buchan, his secretary *cum* artist Herman Spöring and Daniel Solander.

outside the Great Cabin and the three cabins around it, as well as the adjacent storage cupboards, had been turned into six tiny cabins. Cook also had to share his own space, since Banks, Solander and the artists would use the Great Cabin as their study.

Once the alterations were complete, Cook sailed the *Endeavour* from Deptford to Plymouth, arriving there on 14 August, three days after Collinson's death. From here he dispatched an express letter to Banks and Solander, asking them to leave London immediately to join the rest of the crew, which was already on board. Banks and Solander received the message on their way to the opera with Harriet Blosset and her family. Rumours had been circulating that Banks had promised Harriet "to expect he would return her husband," but no official announcement of an engagement was ever made. Although Banks had written to a friend doubting that "matrimony" would bring lasting happiness, Harriet, as another guest noticed, was "desperately in love with Mr. Banks" and had apparently promised to live in seclusion until his return. For his part, Banks did not even tell Harriet about Cook's summons because, as he later explained, "[t]he Ceremony of taking Leave I have always considered a Painfull and unnecessary Struggle." Not knowing of her adventurous admirer's impending departure, Harriet enjoyed the musical evening and the entertaining dinner, while others observed how Banks drowned his emotions, and maybe his fears, in alcohol.

A couple of days later Banks and Solander arrived at Plymouth. The *Endeavour* was swaying with activity as ant-like lines of sailors shifted provisions and materials along the gangway on to the vessel. There were timbers, canvas, nails and tools for emergency repairs; boxes of red and blue beads, little mirrors and other decorative items as gifts for the natives in the South Pacific; as well as twenty tons of biscuits and flour, 10,000 portions of salted beef and pork, and almost 8,000 pounds of sauerkraut—pickled cabbage—to fight off scurvy on the long journey. 1,600 gallons of spirits and 1,200 gallons of beer were loaded to keep the crew happy. In addition there were the best telescopes and a collection of the finest equipment for navigation as well as an astronomical clock and a portable observatory, provided by Royal Greenwich Observatory.

Most of Banks's luggage was already on board. It had taken days to load, as it consisted of hundreds of specimen bottles, sheets of paper, presses, portfolios, pigments for drawing, lenses, razors, knives, chemicals for preserving, wax and several sorts of salt in which to keep seeds. Banks had ordered all manner of nets, trawls and hooks for catching seaweed, corals and fish for the endless months on the ocean. The men carried new aquatic microscopes, telescopes and an armoury of guns and pistols as well as a natural history library of more than 100 volumes. All of this was packed into wooden chests with locks and hinges, tin cases, and specially made wainscot boxes, in order to protect it both from prying eyes and stormy weather. For the Great Cabin, Banks had brought a bookcase and a writing bureau. It "almost frighten[s] me," Banks wrote of the paraphernalia that made him the best-equipped naturalist ever to set foot on a ship.

After Banks's and Solander's inspections, Cook again ordered several joiners and carpenters from the shipyard to make some last-minute alterations to the "gentlemens Cabbins." The cabins, their only private space for the next three years, were tiny. Banks, who was the sole heir of several estates, suddenly found himself the occupier of a cabin no bigger than a modern king-size bed without even a porthole. Before the refitting his room had been an internal storage space, and it was still more of a walk-in cupboard than a cabin, being accessible only from the Great Cabin. The other cabins, on the quarter deck—allocated to the astronomer, Solander and Banks's artists—were no larger, but running as they did along the sides of the ship, they at least had the advantage of portholes and fresh air. Underneath, on the lower deck, the ship's master, lieutenants, carpenter and surgeon had their own cabins, while the rest of the crew would sleep in hammocks pitched between the ship's timbers.

On 25 August the sailors opened the topsails to signal to Banks and Solander, who were still lodging ashore, that the *Endeavour* was ready for her departure. Solander dashed off one final letter, expressing his gratitude and friendship to John Ellis, who, together with Peter Collinson, had laid the foundation for his success in England. Barely hiding his excitement, he ended his note by asking Ellis to send his greetings to all the "museum people." An hour later Solander and Banks were on the *Endeavour*. The wind ruffled the swelling canvas. Ninety-four men were heading into the English Channel. They were leaving behind the world they knew and were uncertain if they would ever see their country again. It was only two weeks after Collinson's death, and almost as if they were picking up the baton from Bartram and Collinson, Solander and Banks began a journey that would add another continent to the expanding empire and transform the English garden yet further.

The voyage captured the public imagination. Newspapers reported it; gossip filled the coffee-houses. "Gentlemen of Fortune, who are Students in Botany," discoveries to the "South Seas" and the "secret" mission of exploring *Terra Incognita* were the talk of the gilded drawing rooms of the aristocracy and the meeting places of the learned societies. The Royal Society had already done much to promote the voyage, since the observation of the transit of Venus was one of the first projects to be organised with international collaboration. America, France and Russia all participated, bringing together more than 150 observers at seventy-seven stations. Franklin's and Bartram's original American Philosophical Society (which was revived in the 1760s after its failure in the 1740s) was also involved in the observation of the transit, overseeing the results of twenty-two American sets of data.

Linnaeus, however, the man who regarded himself as the oracle of scientific thinking, had no idea that his former pupil had joined this epic journey nor that the *Endeavour* was also sailing under the flag of botany.

It was Ellis who felt obliged to tell Linnaeus that Solander was on his way to the other side of the globe. "Solander . . . assured me he would write to you and to all his family," Ellis wrote, but knowing that it was unlikely that Solander would keep this promise (another friend had called Solander "the laziest of mortals"), he provided Linnaeus with the proposed route and a full description of the botanical equipment the *Endeavour* carried, asserting that "[n]o people ever went to sea better fitted out for the purpose of Natural History, nor more elegantly." Two months later, in early November, when Ellis received his first letter from Solander, written when the *Endeavour* was at anchor in Madeira, he again passed on the news to Linnaeus, insisting that Solander "begs his kind respects to you."

The *Endeavour*, meanwhile, with a course set for Rio de Janeiro, had crossed the Equator. Banks and Solander avoided the infamous "ducking" ceremony, in which equinoctial virgins were dipped into the ocean, by proffering a large bribe of brandy. Over the past months the botanists had adjusted to life on board the *Endeavour*. They often rowed around the ship in a dinghy to catch seaweed and fish, in search of new species, and their days were filled with small incidents. Banks was particularly pleased when a flying fish, the first he had ever seen, flew into one of the portholes ready for examination. They all laughed when a squall sent the draughtsman flying across the cabin with his paint pots. Banks kept healthy by bathing and drinking a daily six-ounce dose of concentrated lemon juice mixed with brandy as an anti-scorbutic. He was happy. "I promisd myself three years uninterrupted enjoyment of my Favourite pursuit," he wrote.

In November, three months after they had left Britain, they reached Rio de Janeiro and anchored below the Sugar Loaf Mountain. However, to Banks's and Solander's great frustration, their excited plans to explore the local flora were immediately stopped. The colony's Portuguese governor was highly suspicious of the *Endeavour*'s motives, believing it "impossible that the King of England could be such a fool" as to finance an elaborate expedition simply to observe one planet moving in front of another. To his mind they were either spies or merchants trying to steal business from the Portuguese. He refused to allow any botanical excursions. Banks, who was used to getting his way, was furious, firing off a salvo of letters, reminding the Viceroy of his social standing. "I am a Gentleman, and one of fortune," he fumed. But the protest had no effect and they remained confined to the *Endeavour*.

Solander became so bored that he finally wrote letters both to his mother and to Linnaeus—the first in six years. Making no excuses for his long silence, he promised to travel to Uppsala with all the botanical discoveries once he returned from his voyage and hoped that Linnaeus would be "muster master of our recruits." Banks meanwhile "cursd, swore, ravd, stampd," pacing up and down the deck, feeling that he was going mad. Through their telescopes they could see hummingbirds hovering over clambering bougainvilleas which were dripping with pink blossoms,* as well as hedges of brightly coloured flowers and

---

* *Bougainvillea spectabilis* was named after the French explorer Louis Antoine de Bougainville, who had beaten Banks and Solander by just over a year when he "discovered" it in 1767 on his journey around the globe.

juicy fruits dangling from trees. The flora and fauna of South America were laid out like seductive wares in an exotic bazaar, and Banks wrote to a friend that he felt like a "French man laying swaddled in linnen between two of his Mistresses, both naked [and] using every possible means to excite desire." The mood was worsened by the unbearable heat and by Cook's regimented cleaning programme, in which the *Endeavour* was propped up so that, Solander moaned, "we hardly cou[l]d walk."

To see at least some of the plants, Banks and Solander decided to bribe locals to collect flowers and bring them "as greens and sallading." But after a few days they could not resist the temptation any longer and, under cover of darkness, abseiled from a cabin window into a dinghy and rowed ashore. They returned at dawn with more than 200 specimens, including the tree-like cactus *Pereskia grandifolia* (also called rose cactus for its pink petals) and some species of tillandsia, seemingly magical flowers which clung to tree branches and received their nutrition from rain and air rather than soil—hence their common name, "air plant."

On 2 December, almost three weeks after their arrival, the *Endeavour* left Rio de Janeiro, sailing south towards the treacherous Cape Horn, famed and feared for its volatile winds, strong currents and floating icebergs. During the journey they encountered extremes of heat and cold. At times the air was so damp that Banks's black leather portfolios, books and trunks turned white with mould. Even the knives in their pockets rusted. But as they approached Cape Horn temperatures dropped and Cook issued jackets and trousers made of heavy woollen cloth. Banks avoided the itchy uniform by wearing his own warm flannel jacket, an extra waistcoat and thick trousers. Inside the Great Cabin a little stove kept the room cosy, but outside icy gales rocked the *Endeavour*. One night the dancing swell overturned Banks's bureau and bookcase—"a very disagreeable night," he wrote, as the banging of fallen furniture and books, together with the cacophony of creaking timbers, groaning ropes and the "swinging beds" bashing against the cabin walls kept him up. The crew, though, thought that the *Endeavour* sailed out the high waves better than any ship. She may not have been as elegant as a naval frigate but was "so lively and at the same time so easy," they told Banks and Solander.

They celebrated Christmas with an extra allocation of rum, a feast that was enjoyed by everyone on board as "all hands g[o]t abominably drunk," Banks wrote in his journal. But then, at their landfall in Tierra del Fuego—the island group that formed Cape Horn and one of the most inhospitable environments in the world—disaster struck. Banks, Solander and some others were gathering alpine plants on a mountain when they were caught in a sudden blizzard. The afternoon sun had reminded them of a pleasant English summer's day, but suddenly slicing winds blasted snow across the land. The change of weather was so violent that Solander and two of the servants collapsed, refusing to move. Only Banks had the strength to find shelter and rescued Solander. They had survived vicious gales at sea, only to be struck while on dry land. Both servants died in the storm, but less than a week later, Banks wrote in his journal, about the Tierra del Fuego expedition: "No botanist has ever enjoyd more pleasure in the contemplation of his Favourite pursuit than Dr. Solander and myself among these plants." Either he was being completely heartless or the bursting plant presses made him forget the losses and suffering.

A few days later the *Endeavour* entered the emptiness of the Pacific Ocean, one of the first few ships ever to do so. The voyages of their predecessors, Ferdinand Magellan and Francis Drake, conjured up thoughts of reckless adventure and severe hardship. The plan was to observe the transit of Venus from the island of Tahiti, which had been discovered two years earlier by Samuel Wallis, the captain of the *Dolphin*. Wallis had returned to Britain just in time for Cook to learn about the island's position. Meanwhile the French Louis Antoine de Bougainville had landed in Tahiti in April 1768. The *Endeavour* was therefore only the third European ship to arrive at this pacific paradise.

As the *Endeavour* drew near to the island, Banks and Solander stood on deck, peering through their telescope to see the mountains of Tahiti on the horizon. The summits were wrapped in clouds, but the rest of the sky was blue, with sunbeams playing a game of chase with the rippling water. Two days later, on 12 April, they reached Matavai Bay on the northwest coast of the island. They had been almost exactly three months at sea since they had left Tierra del Fuego, during which time they had seen no land. Cook had found the island by following an almost straight course—a masterful piece of navigation. As soon as the anchors were lowered, Banks and Solander were, as always, the first to go ashore. Walking under groves of coconut and bread-fruit trees laden with fruit, they found, Banks said, "the truest picture of an arcadia of which we are going to be kings."

Banks and Solander collected gardenias, orchids and jasmine, and, in a similar manner, picked up Tupia, a Tahitian priest, to take back to England—to "keep him as a curiosity" as Banks wrote in his journal, just as "my neighbours [have] lions and tygers." And though they filled their bags with exotic plant spec-imens, Banks and Solander also found time to sample the island's women. Banks raved about "a very pretty girl with a fire in her eyes," only complaining that the lack of privacy in the open huts did not allow them "of putting their politeness to every test." The only cloud in these first days was the sudden death of Banks's landscape painter, Alexander Buchan, after an epileptic fit. But judging from the entry in his journal Banks was more sorry for the missed opportunity "of entertaining my friends in England" with paintings of Tahitian scenes than the loss of a life: if Buchan had lived just one month longer, Banks wrote, "what an advantage would it had been to my undertaking."

With their natural curiosity and sociable manners, Banks and Solander became the principal market traders between the *Endeavour* and the Tahitians, negotiating prices for coconuts, breadfruit and hogs and picking up a rudimentary knowledge of the local language. Bartering was fun, especially when it involved women selling swathes of cloth. Pirouetting in front of "Tåpáne" and "Toråno,"* the Tahitian names of Banks and Solander, one woman, for example, slowly unrolled the material that covered her body. By the time the wares were on the ground she was naked, "unveiling all her charms," Banks wrote. He liked that the Tahitian women decorated their hair and ears with sweet-smelling white gardenias—*Gardenia taitensis*—and declared the velvety touch of their skin "superior to any thing I have met with in Europe." They were cleaner than any other people, he wrote, adoring the whiteness and perfection of their teeth—unheard of

* Banks named his new Tahitian acquaintances after Greek thinkers, lawyers and gods. The man who Banks believed to be the fairest judge on the island, for instance, was called Lycurgus after the Spartan lawgiver, while another became Hercules in honour of his strength.

in Britain, where a look into a mouth usually revealed decayed and blackened stumps. Harriet seemed to be entirely forgotten.

For the men of the *Endeavour*, Otaheite, as the natives called the island, was like the mythical land of Cockaigne, the land of milk and honey. Slowly they absorbed the arcadian rhythm of island life: beautiful women, fresh food and exotic blooms. Banks meandered along the sandy paths dressed in a white jacket and a waistcoat embroidered with silver frogs, his head covered with "a turban of Indian cloth." Together with Solander he listened to the natives play the flute with one nostril, watched young girls involved in a game in which they lifted their pleated skirts, "exposing their nakedness," and observed the ceremonial and painful tattooing of a girl's buttocks. They ate dog meat, which they declared "a most excellent dish," and admired the Tahitians surfing the crests of the breaking waves while standing in their canoes—a "strange diversion" according to Banks. He also attended a funeral for which he wore only a loincloth and was rubbed with charcoal, having "no pretensions to be ashamd of my nakedness."

Every man on the *Endeavour* was amazed by this land of

Heads of divers Natives of the Islands of Otaheite, Huaheine, & Oheiteroah.

Portraits of Tahitians drawn by Sydney Parkinson,
Banks's artist on the *Endeavour*.

free love, and after the many lonely months at sea made liberal use of it. By early June, almost two months after their arrival, Cook heard that "Vernerial distemper" was spreading across the island and among his crew. Although Cook tried everything to prevent its progress, and had every sailor and officer of the *Endeavour* examined by the surgeon,* he complained that the men continued to indulge, finally blaming "the Women [who] were so very liberal with their favours."

As trust between the sailors and the islanders grew, and bed companions were swapped, it became increasingly difficult to drum together the crew in the mornings. One morning, for example, Banks woke up naked in the canoe of one of his lovers to discover that his clothes and pistols had been stolen. When he tried to find Solander and Cook to help him to retrieve his belongings, he failed because he did not know where they "had disposd of themselves" the previous evening.

Seven weeks after their arrival they observed the black disk of Venus passing across the sun and took their measurements and calculations. With their first mission accomplished, the crew reluctantly began preparing to leave Tahiti in order to embark on Cook's second assignment, to find *Terra Australis Incognita*— the "unknown southern land." Sails were brought back on board, carpenters made last repairs, and the stores were replenished with fresh water and food. The herbarium had gained more than 300 species, one of which was the breadfruit, a plant that would play an important role in Banks's later life and would be part of the infamous mutiny on the *Bounty*. Finally, on 13 July, the anchors were lifted and Banks climbed up the topmast to wave at the canoes that had surrounded the *Endeavour* for their final farewell.†

---

* European vessels brought syphilis and other diseases to Tahiti, and as a result its population shrank from 40,000 in 1769 to 9,000 only sixty years later.
† In three months of peaceful trading, there had been only a few moments of crisis. One was when the midshipman Jonathan Monkhouse ordered that a Tahitian who had stolen a musket be shot. Banks was so shocked by the incident that he said, "if we quarrelled with those Indians we should not agree with angels." Just a few days before their departure, there was a last-minute disaster when two sailors deserted to stay in Tahiti. In response Cook imprisoned the natives who happened to be on board at the time. The prisoners were only released when the Tahitians brought the absconders back to the *Endeavour*.

Cook plotted a course due south and the *Endeavour* sailed into unchartered waters. As days became weeks, and no land appeared on the horizon, life on board became "botanical" again. Banks and his companions had to mount their discoveries, and spent much of their time describing, classifying and drawing. They kept dried specimens only. The strictures of space on board precluded the collection of live plants, but, in any case, since their purposes were scientific rather than horticultural, they did not need them.

Sometimes Banks wished his friends in England could look through a "magical spying glass" to see them working in the Great Cabin. He sat at the table filling the pages of his journal with his brisk handwriting while Solander bent over the pressed specimens, counting stamens and pistils, once in a while picking up a botanical book from the library, talking constantly to Banks. Opposite them was the draughtsman Sydney Parkinson, surrounded by his paint pots and brushes, sketching the specimens. They had turned the Great Cabin into a laboratory of natural history full of dried plants, pickled animals, stuffed birds and pieces of wood covered in barnacles. From the ceiling's timbers large bunches of drying seaweed swayed to the rhythm of the sea.

After months of delicious fresh Tahitian food, the crew returned to their diet of dried meat, biscuits and sauerkraut. Banks, still delighted by his life as an adventurer, could find no fault in anything and even enjoyed the menu, which was composed of ingredients that had travelled for more than a year through icy gales and humid tropics. The meat, he declared, "is by much the best salt meat I have ever tasted," the peas were "excellent" and the "*Sour Crout* is as good as ever." When he saw insects in "thousands shaken out of a single bisket," he just put the bread in the oven in the Great Cabin "which makes them all walk off." Of course, he had taken the time to classify them according to the Linnaean system in advance. The culinary highlight during those endless weeks at sea was a pie made from apples that had grown on one of the trees Bartram had sent to England.

While Banks and Solander journeyed towards a new empire, one of the men who had paved the way for their voyage was losing his. Philip Miller's troubles had started shortly after the

Endeavour had left Tahiti, when a new Garden Committee began making changes at the Physic Garden. Until then the Society of Apothecaries had left Miller to his own devices: the meetings were away from the garden at the Swan Tavern nearby and were boisterous affairs in which alcohol seemed to take precedence over business, judging by the illegible scribbles, stains and general chaos of the Minute Books of that period. They knew Miller had made the Chelsea Physic Garden famous, had praised his "great diligence in settling a correspondence & procuring Seeds & plants from various parts of the world" and had left him in control. But now the new committee descended upon Miller several times a month, demanding a full inventory of the garden, the greenhouse, the library as well as a collection of all plants, seeds and roots for a new *materia medica* for the Society of Apothecaries.

Miller, now almost eighty, was annoyed by the interference but reluctantly conceded. A month later he presented the inventory but claimed the orange trees were his, initiating a feud that would last for more than a year. Then the committee removed most of the books from the Physic Garden's library to the Guildhall in Blackfriars, which provoked Miller's fury, as over the past five decades he had not only assembled many of these books but also used them to research his own publications. He believed they belonged to the garden, insisting that "several Donors had appointed the Books should always remain at Chelsea." But the committee ignored the old man.

The most onerous task for Miller was that the committee insisted in re-cataloguing the entire garden. Every area was to be allocated a letter—from A for Greenhouse, B for Dry Stove to X for Swamp—and then each plant, instead of receiving a name tag, was to be numbered and cross-referenced with a numerical botanical catalogue. Wooden index sticks were ordered in their thousands, painted with letters and numbers. But Miller was not willing to change his own system for this new order and pretended to be ill every time the committee arrived to place the sticks, knowing that they needed him to identify the plants. On one occasion the committee proceeded with the help of the foreman but Miller retaliated by making sure that nobody from the garden staff was there to attend the committee meetings during the following month. His petulance resulted in an official complaint. The conflict escalated when Miller received a written order in April 1770, demanding that he or his foreman were to assist the committee in order to avoid "future disappointment . . . [and] hindrance to the business."

On that very day, Banks and Solander together with Cook and some of the sailors rowed the *Endeavour*'s yawl through the barrelling surf to the land that would come to be known as Australia. Having explored New Zealand for six months, from October 1769 to the end of March 1770, the *Endeavour* had then followed the route of Abel Janszoon Tasman, the first European to land in New Zealand in 1642, towards Dutch Van Diemen's Land (today's Tasmania). But the ship had been driven north of her intended course by gales, and instead of Tasmania, a previously uncharted coastline had appeared on the horizon. While the Dutch had discovered parts of the west coast of Australia and Tasman had followed the northern coast for almost 2,500 miles, no European vessel had ever reached the east of the continent. After first spying tree-covered hills, the *Endeavour* skirted the crescents of golden beaches along the coast for days without finding a sheltered bay to anchor in. Now, finally, they had found a place to come ashore.

The aboriginal people welcomed them with lances and sticks, to which the *Endeavour*'s men responded with muskets and pistols. After a violent exchange, the aboriginals dispersed and remained in hiding whenever the *Endeavour* men approached the beach during the next few days. Believing the natives to be "rank cowards," Banks did not hesitate to venture out with Solander, and the two men embarked on a series of long botanical expeditions. They were astounded by what they found. There were trees that shed their bark instead of their leaves, and straggly shrubs with spindly branches that left them unprotected against the sun. They were the first Europeans to discover many different species of eucalyptus—some of the tallest hardwoods in the world. One species they called "gum tree"—later named *Eucalyptus gummifera*—because of the red resin that oozed out of its wounds; it was also one of the few seeds that survived the long voyage and was successfully introduced to England on their return. They saw a tree that looked as if it were on fire because it was cloaked in bright red blossom and therefore named it flame tree.

As they clambered through the thickets, they discovered colourful climbers such as wonga wonga vine and the dusky coral pea, which wound itself lasciviously around the branches of trees. Nearby hardenbergia, another climber, paraded its mass of purple pea-like flowers. Banks and Solander also found the first species of a genus which would be named after Banks: *Banksia*, a striking large evergreen shrub with upright, spiky cone-like flowers. In the forest above their heads sculptural orchids sat lightly on fragile branches, clinging to their host trees in a scented embrace.

Despite the forest's beauty, it was eerie. Strange new smells enveloped them as they walked under the laced canopy of shady tree ferns. Sometimes it was so silent that only the distant break of the waves could be heard until the cackling laughter of the kookaburra reverberated through the trees. The men shot birds (which they ate in a pie), investigated the footprints "of an animal clawd like a dog," saw a kangaroo and plucked plants. Some plants had developed unique traits which allowed them to survive the ravaging bush fires. The hard fruits of banksias, for example, protected the seeds from the flames, opening only after being scorched. Similarly the crimson-coloured waratah—today's floral emblem of New South Wales—survives because it regenerates from the rootstock.

Banks and Solander also discovered many of the popular Australian plants that are found in British and American gardens today, such as several species of grevilleas, with flowers that mirrored the plant's needle-like leaves in garish colours, and *Callistemon citrinus*, also called bottlebrush after the hundreds of long red stamens on the flower, each with a dot of yellow pollen at the tip—like the psychedelic bristles of a brush.

They found so many new species that Cook honoured the place with the name "Botany Bay."* "Our collection of Plants was now grown so immensely large," Banks wrote on 3 May, five days after their arrival. In New Zealand they had already discovered and described almost 400 new species—becoming as they did so the first naturalists to catalogue the botanical world of the Antipodes. Now they had hundreds more. But they had collected so many specimens so quickly that they used

---

* Captain Cook explained in his journal on 6 May 1770, "The great quantity of New Plants & cᵃ Mr. Banks & Dr. Solander collected in this place occasioned my giveing it the name of *Botany Bay*."

up their remaining stock of paper, and the plants were all squashed together. Unable to dry properly, the specimens were now in danger of rotting, so Banks and Solander laid them out in the sun, spreading them over a sail on the beach. Three days later, when the *Endeavour* left Botany Bay, Banks's portfolios and presses were filled to the brim.

At the end of May, they approached the Great Barrier Reef. Unknown to them they were entering the most treacherous stretch of coastline in the world—almost 1,500 miles of reef and coral islands. Caught between the reef at starboard and the mainland to port, the *Endeavour* sailed into a labyrinth of perpendicular coral walls, sandbanks, vicious currents and razor-sharp spurs—all hidden by a blanket of azure water. The man at the prow threw his line continuously, calling out the fathoms of the water: three, fourteen, eight. For ten days the *Endeavour* crawled her way north through the shoaling sea, carefully tiptoeing her way like a prim girl dancing around puddles. Then, on 10 June, just before midnight, the night-watch woke the crew with a shout: the *Endeavour* was stuck on the reef twenty miles from shore. Each swell beat her so hard against the rocks that the men could hardly stand and, with every wave, they heard the coral grinding the ship's hull. Attempting to make the *Endeavour* lighter and to keep her afloat, the crew threw fifty tons of ballast overboard. In the morning, with the incoming tide, water gushed into the vessel. The men—from the lowest sailor to Banks—worked the pumps continuously, fighting what felt like a sisyphean battle. "[F]ear of Death now stard us in the face," Banks wrote, knowing that "most of us, must be drownd" but adding that death was probably more desirable than spending the rest of one's life with "the most uncivilizd savages perhaps in the world."

Twenty-four hours after the *Endeavour* struck the reef, as the strength of the men waned, the midshipman suggested drawing a sail under the ship, hoping that the suction of the leak would pull the fabric into the hole like a plug. To make the sail thicker they "fothered" it, stitching tufts of oakum and wool on to it. Most of the crew thought this a strange idea, but to their surprise it worked. They dared to believe that they might survive. Bandaged at the bow, the ship slowly moved towards the shore. Eventually they found the mouth of a river, where they hauled up the ship in order to repair it. Grateful for their unexpected luck, Cook called it the Endeavour River.

On the same day that the exhausted adventurers made their narrow escape, the committee of the Chelsea Physic Garden decided they could no longer handle Miller. Over the previous few weeks he had persistently refused to follow their instructions, declaring the committee's method of cataloguing useless and that "he would give . . . no assistance therein." Every time the committee members arrived at Chelsea, Miller had found excuses not to help by feigning illness or announcing that the ground was too wet. Now the Society of Apothecaries ordered Miller to obey the committee "if he values his place." But once again he refused to budge, treating the committee, as they reported, with "the usual rudeness." When he pulled out some of the index sticks and threatened to remove "all the plants from the Perennial Quarter into . . . the Annual Quarter," the committee decided they had had enough of Miller's "refractory" behaviour and would bring his case to the Court of Assistants, the highest tribunal in the Society.*

When, in October, the case was heard, the prosecutor had read his way through pages of minutes from the meetings of the garden committee over the previous five decades in order to dissect Miller's work. And so, in a strange twist, Miller was accused of the very "crime" that had lifted the apothecaries' garden from obscurity: for having corresponded and procured seeds "without any direction from a Superior" or authorization. Unearthing and re-interpreting the rules from 1722, the prosecutor accused Miller of "Transgression of his Duty"—a rather hypocritical allegation against the man who had turned the Chelsea Physic Garden into the foremost botanic collection in the world. According to them, Miller had not so much revolutionised English horticulture as overstepped the boundaries of his job. In revenge Miller ordered his foreman to fell a large plane tree which he had planted himself fifty years previously and threatened once more to pull out all index sticks, to replace them with some "according to his own method."

---

* At the end of August Miller must have realised how desperate his situation was becoming and sent the committee members "a present of Venison," which was refused because it was regarded as what they called an "improper" attempt at bribery.

Meanwhile the *Endeavour* had begun her return journey. Cook had lost hardly any of his men on this epic adventure but now they sailed towards malaria-infested Batavia (today's Jakarta), the deadliest of all ports, crossing the ocean on hull planks that had been ground down by the coral reefs and tropical worms to one eighth of an inch. Batavia had been founded by the Dutch in the seventeenth century and was laid out as a typical Dutch city along canals. But an earthquake had destroyed the water supply and the canals had become stagnant pools, the breeding ground for mosquitoes. Although Cook knew this, it was the only port where he could get the *Endeavour* repaired properly in order to bring her home. Within days of arriving, Banks and Solander were gripped by malaria. Struck down by violent fever, they could neither walk nor crawl, but none of the servants could help since almost the entire crew was battling with death. When they left two and a half months later, everybody was weakened and more people had died than during the entire voyage. One of them was Tupia, the Tahitian priest.

By the time Banks and Solander returned to Britain in July 1771, almost exactly three years after they had left, Miller had been forced to resign from the Physic Garden for what one observer described as his "obstinacy and impertinence." In February, Miller had handed the keys to William Forsyth, the new head gardener, missing by just a few months what would have been the most exciting moment of his career: the arrival of a bag of seeds from the *Endeavour* voyage. By 16 December, Miller was dead, dying just a few days before Solander invited London's leading gardeners and nurserymen to inspect the specimens from the *Endeavour* voyage. It was almost as if the loss of his daily routine of work and study at the Physic Garden had sapped the spirit of Miller's life. Banks, by contrast, felt more alive than ever before, knowing that he had advanced "this enlightened age" by being, "I may flatter myself . . . the first man of Scientifick education who undertook a voyage of discovery."

11

"An Academy of Natural History"

There are few studies more cultivated at present by persons of taste, than Botany, and ... none can be more deserving of regard. Whether we consider the effect of Botany as enlarging the sphere of knowledge, or as conducive to health and innocent amusement, it ought to rank very high in the scale of elegant amusement.

WILLIAM MAVOR, *The Lady's and Gentleman's Botanical Pocket Book*, 1800

When the *Endeavour* returned to Britain, it was Joseph Banks rather than Captain Cook who was courted both by aristocrats and scientists, and celebrated as the real hero of the most successful expedition of the century. "It is his discoveries and disclosures which provide as much, if not more, interest than the deeds of Lieutenant James Cook," one newspaper reported, while others did not even mention Cook. Linnaeus even suggested calling New South Wales "BANKSIA, from its discoverer, as America was from Americus." Banks had not a moment to rest after his arduous journey. On the day of his arrival back in London, Banks wrote, "I am Mad, Mad, Mad. My poor brain whirls round with the innumerable sensations." From then on, he rushed with Solander from one party to another, toured England's learned societies and dined with both Benjamin Franklin and the President of the Royal Society.

They described to spellbound audiences "the many narrow escapes ... from imminent danger," such as Solander's near death in Tierra del Fuego and their perilous escape from the coral claws of the Great Barrier Reef. They conjured up visions of the strangest blossoms in the world, and an animal that was the oddest looking creature since the dodo—the kangaroo. Their listeners were told about the island they called Otaheite, which Banks and Solander described as a terrestrial paradise and haven of free love. They explained to Franklin that the

Tahitians had no idea of kissing with lips, but that "they lik'd it when they were taught it."

While Banks bragged about his heroic and amorous escapades, the woman who still thought she was to be his wife waited patiently in the countryside for word from her wayward lover. Harriet Blosset had spent three years living in seclusion, having deliberately absented herself from social life in London, only to be rewarded with silence. She waited a further week, but then, still having received no news at all, travelled to the capital for some explanation of Banks's strange behaviour. But instead of declarations of love, Harriet received a note in which Banks declared himself to be of "too volatile a temper to marry." London's high society gorged on the story; even two years later a magazine would mock Banks's ungallant behaviour, portraying him as a scientific libertine who explored not only exotic flowers but also the women in every country. Friends found Harriet to be in a "most distressing as well as ridiculous situation," and urged for some kind of pecuniary compensation, which Banks eventually agreed to pay.

Having bought his freedom and cleansed his conscience, Banks continued to tour the drawing rooms of London. At the same time, he and Solander unpacked and sorted their dried specimens as well as Sydney Parkinson's sketches and drawings of plants which Banks intended to have engraved.* They also met the King at the royal garden at Kew, where they presented him with the first seeds from the Antipodes. Everybody, it seemed, wanted to view the exotic plants from the other side of the world, and newspapers were full of stories about Banks's and Solander's horticultural discoveries, and their "[v]ast numbers of Plants." Banks invited eminent botanists, nurserymen and collectors† to his house to "see fruits of his travels." And when, only six weeks after the arrival of the *Endeavour*, some of the seeds and bulbs germinated, gardeners rushed to

---

* Over the following years, eighteen engravers worked on almost 800 plates. It was planned that they should be published in fourteen volumes. But, in the end, it was not until the 1980s, more than 200 years later, that the Natural History Museum in London published a limited edition of Banks's *Florilegium*.
† Banks and Solander invited, for example, Miller's successor at the Chelsea Physic Garden, William Forsyth, as well as James Lee, one of London's best nurserymen, and William Aiton, the head gardener at Kew.

Pub. by W. Curtis St Geo Crescent Sept 1795

Edwards del Sansom sc

Robinia hispida
The bristly locust arrived in 1743 in England. Two decades later Collinson's specimen was loaded so heavily with dangling plumes of pink blossom that he had "to prop up the branches."

Joseph Banks shortly after returning from voyage aboard the *Endeavour*. The writing on the paper he rests his hand upon reads CRAS INGENS ITERABIMUS AEQUOR— "tomorrow we'll sail the vast deep again."

Captain Bligh watching the transplanting of the breadfruit trees in Tahiti.

Erica cerinthoides
This evergreen shrub was the first of the many Cape heaths that were introduced from South Africa by Banks's collector Francis Masson.

Banksia serrata
Named after Joseph Banks, banksias were one of the first plants to be raised from Botany Bay seeds.

Bird-of-paradise *(Strelitizia reginae)*
This exotic-looking flower from South Africa was Banks's favourite plant.

In 1784, twenty-six chests that held Linnaeus's collections sailed from Sweden to their new owner, James Edward Smith, in London.

Gardeners at work.

Callistemon citrinus
Solander and Banks had seen the so-called bottlebrush at Botany Bay
but had only brought dried specimens for their herbarium. The first seeds
arrived in 1788, once the penal colony had been established.

Pandorea pandorana
The wonga wonga was another
Australian plant that was available
in English nurseries before the
turn of the century.

Kennedia rubicunda
Named after the nurseryman
John Kennedy, this Australian
climber (also called dusky
coral pea) was introduced
to England in 1788.

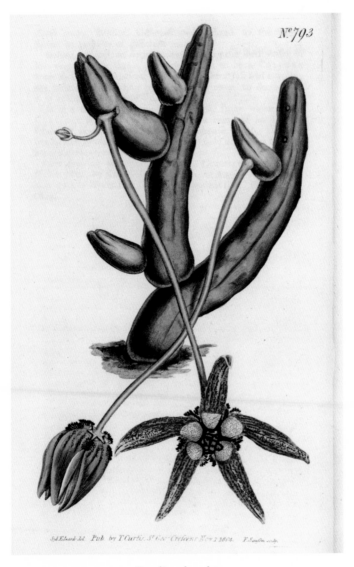

Syd Edward del. Pub. by T. Curtis, S.t Gco. Crescent Nov 2 1804. F. Sansom sculp.

Stapelia pedunculata
The craze for stapelias at the end of the eighteenth century illustrates
to what extent the British had become obsessed with flowers—the plant exudes
a strong smell of rotting meat, hence its common name, carrion flower.

By the end of the eighteenth century, botanising had become a popular pastime.

Battlesden Park *circa* 1818
A garden planted with a large number of the exotics that had arrived throughout the eighteenth century. Flowerbeds pepper the lawn and tender pot plants that were kept in the glasshouse during much of the year were placed outside in the summer.

see them. Their strangeness and beauty conjured up distant countries, dangerous adventures and the riches of the British empire. Much had changed since Bartram's first cargo had arrived forty years previously: Banks and Solander's plants were admired not as a testimony of God's creation but as a manifestation of Britain's successful discoveries, her fertile foreign colonies and brilliant scientific advances.

In total, there were 30,000 dried specimens in portfolios and cabinets: some 3,600 species, 1,400 of which were new to Britain's botanists. There would have been even more if part of their collection had not been destroyed at the Endeavour River, when the ship had been propped up to be repaired and some water had sloshed into the place where Banks and Solander had stored most of their collection. Nonetheless the hoard was an enormous addition to the 6,000 species (the sum of all plants known at the time) which had been catalogued by Linnaeus in his *Species Plantarum.*

Once again, though, Solander forgot Linnaeus, who had been informed by John Ellis that his old pupil was "laden with the greatest treasures of Natural History that ever was brought into any country." Excited by the news, Linnaeus replied within an hour of receiving the letter, saying that if he were not bound by old age and illness to Sweden, he would leave immediately "to see this great hero in botany." Linnaeus also showered Banks with compliments, running the engines of obsequiousness: "Surely none but an Englishman would have the spirit to do what he has done."

But amid the admiration, Banks and Solander were also ridiculed. Horace Walpole, for example, thought the "forty dozen of islands, which Dr. Solander has picked up the Lord knows where, as he went to catch new sorts of fleas and crickets" would only produce more wars, and mocked "that wild man Banks, who is poaching in every ocean for the fry of little islands that escaped the drag-net of Spain."

Such derision seems not to have deterred Banks and Solander, who, by the end of August 1771, a few weeks after their arrival in Britain, were already planning their next voyage around the globe. A month later the Admiralty decided to buy two ships for the new expedition. When Linnaeus heard about these plans he was outraged, believing Solander's time would be better spent classifying the species and working on a publication so that "the learned world may not be deprived of them." The news affected him to such an extent as to "almost entirely deprive me of sleep," Linnaeus wrote. Without a publication the whole *Endeavour* voyage would be worthless to him: Linnaeus fretted that insects might devour the specimens, their store might burn down or Banks and Solander might die before labelling all the plants, leaving nothing for prosperity. John Ellis, obliging as ever, presented to Banks and Solander the letter in which Linnaeus laid out his concerns but they were too busy to acknowledge it. "[T]hey have been so hurried with company that they have little time," Ellis reported. Undeterred, Linnaeus asked Ellis "to persuade Solander to send me some specimens of plants." A month later Linnaeus was growing more despondent and wrote: "If your intercession does not procure me a few plants from my friend Solander, I shall despair." He had treated Solander "like a son," he moaned, but nothing was sent to Uppsala.

Banks and Solander had greater problems to worry about than the complaints of the cantankerous Linnaeus, as their new adventure was suddenly in jeopardy. For this journey the numbers of men in their party had risen to sixteen, including four artists and two musicians—again all paid for by Banks. To accommodate them and their equipment Banks had convinced the Admiralty to alter the ship, the *Resolution,* by laying a new deck and building a "round house" for Captain Cook atop the poop, while converting the Great Cabin into a botanical workshop. As a consequence the *Resolution* had become so top-heavy that she almost capsized when Cook took her on a test run in May 1772. But when Banks heard that the Navy Board had ordered the vessel to be returned to its original state, he reacted like a petulant child. "He swore & stamp'd upon the Wharfe, like a Mad Man," the midshipman reported. Then Banks ordered all his belongings to be removed from the ship, withdrawing from the expedition.* Banks hoped this gambit would persuade the Admiralty to continue with his plan for a reorganisation of the ship according to his original

---

* Captain Cook wrote to Banks on 2 June 1772 that "the Cook & two French-Horn Men" who were already aboard the *Resolution* were allowed to leave the ship. In the end Banks chartered his own ship and went on a much shorter and less dangerous expedition to Iceland instead.

instructions, but he had overestimated his influence and the *Resolution* left without him and his party.

Banks's failure to join the *Resolution* on her circumnavigation of the globe and his clash with the Admiralty were, in some ways, fortuitous. They accelerated his transformation from restless and youthful adventurer to calm and persevering scientist. Under his stewardship, more plants than ever before would arrive in Britain, bringing with them considerable economic advantages for the country, as well as scientific advances. In fact, Banks would do more for the progress and expansion of his country than any other man of his age. His fascination with botany and exploration made him a true man of the Enlightenment. Botany increased the understanding and knowledge of the natural world, while the *Endeavour*'s voyage had made the British citizens "of the whole world." Botany put man in control of nature; exploration put man in control of the globe.

Learning from his mistakes with the *Resolution*, Banks began to control his volatile temper. As he matured over the next few years he became less irascible and impatient, and by summer 1774, a few months after his thirtieth birthday, he had mellowed to such an extent that James Boswell compared him to "an elephant, quite placid and gentle, allowing you to get upon his back or play with his proboscis." Instead of embarking on new adventures, Banks now used his fame and energy to turn Britain into the world's foremost centre of botanical learning. For years gardeners had excelled in practical horticulture, but under Banks's direction Britain would also become the nexus for the scientific and economic exploration of the world's flora— placing plants at the heart of the empire and the nation's psyche.

To achieve this, Banks insisted that the nation needed a storehouse for its plants—a place that could hold as many living plants as possible and provide a research centre where scholars could consult the world's most comprehensive herbaria. Garlanded with praise and highly persuasive, Banks was able to convince George III to transform the royal garden at Kew into the world's largest botanical collection, becoming its nominal director in 1773. One of Banks's first actions in his new role at Kew—where he exercised, as he would later describe, "a kind of superintendence"—was to replace all the numbered plant tags with Linnaean name labels. Banks also decided to systematise Kew's collecting strategy. So, instead of buying the odd plant from nurseries or swapping the occasional specimen with other gardeners, Banks established an efficient supply network across the world. In the name of the King he asked army officers, merchants, diplomats and missionaries to remember Kew on their appointments abroad by dispatching as many plants as they could.

In addition, plant-collectors from across the globe began to send seeds and dried specimens to Banks in the hope of receiving some precious duplicates from the South Sea and the Antipodes in return. Within a few years Bank's original collection was augmented by plants brought back by Cook from his other voyages. And so East Indies specimens were traded for plants from Tahiti, a town councillor from Amsterdam forwarded seeds from Japan in exchange for information on hothouses and plants from Kew, while a West Indian merchant dispatched an annual parcel from Antigua. Banks's contacts ranged from the Governor of St. Helena and the Danish Royal Naturalist in the West Indies to a professor at the Academy of Science in St. Petersburg. Many of his correspondents remained faithful suppliers until their deaths. And unlike other wealthy plant-collectors, Banks had seen thousands of foreign species in their natural habitats for himself, and so was able to request specific plants. Although most of the specimens he and Solander had brought back were dried, he could use his herbarium and its accompanying drawings much like an ordering catalogue. Flicking through the pages, he could remind himself what he needed, as well as showing plant-hunters precisely what to search for.

Slowly Banks became "the first Patron of Botany," and the centre of Enlightenment London. Banks was never tired of new theories, experiments and observations. And because he was open-minded he received throughout his lifetime letters and reports from across the world detailing unusual observations and experiences. They ranged from sightings of "mermaids" to a statement of one Thomas Walker, who swore that he had vomited a "living Toad of about two Inches and an half long." But Banks was also actively involved in experiments himself. When surgeon and anatomist John Hunter, for example, opened a mummy, Solander and Banks assisted and were pleased to find a plant bulb "under one of the soles." In their search for knowledge the pair even entered a "heat chamber," where they remained—fully dressed—for several minutes at temperatures of 100° Centigrade. Banks advised fellow thinkers on book titles, tapped his governmental network to release botanical drawings made in Africa from Customs, and played host to Omai, the Tahitian, whom Cook brought back from the second South Sea voyage in 1774.*

Despite all these wide-ranging activities, Banks remained at heart a field botanist. When he travelled, for example, he stopped his coach whenever a strange weed or unusual flower caught his attention, in order to pluck the specimen. Friends were endeared by his passion and compared him teasingly to "a Newfoundland dog" when he jumped in and out of their boats to fetch aquatic plants during a day trip on a river. But others regarded his behaviour as so strange that one day he was arrested on a common while stuffing his bags full of plants and "carried before a justice as a highwayman." Though some thought it hilarious that Banks had been mistaken for a "highway robber" while crawling through a ditch, for Banks this incident only underlined how the love for natural history was mostly confined to the scientists who gathered in learned societies: why otherwise would he have been arrested, or a fellow naturalist be locked up as a "Madman" when he told his neighbours that he was chasing "the Purple Emperor"† across the village green?

The public's scepticism about natural history seemed to make Banks all the more determined to make it a popular pursuit. "Most think it's not important," he believed, and devoted his life to the promotion of botany—influencing policy-makers but equally teaching the children of friends the rudiments of the Linnaean system by cutting up a cauliflower, insisting that botany "turn'd science into a Sport." Others like Linnaeus had

---

* Banks and Solander looked after Omai for two years, with Banks shouldering all the expenses. By the time Omai left England in 1776 to return to Tahiti, Banks had turned against these kinds of experiments, believing that the medical risks for the natives as well as their emotional distress were too high a price to pay.
† The Purple Emperor is the butterfly *Apatura iris*.

Banks's Library at Soho Square.

laid the foundations and Banks was now building upon them, trying to bring botany to a broader audience, much as Miller had done for practical gardening. This endeavour was undoubtedly boosted by Banks's election to the presidency of the Royal Society in November 1778, a post that he would hold for the next four decades.

Another vehicle for this promotion of botany was his London house. Number 32 Soho Square, which he leased in 1777, became the centre of botanical enquiry—a place which fostered new ideas that reached into the wider world. Every Sunday evening Banks opened his house for informal meetings of thirty or so friends and acquaintances; on Thursday mornings, at ten o'clock, he hosted breakfasts in his library—a tradition that he later extended to every day of the week.*

The panelled room at the back of the house was lined from floor to ceiling with overflowing bookcases holding almost 20,000 natural history titles, while nearly 200 different news-

---

* Horace Walpole dubbed the Sunday soirées, in a letter in March 1791, "Sir Joseph Banks's literary saturnalia."

papers, literary journals and scientific publications from societies around the world were laid out on tables and low shelves. Presided over by the portrait of Captain Cook that hung above the fireplace, Banks assembled the greatest minds of his age, providing a platform for natural history and other scientific subjects that was quite unlike the Royal Society in being, as one visitor described it, "perfectly free from . . . ceremony of any sort." The house was always filled with people, talk and laughter. There was Banks's new assistant and secretary Jonas Dryander—another Swede—several artists and engravers who worked in the workshop at the back of the house, as well as scholars who used Banks's books and collections for their studies. Solander walked every day from the British Museum in Bloomsbury to Soho to help Banks with the library and herbaria. Solander, who was always "cheerful and talkative," enjoyed the endless stream of visitors—"his fondness for company," Banks said, "never allowed him to spend the evening in the Museum."

The move to the large house at Soho Square and his new role as the president of the Royal Society finalised Banks's transformation from adventurer to arbiter of scientific progress and botanical enquiry. He was maturing into this new life with poise and a growing paunch. And, although he had long enjoyed his bachelor life, finding sufficient company among his scientific friends and his mistress Sarah Wells, the now thirty-five-year-old Banks decided that it was time to find a wife. In March 1779, four months after his election to the presidency of the Royal Society, Banks married the rich heiress Dorothea Hugessen. Solander approved: "Mrs. Banks, 20 years & six months old, 14,000£ in money" were her essential characteristics according to him, though he also thought her handsome, chatty and funny, insisting that "our family is much enlivened." Solander was also pleased to report that Banks's mistress had left the scene without a fuss, having "had sense enough to find that he [Banks] acted right." Dorothea quickly found a niche in the male and science-dominated household at Soho Square. She didn't seem to mind that her home was an international scientific research centre and happily participated in the library breakfasts whenever she was in town.

At the heart of what one of the visitors had dubbed "an Academy of Natural History" was the "Herbarium," a room where Banks kept his unrivalled plant collection in Chippendale mahogany "cubes," each with five drawers. In addition to the plants that Banks had collected himself and received from others across the world there were the herbaria of other eminent botanists, which he had bought on their deaths. The first to find its way into his collection had been Philip Miller's, which had been so large that it took Banks and Solander two weeks to get it from Chelsea and to put it in order. Over the years many others followed, including the collection of George Clifford, Linnaeus's first employer in the Netherlands. Banks was particularly keen to assemble the various herbaria on which Linnaeus had worked because they included the so-called "type specimens"—the first plants named according to the new system and therefore the first reference points for identifying a species. Whenever collections were offered to Banks, he wanted to know "if the Herbarium is actually that which Linnaeus used."

Some of the foreign botanists stayed for weeks in Soho Square, or as Banks proudly said "livd almost wholly in my Library." The visitors studied the specimens for their own publications or to identify and classify the species in their own collections. Banks was known for his generosity, allowing botanists to use and examine everything in his collection, because it helped the progress of botany. Visitors rarely exploited this, but if they did Banks was furious. When, for example, one of the botanists dissected several of Banks's most valuable specimens in order to examine their sexual organs, Banks wrote to his assistant Dryander, "[f]or godsake when you are away ... take with you the key of the Cubes." But despite these disappointments Banks maintained an open house in which he welcomed scholars from home and abroad, impervious to wars and revolutions.

In order to further the progress of botany and science, Banks removed himself from all party politics. In his inaugural presidential speech at the Royal Society he had announced himself "free from the Shackles of Politicks." Banks believed that he could influence governmental agencies more effectively if he were not associated with any party. "I have never Enterd the doors of the house of Commons," Banks explained to Benjamin

Banks' herbaria

Franklin, and therefore "I have escapd a Million of unpleasant hours & preservd no small proportion of Friends of both Parties." In the aftermath of the French Revolution, when England's aristocracy feared that the "reign of terror" would spread across the Channel, Banks was even more determined to exclude politics from scientific endeavour. "[Y]ou are welcome to the use of my books," he replied to an enquiry, "but if it is to be any Shape political, I shall hesitate."

On an international level Banks's neutrality proved an equally effective tactic. It allowed him, for example, to stay in contact with Franklin during the American War of Independence, even though Franklin was then an official enemy. Appealing to Franklin as the scientist and founder of the American Philosophical Society rather than as the man who had signed the Declaration of Independence, Banks asked for assistance in granting immunity for Cook, who was on his third voyage across the globe. Franklin responded diligently, and all captains of American war ships received his order to refrain from attacking Cook's vessel and crew as they were travelling "for the Benifit of Mankind in general." "I respect you as a Philosopher," Banks thanked him, and later presented him with a commemorative medal which the Royal Society had issued in honour of Cook.

Banks thought himself "a bird of peace," and every explorer, botanist and scholar knew that he was the conduit for scientific immunity. When the German explorer and naturalist Alexander von Humboldt, for instance, feared for the safety of his French companion Aimé Bonpland, he asked Banks for assistance. Even when France declared war on Britain in 1793, Banks insisted that "the science of two Nations may be at Peace while their Politics are at war," continuing his hospitality to French naturalists throughout the war.* At a time when British laws prohibited correspondence with French citizens, Banks maintained contact "at no small Personal hazard" and canvassed on behalf of the Royal Society, the French Institut National and "other scientific persons" in order to resume communication and the exchange of the societies' publications.

* The botanist and zoologist Pierre Broussonet spent eighteen months at Soho Square during the early 1780s, and the scholar Abbé Gervais de la Rue used the library for many weeks in 1797. On 15 July 1797 Banks wrote that he was pleased that de la Rue "never interferd in politics even in Conversation."

Banks's faith in scientific neutrality was so strong that when the British Navy captured a vessel with the enormous herbarium of Antipodean plants collected by the French botanist Jacques-Julien de la Billardière he did not keep it in Britain. Instead he used all his governmental contacts and royal connections to ensure its safe return to France. And despite the enticing appeal of the specimens, Banks's sense of honour forbade their examination. "I of course did not look them over," he assured the professor at the botanical garden in Paris. "I shall not retain a leaf, a flower or a Botanical idea of his Collection." Instead Banks hoped that one day de la Billardière would come to Soho Square for the two men to compare their herbaria.

Only Linnaeus was excluded from Banks's scientific armistice and global community. Not only had Solander failed to honour the letter he had written from Rio de Janeiro in 1768, in which he had said he would visit Uppsala with their botanical spoils, but neither he nor Banks had sent one of the duplicate specimens they had promised him. Indeed, Linnaeus had heard nothing more of the "ungrateful Solander." Banks, it seems, was wary of seeing his specimens appear in one of Linnaeus's publications before he was ready to publish his *Endeavour Florilegium*—an enterprise that was proving extremely laborious given that, after five years, his engravers had only finished half of the plates. And so it was that, when Linnaeus died in January 1778, aged seventy, the great classifier had not had even a glimpse of a single plant from the collection.

Many of the former "apostles" paid their last respects to their great master and went to the grand funeral in Uppsala—but not Solander. Instead of Solander, Linnaeus's son Carl became his father's successor when he was appointed professor of botany. Yet when Carl asked for duplicate specimens in order to update his father's *Species Plantarum* in winter 1778, Banks replied that he wanted to publish his own discoveries first before the species could be mentioned in any other publication. Carl, though, persisted and travelled to London in April 1781 to work in Soho Square in the hope of gaining access to the *Endeavour* collection. He was treated respectfully by Banks and his assistants, and wrote to Sweden that "the friendliness that Banks and Solander have shown me since I have arrived is more than I could have expected." But despite this, the attic rooms where the *Endeavour* plants were stored remained firmly closed. "Nobody will yet receive any of Sir Joseph's South Sea plants ... least of all Linnaeus," Banks's assistant Dryander wrote during Carl's visit—which was a lie, as many other botanists had been given duplicates over the past decade.

Like his father, Carl would be denied access to the specimens, this time because Banks was fearful of being outdone by Carl's *Supplementum Plantarum*, an addendum to the *Species Plantarum* which would be published before his own book. But this was not the only reason. Carl had refused Banks's invitation to become a fellow at the Royal Society—the highest honour the British scientific community could bestow—because he couldn't afford the fees. Banks was, in Carl's words, "surprised" about this and most certainly deeply offended. Banks might also have been resentful of Carl for refusing to sell Linnaeus's herbarium to him after his father's death. Banks had wanted this collection more than any other in the world. That this treasure was in the hands of Carl, who lacked his father's methodical genius and botanical obsession, was an affront to the entire botanical world—Solander called him "klen Karl"—"weak man." Even Linnaeus had thought that "the greatest collection the world has ever seen" should not be in his son's possession, but his wife had ignored these last wishes. It must have infuriated Banks that the thing he wanted so desperately was denied him by an untalented botanist.

Although Carl was not allowed to see the most valuable specimens, he stayed in London for more than a year, spending much of his time at Soho Square, often joining the philosophical breakfasts and becoming part of Solander's adoring audience. Though more than a decade had passed since the return of the *Endeavour*, Solander still enjoyed recounting the adventures. One morning, in May 1782, he diverted the visitors, including Carl, with the story of his near death in Tierra del Fuego, where he and Banks had been caught in a sudden blizzard. Animated, gesticulating and moving around to make the tale more lively and exciting, Solander described how they had clambered over the tops of an almost impenetrable thicket of chest-high trees, sinking after every step ankle-deep into the bog. But suddenly he stopped in mid sentence, unable to control his arms.

Within minutes, Solander was continuing with his story but onlookers noticed that the left side of his face seemed to have become paralysed, and he was holding his left leg in a strange way. One of the visitors, Charles Blagden, a physician and close friend of Banks, forced Solander to lie down on a couch in the library. Greatly alarmed, Blagden sent for their mutual friend John Hunter, the famous surgeon and anatomist, while Carl went to fetch two of the best physicians in London. But Solander's condition worsened and, in the early afternoon, Blagden sent a messenger to Banks—who was in the country—calling him home because Solander was in "extreme danger."

Banks travelled all night and reached Soho Square in the morning, but the prognosis was not good. Dorothea had prepared a room for the patient and every eminent physician was present, but Solander drifted in and out of consciousness in a feverish state. His agony lasted for five long days, until he died on 13 May aged forty-nine. As befitting to a man who believed in the progress of science through observation, Solander's body was cut open by Hunter in order to investigate what had killed him. When Hunter dissected the brain, he found two ounces of coagulated blood and concluded that Solander had died of a stroke.

12

s good-humoured a nondescript
Otatheitan as ever!"

*The greatest service which can be rendered to any
country is to add a useful plant to its culture.*

THOMAS JEFFERSON, 1800

*He travels and expatiates, as the bee
From flower to flower, so he from land to land;
The manners, customs, policy of all
Pay contribution to the store he gleans;
He sucks intelligence in every clime,
And spreads the honey of his deep research
At his return—a rich repast for me.*

WILLIAM COWPER, "The Task," 1785

Solander's sudden death was deeply mourned by Joseph Banks. For a long time, he refused to talk and write about his closest friend because it was too painful. "[T]o write about the Loss of poor Solander would be to renew both our feelings for little purpose," he said a month after the death. But even years later, he wrote that the loss "forces me to draw a veil over his passing" because he could not bring himself to think about it. Whenever he was reminded of those last dreadful days in which Solander had suffered so terribly, Banks said, he felt "such acute pain as makes man shudder."

In the years that followed, Banks's focus shifted away from studying and collecting plants for the sake of botanical completeness, as he had done with Solander, towards an exploration of their economic value. He became increasingly motivated by a belief that horticulture and botany—especially if applied to the flora of Britain's colonies—would bring improvement and profit to the nation. His botanical knowledge, as well as his understanding of the climate, soil and vegetation of the countries that he had visited with Solander—this vast "book of information"—helped Banks see that useful plants and their efficient exploitation could be the key to the success of a thriving empire.

The first step in this endeavour was to use Kew as an engine of colonial growth. The idea was to find potentially lucrative plants from across the world, test and propagate them in Kew, then dispatch them to other colonies where they might be of agricultural or economic benefit. Kew should become, Banks declared five years after Solander's death, "a great botanical exchange house for the empire." At the same time, the enormous number of plants that arrived at Kew would also find their way to the nurseries and gardens of Britain, feeding amateur gardeners with an unceasing stream of novelties.

Banks was not the first to believe in the economic significance of botany. Already Miller had suggested that the British should grow cacao in the West Indies in order to undercut the French and Spanish; and he had laid the foundation for American cotton production when he sent West Indian cotton seeds to Georgia in 1732. Likewise Linnaeus had suggested—rather unrealistically—growing tea in Sweden. But Banks was able to pursue his plans on a far broader scale and with more tenacity than anyone before him because he was the president of the Royal Society, an explorer and close to King George III, straddling science, first-hand knowledge of the colonies and governmental policy-making. Almost every Saturday Banks walked with the King for hours through the gardens and hothouses at Kew. George III responded enthusiastically to Banks's ideas, acquiring the sobriquet "Farmer George" and financing several plant-collecting expeditions from the early 1770s.

One glance at Banks's dinner table or at his farmers' diet illustrated the benefits the colonies could bring. Potatoes, for example, had been introduced two centuries previously from South America to Europe, by the Spanish. The spice trade had also brought riches to Europe as had cotton, coffee and tobacco. Banks believed that a colony like India, for example, was "by nature intended for the purpose of supplying . . . raw materials" needed to power the factories that were fuelling the Industrial Revolution. If, he argued, the government allocated some of its resources to botanical projects and expeditions, they would be able to locate the countries that had similar climates and transplant useful species such as cotton or tea from one end of the globe to the other. "To exchange between the East & West Indies the productions of nature," Banks urged the Secretary for War, was of the greatest importance, otherwise Britain would forever lag behind the French and Dutch.

Banks was quick to use Britain's economic problems to bolster his arguments. With the defeat in the American War of Independence in 1783 and the loss of the thirteen colonies the Treasury's coffers were depleted. The loss of the large American export market for British goods and the end to the supply of raw materials which were needed for manufacturing centres demanded changes in the regulation of trade policies, and the unprecedented involvement of the state.

As there was no bureaucratic infrastructure to implement the necessary changes, such as a civil service or advisory boards, Banks sensed an opportunity and accepted any request from government for advice, regularly attending the meetings of the Privy Council Committee for Trade and Plantations. Within two years of Solander's death, Banks was widely regarded by politicians, merchants, industrialists and colonial landowners as the expert on botany and its economic benefits at home and abroad.

When, for example, the directors of the East India Company wanted to test the commercial potential of growing Chinese hemp in England, they asked Banks to find the best gardeners. When plantation owners from the West Indies felt neglected by the British government they asked Banks for assistance, believing that his "Knowledge, Philanthropy, Patriotism, & Influence" were "superior to all others." And when the government mounted an expedition to Africa in search of a penal colony, they hoped Banks could recommend a botanist to accompany the vessel in order to assess the plant life that would be the foundation for a self-sustaining settlement. Hurriedly, Banks asked William Aiton, the head gardener at Kew, to choose a suitable man: "For godsake be active and do not let such an opportunity slip," he urged, because "we shall get [dried] specimens for me & seeds for Kew garden."

Banks was extremely wily in the way he used his powerful

connections to further his cause. When the Polish-born gardener Anthony Pantaleon Hove was dispatched to Gujarat in India in April 1787, ostensibly to collect rare flowers for the royal garden at Kew, Banks issued him with a set of "Private Instructions." These were to find the best cotton seeds, and learn about their soils, manure and cultivation. In order that neither the Indian growers nor the East India Company discovered the subterfuge, Hove was to send his observations, written in Polish, to his brother, who would then translate them for Banks. At the same time cotton seeds were to be mixed with the exotic flowers that Hove sent back, with remarks in the letters about their "beauty or curiosity" to justify their place at Kew. By these means, Banks hoped to find a way to overcome the shortage of cotton that had made British manufacturers dependent on foreign sources since the American War of Independence. Banks believed that if Indian cotton plants were transplanted to the West Indies, plantation owners there would be able to supply British cotton weavers—the nation's fastest-growing industry*—in the long term.

It wasn't only Indian cotton that interested Banks. Shortly after Hove had left London for Gujarat, Banks received news that Lieutenant Colonel Robert Kyd had begun to lay out the botanic garden in Calcutta about which he had consulted Banks some time earlier. Banks believed that the garden should be used "solely for the promotion of public utility." The superintendent was to collect plants from throughout Asia in order to dispatch them to London, as well as using the garden to test new fruits and plants on Indian soil in the hope of improving the subcontinent's agriculture. In particular it was to be a nursery for nutmeg and cloves, in an attempt to break the Dutch spice monopoly. With the French introduction of these two spices to their West Indies colonies, so Banks wrote to the Secretary of War a few weeks after Hove's departure, there was no time to lose. "With such example before us," Banks asserted, "nothing but activity is wanting."

As Banks was compiling his plant recommendations for the botanic garden in Calcutta, another of his ideas came to fruition. On 13 May 1787, a month after Hove had left for Gujarat,

---

* British cotton goods consumption grew from 5.3 million pounds in 1781 to 32.5 million in 1788.

the eleven vessels of the First Fleet under the command of Captain Arthur Phillip set sail to cross the globe towards Australia. With him were soldiers, government officials and settlers, but also almost 800 convicts who had exchanged their prison, or even death sentences for seven years' labour in the wilderness of New South Wales. Banks had long been trying to convince the government to establish a settlement in Australia—"my favourite Colony," as he later called it. Already some years previously, when America's Declaration of Independence had deprived Britain of her dumping ground for criminals and radicals, Banks had told the House of Commons that the most suitable place for a new penal colony would be Botany Bay, for its climate, vegetation and "extremely cowardly" natives. Now he was intimately involved with Phillip's expedition, inviting the captain regularly to Soho Square in the weeks before the fleet set sail, and regaling him with Australian tales and precise botanical instructions. When Phillip left Britain he took with him boxes of those seeds judged by Banks to be most suitable for the Australian climate, as well as lists of what plant and seed to collect during the stopovers at Rio de Janeiro and South Africa. When one of Banks's collectors met the fleet at Cape Town, he wrote to Soho Square that the captain's cabin on the *Sirius* was "like a Small Green House." Under the bed and along the walls were pots with cocoa and coffee trees as well as oranges, figs and vines for the gardens of the new colony.

Banks's third great project of that year of 1787 involved a ship called the *Bounty,* under the command of Captain William Bligh, who had served under Cook on the *Resolution*. The purpose of this voyage was another of Banks's ideas: to transfer a crop from the East Indies to the West Indies in order to feed the slave population there. The plant in question was the breadfruit tree, which Banks had seen laden with its large starchy fruits during the *Endeavour*'s stop at Tahiti almost twenty years previously. The fruit, Banks knew, could be baked or boiled, by which process it became like a loaf of bread sufficient for three to four people. And because the tree produced fruit almost all year long, it was one of the highest-yielding food plants in the world.

Again Banks was involved in every aspect of the expedition:

he had conceived the idea together with Solander, corresponded with and met several Jamaican plantation owners (the receiver-general of Jamaica agreed with Banks that it was "a wholesome pleasant Food to our Negroes"), convinced the government, dealt with the administration, suggested the route, advised on the ship, recommended Bligh and found a gardener. He enjoyed being in control and hated it when others meddled with his projects. When the secretary of war, Sir Charles Yonge, for example, dared to make a suggestion after he had seen Bligh and the *Bounty* at the docks in Deptford, Banks appealed to the Undersecretary of State, Sir Evan Nepean. "I wish such dablers as he is could be prevented from interfering in our matters," he complained.

Banks had selected the only gardener in England who had experience of how to tend and protect plants at sea during such a long journey. David Nelson had worked for Banks during Captain Cook's last voyage, collecting seeds and plants for Kew. Having spent some time in Tahiti with Cook, Nelson also spoke the native language, which would help him to negotiate a price for the large number of breadfruit trees he needed to procure. Banks's instructions for Nelson were detailed and precise, covering everything from his conduct on board to pest control. The whole enterprise, Banks reminded Nelson, depended entirely on his diligence and care. "One day, or even one hours negligence may at any period be the means of destroying all the Trees and Plants," and despite—or maybe because of—his own past enjoyment of Tahitian women and feasts, Banks warned "against all temptations of idleness and Liquor." Nelson was to be under the command of the captain, but, Banks insisted, Bligh should "never interfere in directing the manner in which a tree is to be planted." In addition Nelson could demand as much water as he needed to feed and wash the saplings, nor should parrots, dogs or cats be allowed as pets for the crew because they could harm the plants.

Unlike any other ship that had previously set out on such a long sea voyage, the *Bounty* was to be equipped entirely for the transport of plants. Banks turned it into a floating garden, ordering the hull to be packed with barrels of water, the only cargo the *Bounty* was to carry, to feed the breadfruit trees and cleanse their leaves of the salty spray. "[T]he Master & Crew," Banks instructed, "must not think it a grievance to give up the best part of her [the *Bounty's*] accommodations." Bligh, for example, had to relinquish his Great Cabin so that it could be transformed into a greenhouse. Instead he would sleep in a small, windowless cabin—similar to the one Banks had occupied on the *Endeavour*—and could only invite his officers to share his table in a cramped pantry.

The Great Cabin, on the other hand, was given even more windows. Carpenters installed several skylights, allowing light to flood in, as well as putting in false floors. These had more than 600 holes cut into them, to hold the empty plant pots while they awaited their precious contents from Tahiti. Remembering the icy gales and drops in temperature when they had rounded the treacherous Cape Horn in the *Endeavour*, Banks also requested that a stove be installed. Nelson would be responsible for the health of the breadfruit saplings during the ocean

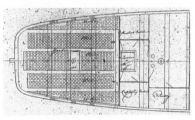

The *Bounty's* lower deck with the Great Cabin filled with empty plant pots awaiting the breadfruit trees—note Bligh's tiny cabin next to the Great Cabin.

crossing and, to ensure that no one else interfered, the Great Cabin was to be locked and the key was to be kept by the gardener. In agreeing to give up the only room on the ship that commanded dignity to a gardener and to empty plant pots, Bligh lost the most important symbol of authority on a ship— a decision that would have fatal consequences.

The *Bounty's* notorious voyage began on 23 December 1787 when the ship sailed into the Channel after being delayed for several weeks at Spithead. Two months into the expedition, Bligh reported that the crew was cheerful and he had "no cause to inflict punishments." The next Banks heard of the ship was a letter he received from Bligh at the end of October 1788. It had been written from Cape Town several months earlier. After several failed attempts to round Cape Horn, Captain Bligh had been obliged to sail around the world in the other direction, adding 10,000 miles to the voyage. After Cape Town he would successfully reach Tahiti. But Banks didn't know this. The Cape Town letter was the last he received. After that the *Bounty* disappeared into the emptiness and silence of the Pacific Ocean.

While Banks waited for news from the *Bounty*, he continued to canvass for more support—financial and political—for imperial botany. One avenue he was investigating was the transplantation of Cantonese tea to the Indian states of Assam and Darjeeling because the climate and soil matched. Tea, according to Banks, was "an article of the greatest national importance" and, if the transplant was successful, it would break the Chinese monopoly on the trade. Banks wrote to the director of the East India Company, suggesting that a few Chinese planters should be persuaded to pack up their tea shrubs and tools in order to move to India. At the new Botanic Garden at Calcutta twenty acres of fertile land were already allocated to propagate the plants. Here the natives would be taught cultivation techniques, an activity that "suit[ed] the Hindoos most perfectly," an acquaintance of Banks agreed, because of "[t]heir patient industry & pliable fingers."*

Slowly, Banks's continuous endeavour was yielding results. Ships packed with plants for the new colony in Australia were sailing across the globe, and Banks received news from a plantation owner in Barbados, who reported that the cinnamon tree that Banks had sent was already more than seven feet. By early 1790 Kew's collection had grown so much that the *Monthly Review* ran an article on "the numberless plants now raising in the gardens at Kew, which never flourished there before; many of them totally unknown." To celebrate these achievements and to demonstrate Britain's horticultural superiority to the rest of the world, the gardener at Kew, assisted by Banks and his workforce at Soho Square, published the *Hortus Kewensis*—a catalogue of 5,600 plants, all cultivated at Kew.† At the same time Hove returned from India with twenty-three varieties of

---

* The project failed, and tea production in India began only when an Indian variety (*Camellia assamica*) of the Chinese tea shrub (*Camellia sinensis*) was found at the beginning of the nineteenth century.

† This first edition of the *Hortus Kewensis* had been in the making for many years—both Daniel Solander and Jonas Dryander had worked on it extensively.

the finest cotton, and Banks received a sample of cotton that had been produced in the West Indies from the seeds he had sent previously—enough to tailor a waistcoat for him.

But just as Banks must have been feeling pleased with the success of his projects he received devastating news. In early 1790 he received a letter from Bligh, who was in Batavia, the malaria-infested port on Java where Banks and Solander had almost died on the *Endeavour*'s return. "I am now so ill that it is with the utmost difficulty I can write to you," Bligh wrote, continuing, "I feel very sensibly how you will receive the News of the failure of the expedition that promised so much."

After the *Bounty*'s arrival in Tahiti in October 1788, all had gone to plan. For five months, Nelson had planted hundreds of breadfruit saplings into pots, nurturing them in a makeshift nursery on the island to make them strong enough for their long journey. In early April 1789, the ship had left the island, its Great Cabin filled with 1,015 breadfruit tree saplings planted in almost 800 pots. Only three weeks later, however, after arguments with their captain, part of the crew including the gardener's assistant had staged a mutiny. On 28 April, at the break of dawn, Bligh had awakened to a nightmare, finding his boat in the hands of the mutineers. With no marines on board to protect the ship and the Great Cabin full of plants, Bligh had neither the physical power nor the symbolic authority he needed to suppress the rebellion. Ignoring Bligh's empty warnings and threats, the mutineers, under the command of master's mate Fletcher Christian, had put him (naked but for his nightshirt) into the *Bounty*'s launch, along with the gardener Nelson and sixteen other men, and abandoned them in the middle of the South Pacific.*

In an overloaded twenty-three-foot open boat with no charts and hardly any food, Bligh had navigated almost 4,000 miles in forty-eight days, skirting the northern tip of Australia, to Timor in the Dutch East Indies. In Timor, Bligh had chartered a schooner which had brought them to Batavia, minus Nelson, who having survived the danger and starvation of their ocean

* After going back to Tahiti to stock up on women and food, the mutineers zigzagged the South Sea for several thousands of miles in search of a suitably remote and uninhabited island until they finally settled at Pitcairn, where their descendants still live today.

crossing had died in Timor of "a Fever." All the plants, however, were lost with the *Bounty*.

The mutiny on the *Bounty* was not the only bad news to reach Banks that winter. Banks's second plant transport to Australia (after the departure of the First Fleet) had also failed when, in December 1789, the ship had hit an iceberg some 1,200 miles south-east of Cape Town. Only a few of the crew and convicts survived and none of the plants reached their destination. Banks, however, remained optimistic, continuing to canvass for colonial botany and influencing government policies accordingly. Only when dealing with the East India Company did Banks's old impatience surface occasionally. "I am tired of promoting projects with a fluctuating body who are sure to be changed by the time I have convinced the first set of the propriety of any measure recommended by them," he wrote to the superintendent of the Botanic Garden in Calcutta regarding some more tea experiments. The prolonged formalities of the East India Company infuriated Banks: as many plants died in their care as were killed by violent storms during the transport across oceans. When one particularly valuable collection from India—including the first thriving nutmeg tree to arrive in Europe—was seized by the East India Company and consequently perished in their warehouse, Banks was furious and "cursd the Chairman," calling him a thief. But Banks refused to be deterred. Horticultural casualties were sent to a special hothouse at Kew—"our Kew Hospital," as Banks called it. And when, in March 1790, Bligh arrived empty-handed in Britain, the next breadfruit expedition was immediately planned. A little more than a year later (after a court martial had acquitted him of losing the *Bounty*), Bligh was commanding the *Providence*, sailing once more towards Tahiti to procure breadfruit for the slaves in the West Indies.

Despite the setbacks Banks's counsel was much sought after. When the East India Company received a gum tree from India called Babul—*Acacia arabica*—they sent a sample to Banks asking whether, in his opinion, the tree would be a useful import to Britain. And when the first consignment of tea plants from China failed to thrive in the Calcutta Botanic Garden, Banks suggested that the procedure be tried again. He gave advice throughout the world, continuing his extensive international correspondence, requesting and providing information on a vast array of subjects. Having admired *Phormium tenax*—New Zealand flax (one of the most popular Antipodean plants in Britain's gardens today)—and its use for Maori cloth production during his time in New Zealand, Banks devised a scheme for a linen industry in Australia. He also insisted on being kept updated on the plant collection in the botanic garden in Calcutta, requesting "a quarterly statement of all new plants." When abolitionist William Wilberforce and the Sierra Leone Company wanted to send a botanist to Africa, Banks found the right man in less than two weeks.

On 3 August 1793, almost two years to the day from when Bligh had left Britain on his second breadfruit expedition, a letter was delivered at Soho Square that repaid all the failures and patient waiting. It was addressed in Bligh's handwriting to "Sir Joseph Banks . . . To be forwarded immediately if not in town." The *Providence* was already in the Channel, Bligh wrote, with a cargo of 703 pots full of exotic plants which the gardeners had collected and nurtured during the long voyage. More than 600 thriving breadfruit trees had been disembarked at the islands of St. Vincent and Jamaica—about a third of what they had procured in Tahiti having survived the ocean crossing. They had been in such excellent condition, Bligh boasted, that "there can be no excuse for their dying in the West Indies."

When the *Providence* anchored in Deptford four days later, Banks was there to inspect the treasure. Stepping into the crowded Great Cabin, he entered another world. The gardeners had plunged potted mango and banana trees into every hole in the false floor, and dangling from the ceiling was a vast variety of orchids still attached to the branches of their host trees—many sorts of epidendrums and the air plant tillandsia, as well as the first two species of the showy orchid genus *Oncidium*. For the first time since Banks had left Tahiti, more than two decades previously, he saw his herbarium brought alive. There were four breadfruit trees, a memento of the prosperous voyage, and the fragrant *Gardenia taitensis*, which Banks had so adored when woven into the hair of Tahitian women. Banks could hardly believe the sight, for he had "never before seen plants brought home by Sea . . . in so flourishing state."

He had not been able to bring back a single living plant from his voyage on the *Endeavour*, but now he was standing among more than 1,000 trees, shrubs and flowers from the other side of the world. Species which no one had ever seen in Britain would soon be on display in the greenhouses at Kew. The expedition exceeded everything he had dreamed of, Banks wrote to the prime minister William Pitt: not only had the breadfruit trees reached the West Indies but Bligh had also returned with the largest living collection of plants that had ever been brought into England.

A few months later Banks received the first letters from the West Indies describing the progress of the breadfruit trees. Bligh had delivered not only plants from the East Indies to Jamaica but also a Tahitian who was teaching the gardeners how to cultivate the trees. In addition, one of the gardeners on the *Providence*, James Wiles, had accepted the position of superintendent at the new botanic garden at Bath, Jamaica, bringing with him the latest English horticultural knowledge. Nine months after his arrival at his new home, Wiles told Banks that the breadfruit trees had been distributed across the island

breadfruit trees

and were flourishing. A year later, a plantation owner and member of the House of Assembly in Kingston informed Banks that the trees were "thriving with greatest luxuriance." The largest specimens had reached a height of fifteen feet and were about to fruit for the first time—"there is every reason to expect complete success." From St. Vincent came similarly good news when the head of the botanic garden there reported that the breadfruit "thrives (if possible) better than in its native soil." In 1801, less than a decade after the arrival of the *Providence*'s nutritious cargo, Wiles wrote that the breadfruit was "perfectly naturalised."

Once Banks had successfully overseen the transfer of plants from one hemisphere to another, he extended his idea to other aspects of agriculture. As a Lincolnshire landowner and sheep farmer, Banks had suffered from the closure of the American market during the war and had investigated ways of improving Britain's wool production in order to target the European cloth-iers, who preferred the finer wool from the Spanish Merino. Smuggling sheep out of Spain, he had cross-bred them with British species, producing a breed that would become the

Australian Merino. These sheep, when dispatched to Australia years later by Banks, formed the basis of the Antipodean wool industry which would bolster the British economy over the next century.*

Banks was now in his fifties, and though he had become corpulent and battled with regular attacks of gout he remained active within the Royal Society, the government and his botanical circle. In spring 1797 his achievements were recognised when he was made a member of the Privy Council, appointed "for the Consideration of all Matters relating to Trade and Foreign Plantations." An acquaintance had once called him "His Majesty's Ministre des Affaires Philosophiques," and now it was as if this sobriquet had become his official title. His friends adored his steadfastness and modesty, as he never boasted about his achievements—when he received the Red Ribbon of the Order of the Bath, for example, one remarked that the honour had made no difference to Banks's behaviour as he still "sprawls upon the Grass kisses Toads and is just as good-humoured a nondescript Otatheitan as ever!" Others continued to complain that his "attention is very much confined to one study, Botany." But whatever people thought, Banks remained one of the most influential men in Britain. Until his death in 1820, Soho Square remained the nerve centre for empire policy-making. Working breakfasts were now held every day and one visitor reported that Banks gave out more than 200 invitations to his Sunday soirées.

During his lifetime, Banks received seeds and living plants from more than 120 plant-collectors in addition to those he garnered from the twenty-one especially commissioned plant-hunters whom he had dispatched across the globe at either his own, the King's or the country's expense.† The collection at Kew remained his "favourite project" because, Banks said, it

---

* In 1821, one year after Banks's death, there were 290,000 sheep in Australia and by the mid-nineteenth century the annual production of raw wool was more than 10 million pounds.

† Banks's usual employment terms were an annual salary of £100 plus up to £200 in expenses for plant-hunters or a piece rate of sixpence per collected species with either fruit or flower, or one shilling if both could be procured. Banks liked to remind the collectors that they should not travel as gentlemen but as servants, and in general he preferred to employ Scottish collectors for their "habits of industry attention & Frugality."

did "honor to the Science of the Countrey, Promotes in Some degree its Commerce [and] aids its Population." Agreeing with Adam Smith's tenet that "commerce and manufactures gradually introduced order and good government, and with them, the liberty and security of individuals, among the inhabitants of the country," Banks had steered—stoically and with perseverance—the country in a direction that would shape Britain's empire, economy and society for the next century. Plants not only changed the English landscape but the very fabric of the nation, contributing to the country's global dominance and imperial strength.

13

"Loves of the Plants"

Flora's treasures from all over the world
are brought here as if by magic.

JOHANNA SCHOPENHAUER, 1803

ATTEND, ye swarms of MODERN TOURISTS
Yclept, or Botanists or Florists:
Ye who ascend the cloud-capt Hills,
Or creep along their tinkling Rills;
Who scientifically tell
The Wonders of each COCKLE-SHELL;
And load the Press with Publications,
With useless, learned DISSERTATIONS.

ANONYMOUS, "Transmigration," 1778

On a crisp winter morning in 1783, Joseph Banks received a letter that would cause great excitement among botanists. Linnaeus's son Carl had died unexpectedly, and his executors were offering Banks the opportunity to purchase Linnaeus's entire collection. It consisted of 3,000 books, thousands of minerals, insects and shells, and most importantly some 19,000 dried and mounted plants—the herbarium, which was the base reference of all botanical enquiry because Linnaeus had used the plants as type specimens for his reclassification of the natural world. For 1,000 guineas it could be Banks's. Ironically, for the first time in his life, Banks was short of money. The war with America had put an end to the export of wool to the American colonies, and it was this that had underpinned his family's fortunes. To make matters worse, Catherine the Great of Russia, a keen plant-collector herself, had expressed interest in acquiring Linnaeus's collection, as had botanists in Denmark, Holland, France, Switzerland and Sweden, including the professors and students at Uppsala as well as King Gustav himself.

Serendipitously, the morning the letter arrived, Banks was

having breakfast with James Edward Smith, the son of a wealthy merchant. Smith had just finished university and was a passionate botanist. Having read the letter, Banks passed it across the table and urged the young man to buy Linnaeus's collection. Its possession would give Smith "honour," Banks said, and place him at once at the vanguard of botanical enquiry. Smith was easily convinced. The next day, he wrote home asking for the money. His father, after warning his son against overexcitement, released the requested funds and Smith struck a deal with the executors of Carl's will.

Ten months later, twenty-six enormous chests arrived from Sweden. There had been a last-minute panic when King Gustav had imposed a ban on their export, but fortunately it had been too late: the ship had already left Swedish jurisdiction. In order to display his purchase, Smith leased an apartment in Paradise Row, near the Chelsea Physic Garden. Banks was the first visitor, poring over the specimens in great excitement. And over the next few months he and his assistant often went to the Chelsea apartment to help Smith unpack the lavish collection.

With Linnaeus's herbarium in Chelsea, Banks's collection at Soho Square, Sloane's bequest at the British Museum and the living plant entrepôt at Kew, London had become the botanic centre of the world.* Nowhere else was there such an accumulation of foreign plants—dried and living—as well as of botanical knowledge. The purchase of Linnaeus's collection, one of Smith's friends wrote, "most decidedly sets Britain above all other nations in the Botanical Empire."

With London the home of botany, the love for plants and classification was also percolating into the boudoirs of the upper and middle classes. "Botany Flourishes here most abundantly," Banks wrote to Smith, explaining that even the Queen was studying it diligently. Women, in particular, enjoyed the study of plants as a pastime. Unlike the collecting of insects, for example, it did not involve cruelty, but combined education with aesthetics, making it a healthy and innocent pursuit for ladies. What was more, the sexual system was so easy to master—"Yes, even for Women themselves," Linnaeus had written a few years before his death—that

* In 1788 Smith co-founded the Linnean Society. Banks was made an honorary member.

women were able to classify plants without any scientific training. They now went "botanising" with portable presses to dry the plants they picked, entertaining themselves by identifying their finds.

To serve this new army of amateur botanists, books, pamphlets and articles were written for the "fair daughters of Albion." In *The New Lady's Magazine*, for example, the story of a romantic stroll through a powdered wintry garden was a thinly veiled botanical lesson, with a young woman exclaiming passionately, on finding a lonely snowdrop, "It belongs, I believe, to the sixth class of the Linnean [*sic*] system, called Hexandria." One of the most popular books of the time was Jean-Jacques Rousseau's *Letters on the Elements of Botany addressed to a Lady*, which included a description of the therapeutic values of plants and nature. Rousseau also insisted that an interest in Linnaean botany and the study of nature "prevents the tumults of passion."

However, the book that cemented botany's place in England's drawing rooms was a charming poem entitled *Loves of the Plants,* written by Erasmus Darwin—physician, inventor, promoter of science schools for girls and grandfather to Charles Darwin. While Banks and Smith negotiated the purchase of Linnaeus's herbarium, Darwin was turning Linnaeus's sexual system of plant classification into "the most delicious poem upon earth," as Horace Walpole declared it.

By the 1780s, Erasmus Darwin had caught the botany fever, having previously dabbled also in geology, zoology and chemistry. For Darwin, botany was not only enlightening, but inspiring of literary endeavour. "The Linnaean system is ... an happy subject for the muse," he insisted, and he made the most of the time he spent travelling as a practising physician (several thousands of miles every year) to write a poem about it. He had converted his brightly painted carriage into a bespoke study: a skylight in the roof ensured enough illumination, a bolted chest contained paper and writing utensils, and books were piled from floor to window.

Erasmus Darwin's *Loves of the Plants* invited the reader to enter an "INCHANTED GARDEN"—a magical place inhabited by glow-worms, horned snails, fluttering butterflies, gnomes and nymphs. Here, silver moonlight, murmuring waters, rustling leaves and swaying oaks set the scene for the "Beaux and Beauties" of flora's realm to make gentle love on "moss-embroider'd beds." Reversing Ovid's transformation of humans into flowers in *Metamorphoses*, Darwin turned plants into people, writing a poem that was populated by lovesick violets, jealous cowslips and blushing roses.

Each verse of the poem is a little vignette—or, as Darwin described it, "little pictures suspended over the chimney of a Lady's dressing-room"—that introduces a plant according to the order in which Linnaeus classified them. With one stamen and one pistil, the monogamous and virtuous canna enters the scene first:

> First the tall CANNA lifts his curled brow
> Erect to heaven, and plights his nuptial vow;
> The virtuous pair, in milder regions born,
> Dread the rude blast of Autumn's icy morn;
> Round the chill fair he folds his crimson vest,
> And clasps the timorous beauty to his breast.

There is the tulip who in a motherly gesture "folds her infant closer in her arms" and the "chaste MIMOSA." As the number of stamens and pistils increases, the love life of the plants becomes more salacious. With the genista, "ten fond brothers" woo one "haughty maid." *Gloriosa superba* seduces the "blushing captives of her virgin chains" and *Colchicum autumnale* sports three "maids" and "six gay Youths." Yet, despite such behaviour, the poem retains its gentility by making the plants tremble, blush or modestly turn their heads—though the plants might be depicted as involved in indecent relationships, at least their cheeks are always flushed and their eyes filled with tears.*

---

* Darwin also digressed from the flowers. Thistle plumes which were carried away by the wind, for example, inspired a verse about balloonists—one of Darwin's favourite topics, as he had been the first Englishman to fly in a hydrogen balloon. Arkwright's industrial textile mills found their way into the cotton verse, and when Darwin asked James Watt for detailed information about steam-engines, which he also included, Watt teased: "I know not how steam-engines come among plants; I cannot find them in the Systemae Naturae ... However if they belong to *your* system, no matter about the Swede."

FLORA attired by the ELEMENTS.

Published June 1st 1791 by I. Johnson, in St Pauls Church Yard London.   Designd by H. Fuseli. Engraved by Anker Smith.

Flora Attired by the Elements, drawn by Henry Fuseli for
Erasmus Darwin's *The Botanic Garden*. The poem *Loves of the Plants*
was published as the second part of *The Botanic Garden*.

When it was published, in April 1789, *Loves of the Plants* became the most talked-about poem of the day, with Darwin being hailed as "sublime" and "divine," the successor to Milton and Shakespeare. Anna Seward, a friend of Darwin and a poet herself, said the poem "shed lustre over Europe in the eighteenth century." George Crabbe and William Cowper adored it, while William Wordsworth praised "the dazzling manner of Darwin." Samuel Taylor Coleridge called Darwin "the first *literary* character in Europe" and Walter Scott later ranked him as a poet "of the highest class." Though Walpole thought it strange that Darwin was inspired to write such lyrics "by poring through a microscope," he nonetheless declared that he could "read this over and over again forever."

What made Darwin's poem so mesmerising was that science suddenly seemed effervescent and delightful. For the first time, botany reached an audience that had no explicit interest in natural history. And this audience was huge. The poem, with its blushing girls and dashing admirers, was devoured by readers of popular romances, and could be found in many a middle-class home. The Linnaean system of classification, once considered an unsuitable topic of conversation for women, had been made genteel.

Erasmus Darwin also made Linnaeus's ideas accessible with his *System of Vegetables*, the first translation of the *Systemae Naturae* from Latin to English (having discussed the minutiae of the linguistic terminology with Samuel Johnson, as well as Banks and other botanists). Using Linnaeus's Latin as his basis, Darwin invented many of the English botanical terms that are still used today, such as "stamen," "pistil," "bracts" or "floret." And so, while John Bartram had needed to ask his learned friends to translate Linnaeus's publications for him, now anybody, provided they were literate, could use the Linnaean system without any knowledge of Latin.

Darwin dedicated the *System of Vegetables* to Banks, for his support throughout the project. As well as sending books from Soho Square to the Midlands, Banks had offered his botanical advice and commented on sample pages. Darwin had sent glossaries of translated terms, asking Banks "to cast your eye over it." Banks, as Britain's chief promoter of botany, Darwin knew, could "much encourage or retard the progress of the work"; his opinion was of the greatest importance.

Together, Darwin's scholarly and poetical writings on botany animated the new national passion. When his publisher had offered him an "immense price" for his botanical poetry, Darwin boasted to a friend, "as every line puts ten shillings in my pocket, I shall go on *ad infinitum*." Others were also keen to capitalise on the new fashion. By the end of the eighteenth century, the number of publications dealing with botany and horticulture increased by thirty-fold, culminating in the launch of *The Botanical Magazine* (subsequently *Curtis's Botanical Magazine*)—the first of its kind—for those who "wish to become scientifically acquainted with the Plants they cultivate." Every month, from 1787 onward, the magazine described and illustrated thirty-six of "the most Ornamental Foreign Plants" cultivated in Britain—because, so the magazine proclaimed, "several Ladies and Gentlemen" were in search of a reference book in which botany and gardening "or the labours of Linnæus and Miller, might happily be combined." The success of *The Botanical Magazine* encouraged rivals, including *Exotic Botany*, whose publisher thought that other publications could not "keep pace with the botanical riches flowing in upon us."

Botany had become so popular that the demonstrator of the Chelsea Physic Garden, who had traditionally explained the garden's plants to apothecaries and their apprentices, was now able to supplement his income by taking budding amateur botanists on Monday mornings to either Hampstead Heath or the meadows at Battersea on "herbarizing excursions," charging two guineas per person. And for those who could not afford this, or preferred to botanise on their own, there was the *Botanical Pocket Book*, which listed plants according to Linnaeus's classes, orders and species, as well as providing plenty of space to add one's own observations on location and habitat. Small enough to be kept in the folds of a skirt or the pocket of a jacket during excursions, this was the perfect publication for the increasing number of people who spent their leisure time searching country lanes, hedgerows and estates for plants.

"*Natural History* and *Botany* are the fashionable and

favourite studies of the polite and the learned in Europe," an American botanist wrote to John Bartram's son William, and the classification of plants became so popular that several garden owners laid out their pleasure grounds according to the Linnaean system. Erasmus Darwin, for example, planted his picturesque garden outside Lichfield according to Linnaeus's orders and families, uniting, as one visitor said, "Linnean [sic] science with the charm of landscape." Another garden owner wrote—to tempt a friend to visit his estate, Bush Hill in Edmonton—"I have specimens of all the 24 Classes, & of most of the Orders, in the Garden." At Woburn Abbey—where Miller had moonlighted in the 1740s and Bartram had sent many of his early boxes—the most fashionable landscape designer of the time, Humphry Repton, designed "a botanical arrangement of all the grasses" as part of a series of specialist gardens, which also included an "American garden." And at Painshill, visitors admired the "Botanical Walk" which snaked through a complete collection of evergreens—a true botanical encyclopaedia—the forerunner of Victorian arboretums and pinetums.*

Because gardening and botany had become a national craze, foreign tourists included a tour of England's gardens in their itineraries, filling their journals and letters with lengthy descriptions of this gardenmania. Everything, from the "great wages" that English gardeners received to the fragrant shrubberies, was noted and admired. The most surprising aspect of the English garden was, as one German tourist commented, the number of "foreign plants in abundance." By the 1780s, many of the American species that had started off as expensive rarities fifty years previously were now affordable. The fringe tree, for example, was now available for 3s., southern catalpa for 2s., *Robinia hispida* for only 1s. Balsam fir, which Bartram had plucked from the Catskills, was even more common: the price had plummeted from 5s. for a single specimen at the Thorndon sale in the 1740s to 6s. for 100 saplings in 1786. Even regional nurseries were now able to offer a vast range of plants. Telford in York listed more than 300 flowering shrubs and evergreens

---

* In the nineteenth century Joseph Paxton's arboretum and pinetum at Chatsworth, as well as John Claudius Loudon's arboretum in Derby, were laid out according to the botanical classification of the trees.

in their catalogue, while a Norwich-based business offered thirteen species of pines. Even the choicest of Bartram's plants, such as magnolia, kalmia and rhododendron, could be purchased from regional nurseries in the North for between 7s. and 15s. depending on their height.

As a result of their availability, Bartram's plants were also included in the many manuals that gave instructions as to how to lay out a shrubbery. A kind of horticultural painting-by-numbers, these manuals allowed amateur and suburban garden-owners to re-create the arrangements of trees and shrubs that had previously only been found on the greatest estates.* And so, graduating rows of native trees and shrubs, intermixed with American species such as balsam fir, witch hazel and southern catalpa, were planted everywhere. A small garden in Upper Gower Street in London featured a shrubbery that included a tulip poplar, as well as several kalmias, magnolias and rhododendrons, while London's squares were planted with clumps of trees in small irregular thickets so that visitors thought they had found the countryside in the town—*"rus in urbe,"* one commented. Banks, who was a member of the Soho Square residents' Garden Committee, oversaw the re-design of the square in the early 1790s, when colourful American shrubs such as azalea and *Euonymus atropurpurea* were mixed with cherries, honeysuckles and roses.

The English garden had become so popular that its plants and designs were exported abroad. In France and Germany, Italy and Russia, gardeners re-created "le jardin anglais," "der Englische Garten" and "il giardino inglese." Whereas English garden designers had for centuries copied foreign styles—playful Italian Renaissance gardens, for instance, or the monumental designs of French and Dutch baroque gardens—now they were leading the way.

When Horace Walpole visited France in 1775, he made fun of the gardeners whose pleasure grounds were "curled and powdered *à l'anglaise,"* writing to a friend, "Shall I bring you a slice of their English gardens?" There were many other examples: in Italy, the English ambassador in Naples, William Hamilton, introduced the new designs, asking Banks to send over a suitable gardener as the Queen of Naples wanted "a specimen

* See, for example, James Meader's instructions in *The Planter's Guide* (1779).

of English pleasure ground" at Caserta. Within a few years, the gardener recommended by Banks had created a typical English garden, with rolling lawns, lakes, a fern-lined grotto and a ruined temple. The groves were planted with American evergreen oaks and Portugal laurels, and on an island in one of the lakes grew tulip poplars, *Magnolia grandiflora* and American persimmon—all dispatched from Kew. There was also a "Botanic garden" with a hothouse, which Hamilton had received from England "in packing cases" because nothing of the same quality was available in Italy. Banks's "beautiful Botany Bay plants," Hamilton wrote, were thriving, and Caserta's visitors delighted at the "British custom of gathering together the rarest gifts of Flora."

Equally, Catherine the Great of Russia was gripped by "anglomania," as she described it to Voltaire. She had sent her landscape architect to England to "visit all the notable gardens and, having seen them, to lay out similar ones here." And while her anglophile lovers had also provided her with information, Catherine hired numerous British gardeners to ensure the authenticity of her gardens. One of them was Bartram's former customer Johann Bush. Catherine had chosen him because, as a nurseryman, he was able to bring with him his "plantations"—the stocks of Bartram's American trees and shrubs that he had cultivated over the years, and which were the most important ingredient of the English garden. In Catherine's garden, Tsarskoe Selo, outside St. Petersburg, Bush laid out an English landscape and built several hothouses for her growing collection of exotics, which was much augmented when Banks dispatched a contribution from Kew that filled sixteen carriages drawn by sixty-four horses.

Even the Americans—despite their hard-won independence—drew inspiration from England. Thomas Jefferson, when in Paris as the American ambassador to France, visited England and with John Adams toured Stowe, Painshill and other landscape gardens because, he explained, "gardening in that country is the article in which it surpasses all the earth." When he saw the groves of American trees and shrubs, he knew that it would be easy to bring the English garden to Monticello, his estate in Virginia. He had only, Jefferson boasted, "to cut out the superabundant plants" in the forest surrounding his garden to achieve the scenery he had admired in England.

"Any of the Nobility will give £100 per Ann for an English Gardener," the gardener James Meader reported from Russia, commenting that the English fashion made foreign landowners "all Gardening mad." Those who did not employ an Englishman sent their own staff across the Channel to tour the new gardens and to procure engravings of the most famous landscapes. Some of these men stayed on to be apprenticed, or worked at Kew to learn the new way of gardening, while some estate owners requested exact instructions from English nurserymen when ordering plants because, as one French garden owner said, they "still had an extremely imperfect knowledge of exotics." Until then, Continental gardeners had not required any specialist botanical expertise. Their gardens, laid out as formal pleasure grounds consisting of straight walks, sharply clipped hedges, turf patterns and repetitive arrangements of topiary, with a few specimen plants, required only a limited horticultural knowledge. By contrast, English gardens needed experts. The painterly composition of the so-called natural landscape required gardeners to be able to contrast the colour of foliage and bark, as well as to arrange plants harmoniously according to height and shape.

For Continentals who could not go to England, there was an increasing number of publications on the subject available. Miller's *Dictionary* had been translated into Dutch, German and French, while Erasmus Darwin's botanical writings were available in Italian, Portuguese, French, German and Russian. Soon original treatises followed in German, French and Italian. These instructed gardeners how to lay out English gardens, as well as providing translated extracts from the works of writers such as Horace Walpole or Humphry Repton. There were even periodicals dedicated to the subject, such as the German *Ideenmagazin für Liebhaber von Gärten, Englischen Anlagen und für Besitzer von Landgütern* (The Ideas Magazine for Lovers of Gardens, English Pleasure Grounds and for Estate Owners).

With the rest of Europe frantically creating landscape gardens, in Britain gardenmania was pushing into new territories. Gardeners now turned their attention to flowers. Many of the flowers that Bartram had sent regularly, such as phlox, rudbeckia and echinacea, had become staples in British nurseries, but Banks's introductions were increasingly available as Kew distributed them to nurserymen in London. One nurseryman wrote that so many flowers had been procured from "every distant part of the globe" that the selection on offer was now vastly increased. For decades, flowers had been relegated to separate areas of the garden, or to the margins of shrubberies, but now the influx of foreign flowers made them a prominent feature. Circular and kidney-shaped flowerbeds peppered lawns like colourful pincushions, with other beds situated below the windows of drawing rooms as the new focus point in the garden. Old-fashioned native species such as honeysuckle or foxgloves were mixed with acclimatised exotics such as red zinnias from Mexico or phlox from North America. Planting manuals advised gardeners to arrange these new flowerbeds like the tiered "seats in a Theatre," with the tallest plants such as sunflowers or Michaelmas daisies—both old favourites from the American colonies—at the centre and low-growing flowers such as violas and pinks mixed with new pelargoniums from South Africa at the edges.

Many of the plants that were first nurtured at Kew still flourish in British and American gardens. One of Banks's most treasured acquisitions, for example, was the first of the Asian peonies to reach England, the Chinese tree peony with its delicate satin-paper bloom. The Chinese called it the "King of Flowers" and admired it as a potted plant for seasonal indoor decoration, much as we now buy poinsettias at Christmas. The first Asian magnolia (*Magnolia denudata*) and hydrangea (*Hydrangea macrophylla*) also made their entry into Britain's garden at this time—the latter causing a sensation because it could change colour from pink to blue. These Far Eastern plants were all the more special as so few reached England, because in China foreigners were confined to the compound of the East India Company in Canton and were not allowed to travel freely.

As the empire expanded, more and more flowers came streaming in: there were showy dahlias and new species of penstemons from Mexico, as well as the first of the shade-loving hostas from Japan. One of Banks's collectors sent from the Canaries the greenhouse ancestor of all garden cinerarias, which was praised as "one of the most desirable" of all ornamental greenhouse plants. And new trees came too. A Kew collector brought, from Chile, cones of the strange monkey puzzle tree, which would become one of the Victorians' most admired conifers. Banks was so passionate about these introductions that whenever a special consignment of rarities arrived at Kew he instructed the head gardener not to move anything before he himself became "acquainted with the plants." When, for example, a fuchsia arrived from South America, Banks reputedly carried it to the greenhouse on his head, "not choosing to trust it to any other person."

Meanwhile, Banks had continued to add to the Australian collection. "[W]e have this summer had vast additions to our Exotic Gardens Principaly from Botany Bay," he wrote in 1789. There was pittosporum, a large bushy evergreen with glossy leaves, as well as tree ferns and the orange-flowering *Helichrysum bracteatum*—all popular garden plants today. Also important for today's garden were two New Zealand species, the pale mauve hebe and the New Zealand flax, with its imposing lance-like leaves.

By this time, the plants that Banks had gathered on the *Endeavour* voyage were maturing. *Sophora microphylla*, for example, had grown into a large shrub with tiny leaves that created a lacy pattern against the English sky. But, with the increasing availability of Australian plants in the nurseries around London, Kew was no longer the only place to view such treasures. Private gardens were catching up quickly. Appropriately, the banksia, named in honour of the patron of Antipodean botany, was one of the first Australian plants to be available in London's specialist nurseries.* Climbers such as wonga wonga vine, hardenbergia and coral dusky pea (named *Kennedia rubicunda* after the nurseryman John Kennedy who sold them first) were also soon on sale. They were easily raised from seed, and hardenbergia was praised as "one of the most ornamental" for its blossom, which lasted from February throughout the summer. Grevilleas were also "common," one garden writer said, as was crimson bottlebrush. The Duke of

---

* The first Australian plant that gardeners could buy in a British nursery was *Eucalyptus oblique*—it was available in William Malcolm's nursery in Kensington from 1778. The first banksia seeds were given to the nursery Lee & Kennedy in Hammersmith in 1788.

Northumberland's gardener was one of the first to raise its fibre-optic blooms at Syon House, near Kew. And because it was easily propagated from seeds and cuttings, young plants could be bought "in most of the Nurseries near town" by the mid 1790s.*

Banks prided himself on the introduction of "an infinity of Plants from China the Cape of Good hope & New Holland [that] come Every day to us." In 1789, when the first edition of the *Hortus Kewensis* was published—the plant catalogue of all species cultivated at Kew—already 5,600 species were listed but within two decades that count rose to more than 11,000. Some 300 of these were Australian species, and more than 1,000 were from South Africa. Since Banks had returned from the *Endeavour* voyage, he and his collectors had brought almost 7,000 species to Kew, most of which were new to England. As Banks proudly wrote to his collector in China, "our King at Kew & the Emperor of China at Jehol solace themselves under the shade of many of the same trees."

There was such a "vast variety of exotics" that Kew's head gardener no longer had time to help the scores of amateur gardeners who requested duplicates. The plants made "a most elegant addition to the Gardens," Banks said. His personal favourite was bird-of-paradise, which unfolded its bright orange petals from a beak-like bud like a cockscomb. Others favoured even more exotic blooms. In fact it seemed that the brighter the colour and stranger the species, the more sought-after the flowers became. Witness to this widespread passion was the weirdest of all plants: the stapelia. This succulent has a striking blossom: large, starfish-shaped flowers, often intricately patterned like marble. Its most distinctive trait, however, is its odour—it exudes the smell of rotten meat, hence its common name, carrion flower. By the 1790s, one of Banks's collectors, Francis Masson, had introduced more than forty species of stapelia from South Africa into Britain (almost the complete genus), filling the country's greenhouses with the highly sought-after smell of putrefying flesh.

* The fashion for Australian plants also hit France when Napoleon's wife, Josephine, became so obsessed that she reputedly spent £2,600 in one year alone at one of the specialist nurseries in London, as well as receiving many seeds from Banks.

The stapelia was not alone in offering gardeners a wide choice of species. There were, for example, more than 100 species of South African pelargonium available in Britain. When Miller had begun collecting the pelargonium for the Chelsea Physic Garden in the early 1720s he had kept his specimens in the greenhouse. By the 1780s they were so common that gardeners placed them outside. In fact, forty years after their mass arrival at Kew, one garden writer observed that "every garret and cottage window is filled with numerous species of that beautiful tribe."

Another African genus that obsessed the English was the large tribe of the so-called Cape heath or erica. In just three decades almost 200 species were introduced, providing a sweeping palette of colour, ranging from porcelain white to deep crimson. They were so numerous that one garden writer bemoaned that it was "beyond the power of Botanists to number up their tribes." Over the next three decades, one nurseryman added yet another 285 varieties to the existing species by crossbreeding them—the first systematic hybridisation programme since Thomas Fairchild had created the Mule in 1716.* By the 1790s botanical magazines reported that ericas were "found in most green-house collections near town," and, whatever the time of year, at least one species would be flowering, since the advantage of a large genus was that there was "scarcely any period of the year in which some of them may not be found to delight the eye with their blossom."

The demand for new plants was so great that some of Banks's plant-hunters took to subsidising their salaries by selling seeds and cuttings to nurserymen, instead of sending everything to Kew. To avoid this, Banks added a clause to his employment contract which prohibited the selling of any parts of the plants or seeds. But the problem persisted, and as late as 1814 Banks felt it necessary to give his employees a strict warning: if any

---

* When Erasmus Darwin published *Phytolagia* in 1800, he declared that the sexual reproduction of plants was the "chef d'oeuvre, the master-piece of nature" and formed the basis of evolution. He was influenced by his observation of plant breeding, as well as of fossils which proved the extinction of species, and built up a theory of evolution which would later be famously developed by his grandson.

plant that was meant for Kew appeared in another garden, an enquiry would be "immediately set out in what manner it was procured."

Unlike the inhabitants of other European cities, many of whom lived in apartments, most Londoners had gardens backing on to their houses. Thus every dwelling from the small terraced houses in Bethnal Green to the grand mansions in the West End had some outside space, providing an incentive to their inhabitants to cultivate it. And of course, England was blessed with a mild and wet climate that made gardening easier and more satisfying than in many other countries. The French, for example, Horace Walpole teased, would never have as verdant and lush gardens as the English "till they have as bad a climate." Envious foreigners complained that their gardens would "never come to perfection of those in England . . . for we want the English climate, which on account of its moisture, keeps the gardens, and the fields, almost the whole year round, in a constant verdure."

How important gardens had become for the British was evident in the way in which they were advertised in sales particulars of houses. An "extensive shrubbery" made a property more desirable, as did a collection of mature trees. Throughout the 1780s and '90s, *The Times* contained announcements of "Sale by Auction" for properties in and around London that highlighted "Rare Shrubs, and Exotics" or a "capital collection of exotics, and other curious plants." Ornamental shrubs and flowers—rather than useful vegetable plots or fruit trees— were most popular. A "London garden," as one garden lover wrote, "must be made to please the Eye only." For an English gardener the "main criterion" for choosing flowers was, so one German tourist observed, "that a plant should come from a distant corner of the globe."

Banks's introductions would become staples in Victorian gardens, much as Bartram's trees and shrubs had become part of the English landscape, but with one important difference. While most of the specimens introduced from North America thrived in Britain's temperate climate, many of the plants that hailed from the warmth of Australia, Africa and Asia required the protection of greenhouses during the winter. Soon the hothouse became as widespread as the plants it protected. Once seen as a luxury of the aristocracy, it was now a sought-after accessory for the middle classes.

Again foreign visitors were amazed to see that English gardeners had hothouses "in most of the gardens in town." Humphry Repton—the self-declared successor to garden designer Lancelot Brown—designed many for both the town and the country. Unlike those of previous generations, these were to be placed adjacent to the house, linking the inside to the outside world. The introduction of "the numerous tribes of geraniums, ericas, and other exotic plants," Repton wrote, "have caused a very material alteration in the construction of the greenhouse." Rather than the brick building advised by Miller, Repton devised glass-roofed "flower passages" and greenhouses that flooded light on to the tender exotics—the first step towards the modern conservatory. By the turn of the century a greenhouse had become, as one writer declared, "an appendage to every villa, and to many town residences."

But it was not only the wealthier middle classes with their hothouses who were becoming converts to gardening. The greater supply and availability of plants drove plant prices down and turned gardening into a pastime that permeated society's different strata. England's bookshelves were filled with cheap garden calendars and horticultural manuals teaching men and women of all classes how to garden. More than 200 nurseries flourished in London alone—compared to only fifteen at the beginning of the century. Small tradesmen and artisans took to crossbreeding and competing for prizes at horticultural shows, and a century after Thomas Fairchild's sacrilegious experiments, hybrids started to become standard border plants.*

This craze for flowers was also manifested in other facets of daily life. Dresses were embroidered and printed with floral ornaments; gems and precious metals were fashioned into naturalistic bouquets to adorn the pale décolletés of society women and flower-arranging and painting were popular female pastimes. Even parlour games were infused with floral references, with playing cards featuring engravings of plants and

* The extensive hybridisation of ericas marks the beginning of the crossbreeding of ornamental plants on a large scale, which accelerated in the 1830s with gladiolus, pelargonium, verbena and petunia.

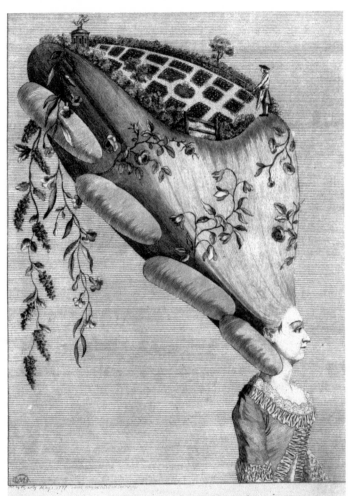

THE FLOWER GARDEN.

Eighteenth-century satire on flower and gardening fashions.

botanical questions. The drivers of carriages that could be hailed on the streets wore a small "bunch of flowers" pinned to their coats, while the wealthy flocked to the masquerades at Carlisle House in London, where, for the princely sum of two guineas per head, visitors could admire a floral extravaganza. Here the rooms were "metamorphosed . . . into gardens," with hedges, grottoes and arbours—with every inch filled with scented greenhouse plants and enormous potted shrubs.

The garden obsession filtered into interior design. Fashionable houses were filled with all manner of floral displays, and to capitalise on this, entrepreneurial London gardeners and nurserymen offered contracts whereby flowering plants could be hired for the week by clients who did not live permanently in the capital. And though few could match the masquerades at Carlisle House, many Londoners rented potted plants for dinners and balls. Even Bartram's rhododendrons and kalmias were available for nightly hire.

In the midst of this flower frenzy, Banks gave his wife a piece of dried moss to wear as a brooch. While he thought it a gorgeous botanical specimen, his wife found it boring and unsightly. When she refused to pin it to her blouse, Banks called her a "Fool that She Likes diamonds better, & Cannot be persuaded to wear it as a botanists wife Certainly ought to do." Other women were more taken by botanical and horticultural themes. When the French queen Marie-Antoinette made the *pouf* fashionable before the French Revolution, English women used these elaborate wigs to create mobile gardens. Made of wire, cloth, horsehair, and dusted with copious amounts of flour, the *pouf* provided the perfect substructure for a model landscape. Marie Antoinette herself wore a wig that depicted "a whole *jardin anglais*," while English women decorated theirs with festoons of silky blossoms, miniature groves and flowerbeds complete with figurines of gardeners. "Eleven damsels," one onlooker observed at a dinner-party, "had, among them, on their heads, an acre and a half of shrubbery, besides slopes, grass-plats, tulip-beds, clumps of peonies, kitchen-gardens, and green-houses."

As the new century dawned, ownership of a garden came to be seen as an essential prerequisite for happiness—and perhaps Englishness itself. "[S]carce a person, from the peer *plant renting! garden "pouts"* (handwritten annotation)

to the cottager," one horticultural writer observed, "thinks himself tolerably happy without being possessed of a garden." The expanding empire flooded the English garden with plants, and to coax these strange flowers into blossom, to be the first to see a rare plant that conjured up the smells and colours of a distant country, was all part of the joy of gardening. At the same time, in a world that was becoming increasingly dominated by speed, technology and the churning of factories, where the urban inhabitants were alienated from the day-to-day experience of cultivating the land, the garden became a retreat from modern life, a place to re-engage with nature.

Without the achievements of Miller, Collinson, Bartram, Linnaeus, Solander and Banks, England would not have become such a nation of gardeners. Miller taught his fellow countrymen practical horticulture with his matter-of-fact advice in the *Dictionary*—making it the model for all future plant encyclopaedias. Collinson and Bartram enabled plant-lovers to translate their ideas about the "natural" Arcadian landscape into reality—incidentally nurturing the commercial seed trade and nurseries in England. Linnaeus and Solander transformed botany from the scholarly pursuit of a few educated men to a common pastime, for without the standardisation of plant names there would have been chaos and confusion, making it impossible for people to share botanical knowledge and research. Banks built on these achievements when he consolidated practical horticulture, systematic botany and imperial expansion into a coherent enterprise. As President of the Royal Society, member of the Privy Council, confidant of King George III and founder of the Horticultural Society, he, more than anyone before or after, saw how the three elements could bring pleasure and prosperity to the nation.

Today's flowerbeds are crowded with the suburban descendants of plants brought to Britain from North America, South Africa, Australia and the Far East. They are the offspring of species such as the fuchsia that Banks carried on his head into the greenhouse at Kew, or Bartram's carefully packed seeds. Even England's rolling parkland—the embodiment of the "green and pleasant land"—is made up of foreign introductions. Without these, we would be deprived of the spectacular effect of glossy evergreens, dazzling autumn foliage and colourful blossom. From Bartram to Banks, the brother gardeners left a lasting legacy. The garden revolution of the eighteenth century is still alive in the English landscape, and ingrained in the nation's psyche today.

Epilogue

*We have discovered the point of perfection. We have given the
true model of gardening to the world; let other countries mimic
or corrupt our taste; but let it reign here on its verdant throne,
original by its elegant simplicity, and proud of no other art than
that of softening Nature's harshness and copying her graceful
touch.*

HORACE WALPOLE, *The History of the
Modern Taste in Gardening*, 1771

My morning started abruptly at precisely 8 a.m., when the
blackout blinds shot open automatically in the *Privatzimmer*—
the East German version of a bed & breakfast. In the dark,
the themed décor of the room had been invisible. Now the
hideous plastic flower arrangements on the walls leapt out at
me, and through the open bathroom door I could see gaudy
garden scenes painted on the tiles. There was no doubt that I
was in Germany. Suddenly my trip to Wörlitz, near Dessau, to
see for myself how English ideas about the garden had been
exported, seemed like a bad idea.

My spirits improved when I entered the Wörlitz gardens
themselves. Evidence of English influence was everywhere.
Although I had read eighteenth-century diaries and letters
describing "der Englische Garten" in Germany, it was only
when I followed Wörlitz's serpentine paths through thickets
fringed with the same colourful and perfumed shrubs that
adorned the English landscape garden that I could picture
what they meant.

The first German publication to discuss landscape gardens
had told its readers in 1770 that "[i]t is now a fashion in
England to create plantations which consist almost entirely of
seeds from America." Other horticultural writers recommended
John Bartram's trees and shrubs to garden lovers across Europe:
fast growing white pine, scented Carolina allspice and magnolia,

evergreen kalmia and *Thuja occidentalis*, and colourful callicarpa, azalea, hydrangea and rhododendron. I had devoured
these documents in the archives. But to see the plants growing
in that flat German landscape beside the river Elbe made what
I had read about it all the more tangible.

As I wandered around the lakes, once in a while crossing
little bridges to reach the different parts of the garden, it became
evident to me that "my" protagonists had shaped this landscape. The white stems of the American paper birches against
the darkness of yews and other evergreens were proof that
Petre, Collinson and Miller's tenet of painting with foliage,
bark and shape had been preached across Europe. Outside the
Gothic House the fluttering light green leaves of the black
locust were set against the dark tufts of white pine while, in
the same clump of trees, the blossom of the tulip poplar was
preparing to take over the show from the virginal white petals
of flowering dogwood. A little further on I saw evidence of
Miller's advice on "painting" with different shapes of tree. The
vertical brushstrokes of Eastern hemlock and *Juniperus
virginiana* contrasted with the looser and rounder shapes of
the rhododendrons and other shrubs.

Elsewhere were Collinson's elusive kalmia, as well as different
sorts of magnolia and the fringe tree. Wherever I turned, I
found Bartram's American species amid other trees and shrubs—
southern catalpa, honey locust and staghorn sumac. It was
early summer but I pictured what the gardens would look like
in autumn—a kaleidoscope of colour, from the orange of maple,
the blood-red of scarlet oak, yellow tulip poplar and aubergine
liquidamber. All this set against the varied greens of the conifers
and evergreens. I imagined Wörlitz's creator, Prince Franz of
Anhalt-Dessau, poring over engravings in order to re-create an
English landscape garden in his tiny principality. It brought
home to me just how powerful the influence of the English
garden revolution had been.

For many eighteenth-century thinkers, landowners and
writers, visits to Britain were like a journey into the future—
unlike the traditional Grand Tour to Italy, where sightseeing
took place among the ruins of the past. Britain's burgeoning
industrialisation and agricultural reform were seen as emblem-

atic of progress and the Enlightenment. The English garden, with its trees and shrubs set out according to the patterns of nature, was the visual expression of a country that was not ruled by absolutism and was known throughout Europe as "the seat of liberty." At Wörlitz the anglophile owner's beliefs are indelibly stamped on to the landscape. Hailed as "the secret birthplace of the Enlightenment in Germany," the English garden at Wörlitz was a political and philosophical statement.

In many ways, walking through Wörlitz was like walking through English landscape gardens such as Stowe or Stourhead. Most of the little temples and bridges were replicas of English originals, with "Der Englische Sitz" (The English Seat) copied from Stourhead, and a Temple of Flora modelled on William Chambers's "Casino" at Wilton House in Wiltshire. From Kew, Prince Franz copied Chambers's White Bridge and the Pagoda (albeit a few storeys lower), while the Temple of Venus was inspired by John Vanbrugh's Rotunda at Stowe. At one time Wörlitz had even had a "Garten Bibliothek," a public garden library, which would have been filled with English publications such as Philip Miller's *Dictionary* and Horace Walpole's *History of the Modern Taste in Gardening*. And besides the American trees and shrubs, there was yet more evidence of the impact that the British empire had on the garden. In a pavilion next to the garden library, for example, Prince Franz kept his collection of South Sea artefacts brought back from Captain Cook's second voyage, while the neo-gothic hothouse, a short walk north of the "Otahaitisches Kabinett," would have held many of the tender flowers that English plant-hunters sent back to Kew.

But the longer I spent at Wörlitz, the more I thought that there was something strange about this garden. Somehow Prince Franz's landscape felt like a faded copy. It lacked the beauty of the real thing. I walked back to the *Privatzimmer* pondering why this might be. Maybe it was because the landscape was that little bit too flat, and the brooks just slightly too straight, or maybe it was because I knew that Prince Franz's passion for gardening had never crossed the threshold of his own class, remaining an exclusively aristocratic pursuit. Suddenly I longed to be back in Britain, walking along the meandering paths in Stourhead or standing below the towering trees at Painshill. Wörlitz gives a snapshot of that moment at the end of the eighteenth century when anglomania arrived in Europe. But only gardens like Painshill and Stourhead allow you to experience, in its true glory, that revolutionary time when English gardeners dug up their topiary and planted a new world.

Glossary

The information on plants, their different names, and their introduction to Britain is based on the correspondence of the protagonists, the different editions of Philip Miller's *Gardeners Dictionary*, the two editions of William Aiton's *Hortus Kewensis*, Linnaeus's *Species Plantarum*, *The Botanical Magazine*, eighteenth-century nursery catalogues, plant lists and other plant catalogues listed in the Bibliography.

Latin Name	Common Name	Introduction	Additional Information
Abies balsamea	balsam fir	1696	John Banister sent this conifer to England from North America to Bishop Compton for the garden at Fulham Palace, but it remained a rare tree. Bartram found it first in 1741 in the Catskill Mountains and continued to dispatch what he called "Balm of Gilead Fir" over the following decades. In the Thorndon auction of 1742 trees sold for 15s. each, but by 1775 one-foot high saplings were available for only 6d. in a Yorkshire nursery.
Acer saccharum	sugar maple	1725	One of Collinson's acquaintances, Charles Wager, was the first in England to grow this deciduous North American tree, which was much admired for its orange and red autumn foliage. Bartram always sent seeds in his boxes, and in the 1752 edition of the *Dictionary* Miller wrote that the sugar maple "is now become pretty plenty." Red maple (*Acer rubrum*) was introduced by John Tradescant the Younger and grew in his garden in South Lambeth by 1656.
Aeculus hippocastanum	horse chestnut	1616	Originally introduced from Macedonia and Albania, the horse chestnut quickly became one of the most popular park trees in Britain. For years Bartram tried unsuccessfully to raise it in America from the seeds that Collinson had sent. When Bartram's son Moses visited Collinson in London in 1751, he was given seeds to take home to Philadelphia. Eventually, in May 1763, Bartram was able to report that "my europian ... horse chestnut is in full bloom."

Aesculus pavia	red buckeye	1711	Fairchild introduced this small tree with red blossoms from North America to England. It flowered for the first time in his nursery in Hoxton in 1722. Bartram collected it in Virginia as well as in North and South Carolina. By 1759 the nurseryman Christopher Grey sold the "scarlet flowered horse chestnut" for 2s. 6d.
Ailanthus altissima	tree of heaven	1751	Father D'Incarville sent seeds of this deciduous tree from China to England in 1751. When Collinson received them, he passed them on to the nurseryman James Gordon, who succeeded in raising the first saplings. Within a few decades the tree had arrived in America, where it grows in plenty because of its resistance to heavy pollution. Today it is regarded as an invasive species. Aiton listed it as *Ailanthus glandulosa*.
Amaryllis belladonna	belladonna lily	1712	This South African flower was probably introduced by the royal gardener and nurseryman Henry Wise. Before Linnaeus named it *Amaryllis*, the flower had been classified as "Lilio-Narcissus."
Araucaria araucana	monkey puzzle tree	1796	This unusual-looking coniferous tree from Chile was introduced by Archibald Menzies, one of Banks's collectors. Aiton listed it as *Araucaria imbricata*. It became one of the most popular Victorian specimen trees.
Arbutus andrachne	strawberry tree	1714	This small evergreen tree with red strawberry-like fruits was introduced by William Sherard when he was consul in Izmir, but it initially failed in Britain and was reintroduced in the 1750s. Collinson noted in his Commonplace Book that his friend John Fothergill gave the valuable seeds to nurseryman James Gordon, despite having fifteen gardeners himself, for only Gordon could coax them to life. Gordon sold one for £2 2s. in May 1761, but it was in Fothergill's garden at Upton in Essex, in May 1766, that the tree flowered for the first time. Collinson also had a tree in his garden in Mill Hill.
Aristolochia macrophylla	Dutchman's pipe	1763	Bartram introduced this vigorous North American climber to England. Gordon sold it in 1765 as "Bartram's Aromatick Vine a new species of Aristolochia from Canada" to the Duchess of Portland at Bulstrode, and Collinson also grew it in his garden in Mill Hill.

Aster grandiflorus	large-flower aster	1720	Catesby had sent the seeds of this flower from Virginia and it flowered two years later, in 1722, at Fairchild's nursery, where it was sold as "Mr. Catesby's new Virginian Starwort."
Banksia serrata	saw banksia	1788	This striking large evergreen shrub with upright, spiky cone-like flowers was named after Banks, who had seen it at Botany Bay in 1770. Banksias were one of the first plants to be raised from Botany Bay seeds and were available from the nursery Lee & Kennedy in Hammersmith.
Betula nigra	river birch or red birch	1736	This North American birch with peeling bark was introduced by Collinson and Bartram. In 1768 Miller wrote that "many of the plants have been raised, which thrive very fast here."
Betula papyrifera	paper birch	1750	In the 1752 edition of the *Dictionary* Miller wrote that this tree "has been raised in the Gardens lately." Bartram collected these coveted trees in the Catskills and in Pennsylvania. English gardeners adored them because they could contrast the white bark with the darker stems of other trees, "painting" with the trees.
Bignonia capreolata	cross vine	1710	The stone columns of Bartram's porch were overgrown with this vigorous North American climber with orange-red flowers. Collinson also grew it in his garden, as did Charles Hamilton in Painshill.
Bletia purpurea	modest orchid	1725	This was the first tropical orchid in Britain. It was probably collected by Catesby in the Bahamas and then sent to Collinson, who was the first to bring it to flower in his garden in Peckham.
Brugmansia arborea	angels' trumpets	1733	Miller received a parcel of seeds from Chile, sent by William Houstoun, but they had been almost entirely eaten by insects. Nevertheless, he still managed to raise two plants of what was then called *Datura arborea*. Lord Petre also raised a couple from the same batch of seeds in his stove at Thorndon. Their fragrant white blossom dangled as six-inch-long trumpet-shaped flowers from their upper branches.
Callicarpa americana	beauty berry	1724	Catesby introduced this shrub from North America but most of the plants died in the severe winter of 1739/40. Collinson also received specimens of what was then called *Johnsonia*

Callistemon citrimus — crimson bottlebrush — 1788

Banks had collected dry specimens of this Australian shrub in 1770 and introduced the first seeds once the penal colony had been established in Botany Bay. The blossom, with its hundreds of long bright red stamens, each with a dot of yellow pollen at the tip, gave it its common name: bottlebrush.

Callistephus chinensis — China aster — 1731

The first of these fast-growing annuals arrived in England when Miller obtained the seeds from France, where they had originally been sent to from China by French missionaries. Bartram had received them in one of the first seed dispatches from Collinson in 1735 and they grew annually in his garden.

Calycanthus floridus — Carolina allspice — 1726

Catesby introduced this strongly scented North American shrub, which he described as looking like a "starry Anemone," but it remained rare in England until Bartram sent it in greater numbers from Carolina. Collinson wrote in 1756 that it flowered abundantly every year in his garden. It was listed in Miller's *Dictionary* as *Basteria*.

Camellia japonica — camellia — 1739

Lord Petre was the first to grow camellias in England. He kept them in his warmest hothouse, which was the cause of their speedy demise, but his gardener, James Gordon, seems to have managed to take some cuttings, for he sold camellias after Petre's death in his nursery in Mile End. By 1775 they were available for 4s. each.

Campsis radicans — trumpet vine — 1640

John Parkinson mentioned this deciduous North American climber with showy trumpet-shaped flowers in 1640, but it remained rare in England. Bartram sent it to Collinson who grew it up the outside of his greenhouse in the mid-1750s. It was known as *Bignonia radicans*, and in 1756 Miller wrote in his *Figures of the most Beautiful, Useful, and Uncommon Plants* that the plant at Chelsea was older than fifty years. It became one of the most popular Georgian shrubbery plants.

americana from Bartram, who collected them in South Carolina, but in England the shrubs did not flourish in the open. In the 1768 *Dictionary* Miller wrote that they had flowered, but to the great frustration of gardeners had failed to produce the desirable bright purple berries. Today's garden centres in Britain offer oriental callicarpas because they are hardier than their American cousins.

Catalpa bignonioides — common catalpa or southern catalpa — 1722

This tree, with its unusually large leaves and abundant white summer blossom, was introduced from North America to England by Catesby. Bartram collected it in North and South Carolina. It was classified as *Bignonia*, and Miller wrote in 1752 that it was "propagated pretty commonly in the Nurseries near London, and sold as a flowering Tree to adorn Pleasure-gardens." The nurseryman Christopher Gray sold them for 1s. each in March 1759.

Cedrus libani — cedar of Lebanon — 1638

This conifer originates from Lebanon and became one of the most majestic parkland trees in England. Some specimens were planted at the Chelsea Physic Garden in 1683. When they—under the care of Miller—produced cones with ripe seeds for the first time, it caused such a sensation that the president of the Royal Society, Hans Sloane, presented one branch with nine cones to the fellows in 1729. Bartram received the first seeds in 1735 from Collinson but repeatedly failed to bring them to maturity in America. In 1755 Horace Walpole wrote that Christopher Gray sold them for half a guinea, but by 1775 the price had dropped to 4s. for two-foot saplings.

Chamaecyparis thyoides — white cedar or Atlantic white cedar — 1736

This narrow conical tree from North America caused much excitement when Collinson received the first batch from Bartram. Nobody had grown it in England before, and it quickly became one of the most expensive trees. Only five years later one of Bartram's young trees was sold for £2 2s. When Collinson discovered that the tree was easily propagated from cuttings, he promised to keep it a secret so as not to spoil Bartram's business. It is also sometimes called Atlantic white cedar, and Linnaeus listed it in his *Species Plantarum* in 1753 as *Cupressus thyoides*.

Chionanthus virginicus — fringe tree — 1736

With its dangling white plumes of fragrant flowers, the fringe tree was a coveted addition to the English garden. Collinson introduced it when he received it from his Virginian correspondent John Custis in 1736. The following year it flowered for the first time in his garden in Peckham. Bartram regularly sent seeds to his customers because none of the plants fruited in England. In 1766 John Hope saw a specimen in Collinson's garden in Mill Hill that was seven feet high. By 1768 Miller reported that they were "now more common" and in 1775 they sold for 3s. each.

Cinnamomum camphora	camphor tree	1727

Miller received this tree from tropical Asia from one of his correspondents. In the 1731 edition of the *Dictionary* he wrote that it was "at present very rare in Europe." In 1753 Linnaeus listed it as *Laurus camphora* in his *Species Plantarum*. Banks sent it to a plantation owner in Barbados, who wrote in December 1787 that it was thriving in his garden.

Colutea orientalis	bladder senna	1710

Tournefort brought this shrub with voluptuously shaped petals from the Levant and supplied several gardeners in England. In 1752 Miller wrote that it "is now become common in several Nurseries near London." By the end of the eighteenth century it had become a popular plant for Georgian shrubberies.

Cornus florida	flowering dogwood	1722

In 1722 Fairchild listed the species in his "List of those Plants which flower every Month in my Garden." Catesby had sent it to him from North America and wrote in 1730 in his *Natural History* of "the white Flower Mr. Fairchild has in his Garden." Bartram always packed seeds into his boxes. The shrub-like tree was adored for its profuse alabaster flowers in spring and red foliage in autumn. Collinson had to wait until May 1761 to see the tree flower in his own garden.

Cypripedium calceolus var. *parviflorum*	lady's slipper orchid	1731

Miller listed these North American lady's slipper orchids in the Appendix of his 1731 *Dictionary* because they had only just been introduced. They were one of Collinson's favourite flowers and he constantly ordered them from Bartram.

Delphinium grandiflorum	Chinese delphinium	1741

The professor of botany in St. Petersburg, Johannes Amman, sent this stately perennial from Siberia to Collinson, who was the first to grow it in Britain. John Hope noted in 1766 that Collinson cut it after it flowered so "that it may flower twice in one season."

Dionaea muscipula	Venus flytrap	1768

In 1763 Bartram sent seeds of this insect-devouring plant from Carolina to Collinson, who gave them to James Gordon because "his skill Exceeds all others in raising seeds." But Gordon does not seem to have succeeded, because in September 1769 John Ellis wrote to Linnaeus that the plant which Bartram called "tipitiwitchet" was growing in England, but that William Young, not Bartram, had introduced the first plants.

Diospyros virginiana	persimmon	1629	John Parkinson had cultivated this North American tree by 1629, but it remained rare in England. Collinson ordered it from Bartram, who collected it in southeast Pennsylvania, and in 1768 Miller wrote that they "are now become pretty common in the nurseries about London." In the 1790s Banks dispatched seeds to William Hamilton, who created an English garden at Caserta for the Queen of Naples.
Dodecatheon meadia	shooting star	1704; reintroduced in 1744	This clump-forming perennial with dangly dart-shaped magenta-pink petals was originally introduced from North America by John Banister, but then lost. Bartram found it in the Appalachian Mountains of Virginia and it flowered for the first time in Collinson's garden in May 1745. Collinson sent it to Linnaeus and suggested—to no avail—that he name it *Bartramia*. Catesby and Miller listed it under its original name *Meadia*.
Echinacea purpurea	purple coneflower	1699	John Banister introduced this purplish-red flower from North America to England, but it was soon lost again. Collinson had it in his garden, and Miller wrote in the 1752 edition of the *Dictionary* that "it is a scarce Plant, it is generally sold at a good Price by those who deal in curious Plants"—it was classified as *Rudbeckia purpurea*.
Erica cerinthoides	fire heath	1774	This evergreen shrub with bright red flowers was the first of the many Cape heaths to arrive in England from South Africa. It was introduced by Francis Masson, one of Banks's collectors, who over the next years dispatched almost 100 different species to Kew.
Eucalyptus gummifera	red bloodwood	1771	This was one of the few Australian seeds that survived the long voyage on the *Endeavour*. Banks and Solander called it "gum tree" because of the red resin that oozed out of its wounds. The tree still grows along the coast of New South Wales.
Euonymus atropurpurea	burning bush	1744–6	Bartram collected seeds of what he called "Broad Leaved Euonymus" in south-east Pennsylvania and always included it in his five-guinea boxes. William Aiton gave 1756 as its introduction date to England, but Bartram had already sent it earlier, for it was included in the 1744–6 plant list. Later, in the 1790s, when Banks was a member of the residents' Garden Committee of Soho Square, this colourful shrubs was planted there.

Franklinia alatamaha	Franklin tree	1774	This showy shrub is related to *Gordonia lasianthus* and camellias, and named after Benjamin Franklin. The *Franklinia* was only ever found in Georgia in the location where Bartram had seen it in 1765 by the river Alatamaha, but within a few decades it had become extinct in the wild. Reputedly all of today's specimens across the globe derive from the seeds that his son William collected a few years later. British botanists classified it first as *Gordonia pubescens* and later as *G. alatamaha*—its current name was only internationally legitimised in 1925.
Fraxinus americana	white ash	1724	The tree was first raised in England from seeds that Catesby had sent from North America in 1724. Bartram always included it in his seed boxes. In 1768 Miller wrote that, still, no seeds had been produced in England, but the trees were grafted on to the common ash, as "there has been of late years a great demand." In 1764 saplings sold for 1s. and in 1775 for 6d.
Fuchsia coccinea		1788	The nurseryman James Lee apparently saw a potted specimen on a windowsill in Wapping and convinced the owner to let him have the plant. By the following year he had propagated 300 plants, which he sold at the enormous price of one guinea apiece. Banks carried the first plant of this South American species on his head into the greenhouse at Kew. There is some confusion as to whether this might not, in fact, have been *F. magellanica*.
Gardenia augusta		1754	Not since the introduction of *Magnolia grandiflora* had English gardeners been so excited about a plant. The naming of the showy Asian evergreen with ivory flowers escalated when Miller insisted that it was a species of jasmine and not a new genus. James Gordon reputedly made a £1,000 profit in 1759 from its sales. In 1766 he sold them for one guinea apiece but by 1775 the price had dropped to 7s 6d.
Gleditsia triacanthos	honey locust	1698	Because of its acacia-like leaves and spiky thorns, this tree was also called "3 thorn'd acacia." It was introduced from North America in the seventeenth century by Bishop Compton, and in 1734 the nurseryman Robert Furber was able to deliver two twenty-foot trees to the Prince of Wales at Carlton House for £6 6s. Bartram always packed the seeds in his boxes and by 1760 it had become quite common—saplings could be bought for 1s. each. Today the tree lines many streets in American cities.

Gordonia lasianthus	loblolly bay	1739	Bartram collected this evergreen shrub with fragrant white flowers in North and South Carolina. It was named after the nurseryman James Gordon, who, in 1763, after twenty years of trials, succeeded in raising it from seed. Bartram had found plenty during his Florida expedition and sent some to Collinson. Linnaeus had originally classified it as *Hypericum lasianthus* in 1753.
Grevillea		1790	Banks and Solander collected the first plants of this genus in Botany Bay, bringing them back to England as dried specimens. The first living specimens were available at Lee & Kennedy's nursery, and by the early nineteenth century Aiton was able to list five species growing in Kew.
Halesia carolina	silver bell or snowdrop tree	1756	This tree was admired for its drooping clusters of white bell-shaped blossoms in spring. Bartram collected it in North and South Carolina, and Collinson mentioned it for the first time in 1764. It was included in Bartram's plant list of 1769.
Hamamelis virginiana	witch hazel	1736	This North American shrub was grown for its winter blossom. Bartram had found it at the source of Schuylkill River in Pennsylvania and introduced it to England when he sent it in his five-guinea boxes. In 1759 the nurseryman Christopher Gray sold them for 1s. each and by 1768 they had become so popular and easy to cultivate that Miller could report in his *Dictionary* that "many plants have been raised in the English gardens, where they are propagated for sale by the nursery gardeners."
Hardenbergia violacea	purple coral pea	1790	This climber, which paraded its purple pea-like flowers from February through to the summer, was one of the first Australian species to adorn the English garden and to be sold commercially because it was easily raised from seed. It was praised as "one of the most ornamental." Banks had seen it during the *Endeavour* voyage in Australia, but it was one of his collectors who sent seeds to Britain.
Helichrysum bracteatum	golden everlasting strawflower	1799	The orange strawflower was another popular Australian plant that was introduced to England in the late eighteenth century after the first settlers had arrived in Botany Bay.

Hibiscus rosa-sinensis	rose of China	1731	Miller was the first to document growing this Chinese shrub, which changed colour from white when the buds appeared, to "blush rose colour," and then to dark purple as they withered. It was first listed as *Ketmia* in the *Dictionary*.
Hosta ventricosa	blue plantain lily	1790	This was the first hosta to arrive in England from Japan. It was introduced as *Hemerocallis coerulea* by George Hibbert.
Hydrangea arborescens	wild hydrangea	1736	Bartram regularly sent seeds of this North American shrub which he collected in Pennsylvania to England. In 1746 Collinson wrote "my hydrangia—perhaps the first in England flowered in my Garden." By 1775 they sold for 3d. each.
Hydrangea macrophylla	mophead hydrangea	1788	One of Banks's plant-collectors introduced the first of the Asian hydrangeas to Britain. It caused a sensation because it could change colour from pink to blue. It was then called *Hydrangea hortensis*.
Juglans nigra	black walnut	1629	Though already cultivated in England by 1629, Miller classified these North American trees still as "rarities" in 1731. Lord Petre ordered several thousands from Bartram for his plantation at Thorndon. All of Bartram's plant lists include the species.
Juniperus virginiana	red cedar	1664	The dense conical shape of this North American conifer (at least when it is young) made it the perfect tree to translate the columns of the fashionable Palladian architecture into the landscape. Lord Petre planted more than 1,000 on the mounts at Thorndon. Because of their popularity in Britain, Bartram always included them in his seed boxes.
Kalmia latifolia	mountain laurel	1734	This evergreen shrub with pink flowers that open like mini-umbrellas was introduced to Britain from North America by Bartram, who sent them in his first box to Collinson. Bartram always included the seeds in his boxes, but because they were difficult to raise he often also dispatched living saplings planted in tubs. Miller thought they were "one of the greatest ornaments to the country." In 1775 they still sold for 15s. each. Collinson and Bartram also introduced *K. angustifolia* in 1736, which flowered in Collinson's garden in 1743.

Kennedia rubicunda	dusky coral pea	1788	Banks had seen this climber in the area around Botany Bay but it was one of his collectors who introduced the seeds to Britain. It was one of the first Australian species to be available in nurseries. It was first classified as *Glycine* and then named after the nurseryman John Kennedy. In 1790 and 1803 two more species were introduced, *K. prostrata* and *K. coccinea*.
Leptospermum scoparium	tea tree	1771	This tree-like shrub, which is covered with tiny white flowers, was introduced by Banks from New Zealand. It was one of the few seeds that survived the voyage on the *Endeavour*.
Lilium superbum	American turkscap lily	1727	In early 1736 Bartram sent some bulbs of this lily to England, and later that summer an eight-foot-five-inch plant flowered in Collinson's garden in Peckham. The sculpted petals resembled mini-turbans—hence its common name, the American turkscap lily. Collinson also called it marsh martagon. The celebrated botanical artist Georg Dionysius Ehret painted Collinson's plant in its full bloom.
Lindera benzoin	spicebush	1683	This shrub with upright branches was cultivated for its aromatic leaves. It was introduced to Britain from North America by Banister, and in 1730 the Society of Gardeners wrote in their catalogue that the few trees that were grown in England had produced no seeds yet. Linnaeus classified it as *Laurus benzoin* in his *Species Plantarum*, and Bartram always included seeds of what he called "Benjamin" or "all-spice tree" in his boxes. By the mid eighteenth century it thrived in many landscape gardens, including those at Syon House, Whitton and Painshill. Collinson also grew it in Mill Hill.
Liquidambar styraciflua	sweetgum	1680s	This deciduous North American tree, which was clothed in dark aubergine-coloured leaves in autumn, was already grown as a great rarity in the seventeenth century, and in 1734 the nurseryman Robert Furber sold one thirteen-foot tree for £3 3s. But it was Lord Petre who succeeded in raising thousands of them, and in 1742, at the Thorndon auction, 300 of them were sold for 2s 6d each. Bartram always added seeds to his five-guinea boxes.
Liriodendron tulipifera	tulip tree or tulip poplar	*c.* 1638	This was one of the most coveted North American trees, for its showy blossom and strangely shaped leaves. It had been introduced by John Tradescant the Younger, who had brought it from Virginia. One of Petre's first orders had been for tulip poplars, and over the years

			Bartram sent thousands. By 1768 Miller could report that the tree which he called *Tulipifera* was "now common in the nurseries about London." By 1775 they were available for 2s 6d. each.
Lobelia erinus	edging or trailing lobelia	1752	Today almost every windowbox contains a lobelia, spilling over the edge of its planter like a blue cushion. They are the descendants of the plants that arrived in the eighteenth century from the Cape of Good Hope. Miller grew them in Chelsea.
Magnolia denudata	Yulan or lily tree	1789	This was the first Asian magnolia to arrive in Britain. It was deciduous and admired for its alabaster white flowers. One of Banks's collectors introduced it to England from China and it first grew at Kew.
Magnolia grandiflora	laurel-leaved magnolia or southern magnolia	1734	This evergreen North American magnolia flowered in England in summer 1737 in a garden in Parsons Green, near Fulham. Like a forgotten porcelain teacup on a ruffled tablecloth of green taffeta, one lonely flower clung to its branch as London's gardeners swarmed around it with great excitement. To keep it for posterity, Collinson took Georg Dionysius Ehret to draw it. Miller wrote in the 1752 edition of his *Dictionary* that "almost every Person who is curious in Gardening is desirous to have some of these beautiful Trees in their Gardens." Bartram sent it regularly to Britain, increasing its numbers substantially. By 1775 they had become much more common and were offered in Telford's nursery catalogue for 7s 6d. each.
Magnolia virginiana	swamp bay or sweet bay	1688	This was the first magnolia tree to arrive in England. It was introduced from North America by John Banister, and Catesby wrote in 1730 that it was blossoming in Fairchild's nursery and in Collinson's garden. In the 1739 edition of the *Dictionary* Miller still described it as "very rare." For four decades Bartram sent vast numbers of what he called "swamp magnolia" to England. Specimens sold for £2 2s. at the Thorndon auction and for 15s. in 1775.
Mertensia virginica	Virginia bluebells	1699	Collinson received roots of this clump-forming perennial with beautiful violet-blue or white bell-shaped blossoms from his Virginian correspondent John Custis. It flowered in his garden in Peckham for the first time in 1735. Linnaeus listed it as *Pulmonaria virginica* in the *Species Plantarum* and Aiton stated that it was first grown in Britain in the Chelsea Physic Garden in 1699.

Monarda didyma	scarlet beebalm or Oswego tea	1743	This clump-forming perennial is the parent of most monardas today. Bartram collected the bright scarlet flower during his expedition to the League of the Iroquois in 1743 near Oswego—hence its common name of Oswego tea. It flowered for the first time in Collinson's garden in 1746.
Nerine sarniensis	Guernsey lily	1659	This flower was introduced in 1659, but Fairchild was the first to make it bloom repeatedly every autumn. In 1768 Miller listed it as *Amaryllis sarniensis*. Until Banks's plant-hunter Francis Masson saw them in the wild in South Africa in the late eighteenth century, it had been believed that they came from the Channel Islands.
Nyssa sylvatica	black gum or tupelo	1752 (probably earlier)	Throughout the 1740s Collinson sent orders to Bartram for this deciduous North American tree, which was adored for its fiery autumn colour. Bartram always included them in his seed boxes, but they were difficult to raise. The tree only just made it into the Appendix of Miller's 1752 edition of the *Dictionary*. Miller wrote that there were only "few of these Trees at present in the English Gardens."
Oxydendrum arboreum	sourwood or sorrel tree	1752 (probably earlier)	This tree was also called lily-of-the-valley-tree, for its delicate tiny urn-shaped flowers. Catesby described it in his *Natural History* as "Frutex" and Linnaeus listed it as *Andromeda arborea* in the *Species Plantarum*. Collinson ordered it from Bartram in 1735, but the first mention of its successful cultivation in England was in Miller's 1752 edition of the *Dictionary*.
Paeonia suffruticosa	Moutan or tree peony	1787	Banks introduced the first tree peony to England. The Chinese called it the "King of Flowers" and admired it as a potted plant for seasonal indoor decoration. It perished quickly at Kew, but Banks's collector in China, William Kerr, procured more in 1804.
Pandorea pandorana	wonga wonga	1793	This vigorous climber with bell-shaped flowers was one of the first Australian plants to be commercially available in Britain. It was introduced and sold by the nursery Lee & Kennedy under the name *Bignonia australis*.
Papaver nudicaule	Icelandic poppy	1730	This beautiful flower was introduced by plant-collector James Sherard from Russia, but Collinson was the first to see the blossom in his garden in Peckham.

Papaver orientale	Oriental poppy	1714	In 1768 Miller wrote that Tournefort had sent the Oriental poppy from Armenia to Paris, from where it was dispatched to plant-collectors across Europe. Fairchild was the first to grow it in England.
Pelargonium zonale	geranium	1710	This was one of the first pelargoniums to arrive in Britain from South Africa and it is still one of the most commonly planted in today's garden. It was introduced by the Duchess of Beaufort and first called *Geranium*. Miller grew many species of pelargonium at Chelsea, but it was Francis Masson who brought the largest collection to Britain. By the end of the eighteenth century there were more than 100 species available, forty-seven of which had been introduced by Masson. The distribution to nurserymen and an intensive hybridisation programme made the plants widely available.
Pericallis cruenta	cineraria	1777	Banks's collector Masson collected this flower from the Canaries. It was the greenhouse ancestor of all garden cinerarias (then *Cineraria cruenta*), and William Curtis praised it in 1798 as "one of the most desirable" of all ornamental greenhouse plants. It would become the most popular pot plant in the mid-twentieth century.
Phlox divaricata	wild blue phlox	1746	This North American flower was introduced by Bartram—then called lychnidea. Collinson wrote that all species of phlox in England "came from my Garden where they first Flower'd." This was true of *P. divaricata* and *P. maculata* (introduced in 1740), but Miller was already growing *P. glaberrima* at Chelsea in 1725.
Phormium tenax	New Zealand flax	1772	In 1772 John Ellis found some of the seeds—still in their pods—on Banks's dried and mounted New Zealand specimens from the *Endeavour* voyage and gave them to "our best gardeners." Today the plant, with its lance-like leaves, is one of the most popular Antipodean plants in Britain's gardens. Banks, though, was interested in its economic value. He envisaged a thriving linen industry in Australia after he had seen its use for Maori cloth production during his time in New Zealand.
Pinckneya bracteata	fever tree	1786	Bartram found this tree during his Florida expedition in 1765 and identified it as a "very odd Catalpa." Had he seen it in early summer, when it flowered, he would have immediately seen that it was not a catalpa because of its unusual blossom, in which the sepals parade

in bright pink disguise—similar to the bracts of poinsettias. In 1803 it was named by the French plant-collector André Michaux.

Pinus echinata	shortleaf pine	1736	Lord Petre was the first to grow this pine in England, from cones sent by Bartram. In the north-east and southern states of America it is the most common yellow pine. In England it was adored for its unusual bark of almost square scales. Bartram always sent it in his boxes.
Pinus palustris	longleaf pine	1727	This North American pine was admired for its extremely long and flexible needles, which could reach up to almost 20 inches, growing in feathery tufts which gave the tree an exotic feel. The nurseryman Robert Furber sold it by 1727, and later it became one of the most popular conifers in the landscape garden. Bartram sent it regularly, and Collinson's tree was praised by John Hope, the professor of botany at Edinburgh, as having "a most singular appearance."
Pinus strobus	white pine or Weymouth pine	1705	This North American pine was introduced in 1705 and named after Lord Weymouth, who was the first to grow it in larger numbers in England. Collinson ordered it regularly because "[t]his sort is scarce & Rare with us." In December 1746 Bartram was able to send a whole box full of cones because the harvest had been so abundant—he even had to employ people to help with the gathering. Horace Walpole adored the "clean straight stem, the lightness of its hairy green, and for being feathered quite to the ground." In 1742 at the Thorndon sale they sold for 5s. each, but by 1775 one-foot saplings were sold for only 4d. each.
Pittosporum revolutum	yellow pittosporum	1795	This large bushy evergreen with glossy leaves and fragrant yellow flowers was introduced by one of Banks's collectors from Australia and grown at Kew. Today it is a popular garden plant in Britain.
Platanus occidentalis	American sycamore or buttonwood	1638 circa	American sycamore was introduced by John Tradescant the Younger. It is one of the parents of the London plane (*Platanus x hispanica*). In 1734 the nurseryman Robert Furber sold two twenty-foot specimens for 15s. apiece, and in May 1740 Lord Petre planted 120 on his mounts at Thorndon. Bartram regularly sent seeds, which he collected in south-east Pennsylvania. By 1775 a four-foot specimen cost only 4d.

Polygonum orientale	persicaria	1707	This large flower with pendulous pink clusters of blossoms was sent from China to the Duchess of Beaufort. Collinson dispatched it to Bartram in March 1737. In England it became a popular Victorian annual.
Prunus lusitanica	Portugal laurel	1648	The gardener Thomas Knowlton wrote in 1722 that Fairchild was the first to cultivate the evergreen shrub in England, but Aiton gave its introduction date as 1648. In October 1759 Collinson sent berries to Bartram, who wrote back that they were so valuable to him that he "would not take ten guineas A piece for them (as poor as I am)." The Victorians later trimmed Portugal laurels into neat balls in imitation of orange trees.
Quercus coccinea	scarlet oak	1691	This deciduous tree, which was adored for its autumn foliage, was sent by John Banister from North America to Bishop Compton's garden at Fulham Palace, but it was Bartram who dispatched acorns in large numbers over four decades. In 1775 two-foot saplings were sold for 6d. each.
Quercus phellos	willow oak	1723	This North American oak was grown for its unusual willow-shaped leaves and almost black bark. Thomas Fairchild had succeeded in raising them by 1723—probably from acorns that Catesby had sent. In May 1741 Collinson was able to report that they had raised a "Vast Quantity" from Bartram's harvest of 1740. Bartram continued to send them in all his seed boxes. Today they are popular street trees in the north-eastern and southern states of America.
Rhododendron maximum	rosebay or great laurel	1736	Bartram introduced this North American evergreen shrub to Britain when he sent a number of living saplings in tubs to Collinson. When Catesby published the Appendix of his *Natural History* in 1747 the rhododendron had not yet flowered. Almost ten years later, in June 1756, Collinson wrote to Bartram that his plant had finally shown its first blossom. It was this rhododendron species that Bartram sent most regularly to his customers.
Rhododendron periclymenoides	pink azalea or pinxterbloom azalea	1734	Bartram called this North American shrub with soft pink flowers "swamp azalea." Collinson received the first batch in 1734 but was frustrated with the progress. They flourished at Thorndon, he reported to Bartram in 1742, but not in his garden. In the 1768 edition of the *Dictionary*, Miller wrote that gardeners had failed to produce seeds in England but the

Rhus typhina	staghorn sumac	1629	This was one of the earliest North American plants to be introduced to Britain, mentioned in John Parkinson's *Paradisi in Sole Paradisus Terrestris* in 1629. In autumn the leaves turn fiery red and throughout winter tiny bright red fruits form cones on the branches. Bartram included it in most of his boxes. By 1752 Miller wrote that it had become common in England.
Robinia hispida	bristly locust	1743	This shrub was introduced from North America in 1743 by Sir John Colliton, and twelve years later its dangling plumes of rose-pink petals were on display in Collinson's garden in Mill Hill for the first time. Miller reported that it was still rare in 1768, but by 1791, when Francis Douce bought it for his London garden in Upper Gower Street, its price had dropped to 2s 6d.
Robinia pseudoacacia	black locust	c. 1630	John Tradescant the Elder received this North American tree from France, where it had been introduced more than twenty years previously by the royal gardener Jean Robin (in whose honour it was named). In June 1736 Collinson wrote that it flowered in his garden. Bartram, who called it "Sweet flowering Locust" or "sweet Honey Locust," sent it regularly. In 1768 Miller wrote that it "is become fashionable again, and great numbers of the tree have been raised in most parts of England."
Sarracenia purpurea	pitcher plant	c. 1638	John Tradescant the Younger collected this insect-devouring flower in the swamps around Jamestown in Virginia, but English gardeners struggled to keep them alive. Collinson managed to bring them to blossom every year by keeping them in a perforated pot filled with moss and placed in a tub of water. Bartram sent roots wrapped in moss to keep them moist during the long journeys across the Atlantic.
Sassafras albidum	sassafrass	c.1630	Catesby wrote in 1730 that "several [are] now at Mr. Collinson's at Peckham," and Miller described this North American tree from the laurel family in 1731 as "one of the most difficult Trees to grow with us." In the different editions of the *Dictionary*, he listed it initially as *Cornus* and then later as *Laurus*. In 1753 Linnaeus classified it as *Laurus sassafras*. Bartram always sent seeds in his five-guinea boxes. In 1775 saplings still cost 10s. each.

blossom could be admired in "many curious gardens." It is the parent of many modern hybrids. When Linnaeus named it, he combined the Greek word for rose "rhodon" and tree "dendron"—rose tree.

Solidago	goldenrod	1722	Fairchild was the first to grow this yellow flowering perennial from North America, and in June 1735 Collinson wrote to Bartram that he had some thriving in his garden. By 1768 Miller listed thirty-one different species in his *Dictionary*.
Sophora microphylla	kowhai	1771	Banks brought back the seeds of this evergreen tree with pendulous yellow flowers on the *Endeavour*. Today it is New Zealand's national flower.
Stapelia	carrion flower	1690	The often starfish-shaped and patterned flower exudes the most revolting smell of carrion—hence its common name. The first specimens were introduced from South Africa in 1690 and grew in William and Mary's hothouses at Hampton Court, but of the forty-four species that Aiton listed in 1811, forty-one had been introduced by Francis Masson between 1774 and 1795. They became some of the most popular Victorian greenhouse plants.
Stewartia malacodendron	silky camellia or Virginia stewartia	1742	This upright shrub with rose-like white flowers first blossomed with Catesby in May 1742. By 1768 Miller wrote that it "is at present very rare." It was only from the mid-1770s that it became more widely grown in Britain.
Strelitzia reginae	bird-of-paradise	1773	This spectacular South African flower, which unfolds its bright orange petals like a cockscomb from a beak-like bud, was Banks's favourite plant. Masson introduced it in 1773 to Kew, and by 1790 it had flowered. It was named after George III's wife, Charlotte of Mecklenburg-Strelitz.
Telopea speciosissima	waratah	1789	Banks collected dry specimens for his herbarium in the area around Botany Bay. It is one of the many Australian plants equipped to survive bush fires because it regenerates from the rootstock. It was one of the first seeds to arrive from the new penal colony. Today its red-coloured blossom is New South Wales's floral emblem.
Thuja occidentalis	arborvitae	c. 1536	This was one of the first North American coniferous trees to arrive in Europe in the sixteenth century. Bartram found it in autumn 1738 during his expedition to Virginia and continued to send it regularly to England. In 1775 two-foot saplings sold for 1s.

Tsuga canadensis	Eastern hemlock	1736	Collinson was the first to cultivate this elegant North American conifer in England from seeds that Bartram had sent. Six years after Bartram had dispatched the first batch, seven specimens sold for £2 2s. each at the plant auction at Thorndon in 1742, and by 1775 saplings sold for 2s. each. Today, though, the state tree of Pennsylvania is under threat from an Asian insect—the hemlock woolly adelgid. It has wiped out great swathes of forests in many parts of the eastern United States, leaving only naked, skeletal trees.
Viburnum dentatum	arrowwood	1736	In April 1742 Collinson wrote that he had introduced this Northern American shrub to England. Bartram collected it in southeast Pennsylvania and regularly sent it to his customers. It was called arrowwood because the native Americans used the straight branches as arrow shafts.
Wisteria frutescens	American wisteria	1724	Miller wrote in 1737 that the "seeds were sent by Catesby in 1724 & distributed to several curious Persons in London, from which plants were raised" (he listed it as *Phaseoloides carolinianum* and later as *Glycine frutenscens*). In the 1752 edition of the *Dictionary*, Miller reported that the North American climber was propagated in "several Nurseries near-London." It became one of the most popular plants in Georgian shrubberies.
Yucca filamentosa	Adam's needle	c.1638	John Tradescant the Younger introduced this North American plant with its sturdy lance-like leaves to England, but Collinson claimed that he was the first to make it flower, on 16 July 1747. In 1768 Miller wrote that it was "not so common . . . in the English gardens."
Zinnia elegans		1796	This Mexican zinnia is the parent of today's zinnias. It was introduced to Britain by the Marchioness of Bute, who received it from Spain, where her husband was the British ambassador.
Zinnia pauciflora		1753	This red and yellow flower was the first zinnia to arrive in Britain. Miller received seeds from the gardener of the Royal Garden of Paris, who had procured them from Peru.

265

Bibliography, Sources
and Abbreviations

Abbreviations

BL	British Library, London
CPG	Chelsea Physic Garden, London
DTC	Dawson Turner Collection, Natural History Museum, London
ECRO	Essex County Record Office
EUL	Edinburgh University Library
HMC	Historic Manuscript Commission
Kew JBK	Joseph Banks's Correspondence, Archives of the Royal Botanic Garden, Kew
LHASA	Gutsarchiv Harbke, Landeshauptarchiv Sachsen-Anhalt, Abteilung Magdeburg, Aussenstelle Wernigerode, Germany
LS	Linnean Society, London
MHS	Museum of the History of Science, Oxford
Miller	Philip Miller's *Gardeners Dictionary*
NHM	Natural History Museum, London
NLW	National Library of Wales, Aberystwyth
NMM	National Maritime Museum, Greenwich
PA	The Parliamentary Archives, House of Lords Record Office
RS JBC	Journal Books, Royal Society, London
SLNSW Banks	The Sir Joseph Banks Electronic Archive, State Library of New South Wales, Sydney
TNA	The National Archives, London
WSA	The Worshipful Society of Apothecaries, London
WSRO	Goodwood Estate Archives, West Sussex Record Office

Unless otherwise referenced, all quotes from letters between John Bartram and Peter Collinson are from Berkeley, Edmund and Dorothy Smith Berkeley (ed.), *The Correspondence of John Bartram, 1734–1777* (Gainesville: University of Florida Press, 1992).

Internet Archives

The Linnaean Correspondence, prepared under the aegis of the Swedish Linnaean Society, the Royal Swedish Academy of Sciences, Uppsala

University and the Linnean Society of London, with the collaboration of the Centre international d'étude du XVIIIe siècle (searchable database of Carl Linneaus's letters)
www.linnaeus.c18.net

The Sir Joseph Banks Electronic Archive, State Library of New South Wales, Sydney (searchable online database)
www.sl.nsw.gov.au/banks

National Agricultural Library, United States Department of Agriculture (online edition of *The Botanical Magazine*, Vols.1–26)
http://www.nal.usda.gov/curtis/

South Seas, National Library of Australia (online editions of James Cook, Joseph Banks and Sydney Parkinson's *Endeavour* journals as well as maps)
http://southseas.nla.gov.au/

Newspapers and Magazines

Boston Gazette
Gardener's Magazine
Gentleman's Magazine
Lloyd's Evening Post
London Chronicle
Monthly Review
The Botanical Magazine (subsequently *Curtis's Botanical Magazine*)
The General Evening Post
The Guardian
The London Evening Post
The Middlesex Journal, or Chronicle of Liberty
The Public Advertiser
The Spectator
The Times (London)
The World
Town and Country Magazine

Nursery Catalogues and Other Plant Lists and Catalogues

"A Catalogue of English and Foreign Trees, Sold by Robert Furber" (London, 1727).
"A Catalogue of Forest Trees, Fruit Trees, Evergreen and Flowering Shrubs, Hot-House, Green-House and Herbaceous Plants, Kitchen-Garden and Flower Seeds, sold by John Mackie, Nursery and Seeds-Man" (Norwich, 1790).
"A Catalogue of Forest-Trees, Fruit-Trees, Ever-Green and Flowering-Shrubs, sold by John and George Telford, Nursery-Men and Seeds-Men" (York: A.Ward, 1775), reprinted in Harvey 1972.
"A Catalogue of Garden, Grass and Flower Seeds, Trees, Shrubs, Herbaceous, Green-House and Hot-House Plants, Sold by Russell, Russell,

& Willmott, Nursery & Seedsmen, Lewisham, Kent" (London: T. Plumer, 1800).

"A Catalogue of Hardy Trees and Shrubs, Greenhouse and Stove Plants, herbaceous Plants, and Fruit Trees ... sold by Luker and Smith, Nurserymen and Seedsmen" (London, 1783).

"A Catalogue of Plants and Seeds which are sold by Conrad Loddiges, Nursery and Seedsman, Hackney" (London: C. Heydinger, 1777).

Aiton, William, *Hortus Kewensis* (London: George Nicol, 1789).

———. *Hortus Kewensis* (London: Longman & Co, 1810–13).

"A List of Forrest Trees and Shrubs, gathered in Pennsylvania, East and West Jersey by John Bartram, Botanist, written by John Breintnall, circa 1744–6" (Library of Philadelphia, Papers of Edwin Wolf, 2nd, Box 8, f.12), copy at Bartram's Garden, Philadelphia.

"A List of Seeds, circa 1769, probably in William Bartram's hand and probably sent to Benjamin Franklin in London" (Stanford University Library, Department of Special Collections), copy at Bartram's Garden, Philadelphia.

"A List of Seeds of Forest Trees and flowering Shrubs, gather'd in Pensilvania, the Jerseys and New York, by John and William Bartram, and sent over the last Year to their Correspondents, being the largest Collection that has ever been imported into this Kingdom," *The Gentleman's Magazine*, vol.24 (February 1754).

"An Account of the Introduction of American Seeds into Great Britain, 1766" (Peter Collinson, MSS Col1), Natural History Museum, London.

"Bartram's seeds sent to Charles Polhill, January 1760" (Polhill MS 528/5 Memorandum Book for the Estate, University of London Library), copy at Bartram's Garden, Philadelphia.

"Catalogue of the Collection of Trees and Shrubs Formed by Lord Islay (Later Duke of Argyll) at Whitton, Middlesex," reprinted in Symes et al. 1986.

Fairchild, Thomas. "List of those Plants which flower every Month in my Garden," 1722, reprinted in Bradley 1724, vol.3 and Harvey 1974.

Hill, John. *Hortus Kewensis* (London, 1768).

"List of Plants from the Endeavour Voyage," reprinted in Stearn 1969.

"List of Trees in the Thorndon Nurseries, 1740–42," reprinted in Harvey 1996.

Miller, Philip. *Figures of the most Beautiful, Useful, and Uncommon Plants described in the Gardeners Dictionary* (London, 1755–60).

———. *Gardeners and Florists Dictionary* (London, Charles Rivington, 1724).

———. *Gardeners Dictionary* (London, 1731, and all subsequent editions to 1768).

Society of Gardeners. *Catalogus Plantarum. A Catalogue of Trees, Shrubs, Plants, Flowers both Exotic and Domestic which are propagated for Sale in the Gardens near London* (London, 1730).

Books

Abbot, Charles. *Flora Bedfordiensis* (Bedford: W. Smith, 1798).

Abercrombie, John, and Thomas Mawe. *The Universal Gardener and Botanist* (London: G. Robinson & T. Cadell, 1778).

Adeane, Jane H. (ed.). *Early Married Life of Maria Josepha Lady Stanley* (London: Longmans, Green, and Co., 1899).

Afzelius, Adam. *Linné's Eigenhändige Anzeichnungen über sich Selbst* (Berlin: G. Reimer, 1826).

Alexander, Caroline. *The Bounty. The True Story of the Mutiny of the Bounty* (London: Harper Collins, 2004).

Allen, David Elliston. *The Naturalist in Britain. A Social History* (London: Allen Lane, 1976).

Alston, Charles. *A Dissertation on Botany* (London: Benj. Dod, 1754).

Anon. "Botanical Conversation," *The New Lady's Magazine, or Polite and Entertaining Companion for the Fair Sex* (May 1786).

Archenholz, Johann Wilhelm von. *A Picture of England* (London: Edward Jeffery, 1789).

Armstrong, Alan W. (ed.). *Forget not Me & My Garden. Selected Letters 1725–1768 of Peter Collinson, F.R.S.* (Philadelphia: American Philosophical Society, 2002).

Aughton, Peter. *Endeavour. Captain Cook's First Great Voyage* (London: Phoenix, 2003).

Banks, R. E. R. (ed.). *A Global Perspective* (London: Royal Botanic Garden Kew, 1994).

Barrett, Charlotte (ed.). *Diary and Letters of Madame D'Arblay* (London: Macmillan & Co, 1904).

Bartram, John. *A Journey from Pennsylvania to Onondaga in 1743 by John Bartram, Lewis Evans [and] Conrad Weiser*, first published 1751 (Barre, Mass.: Imprint Society, 1973).

Baybrooke Marschall, John. "Daniel Carl Solander, Friend, Librarian and Assistant to Sir Joseph Banks," *Archives of Natural History*, vol.11:3 (1984).

Beaglehole, J. C. (ed.). *The* Endeavour *Journal of Joseph Banks, 1768–1771* (Sydney: Angus & Robertson, 1962).

———— (ed.). *The Journals of Captain James Cook on His Voyages of Discovery. The Voyage of the* Endeavour, *1768–1771* (Cambridge: Cambridge University Press, 1955–74).

Beauman, Fran. *The Pineapple, King of Fruits* (London: Chatto & Windus, 2005).

Bedford, John Russell, Duke of. *Correspondence of John, Fourth Duke of Bedford* (London: Longman, 1842).

Beer, E. S. de (ed.). *Correspondence of John Locke* (Oxford, Clarendon Press, 1979).

Beer, Gavin de. *The Sciences Were Never at War* (New York: Thomas Nelson and Sons, 1960).

Bell, Whitfield J. "John Bartram: A Biographical Sketch," in *America's Curious Botanist. A Tercentennial Reappraisal of John Bartram 1699–1777*, eds. Nancy E. Hoffmann and John C. van Horne (Philadelphia: The American Philosophical Society, 2004).

Berkeley, Edmund, and Dorothy Smith Berkeley (eds.). *The Correspondence of John Bartram, 1734–1777* (Gainesville: University of Florida Press, 1992).

————. *The Life and Travels of John Bartram, from Lake Ontario to the River St. John* (Tallahassee: University Presses of Florida, 1982).

Black, David. (ed.). *Linnaeus' Travels* (London: Paul Elek, 1979).

Black, Jeremy. *George III. America's Last King* (New Haven and London: Yale University Press, 2006).

Blair, Patrick. "Observations upon the Generation of Plants, in a Letter to Sir Hans Sloane." *Philosophical Transactions*, vol.31 (1720–21).

———. *Botanick Essays* (London: William and John Innys, 1720).

Bligh, William. *The Mutiny on Board H.M.S. Bounty* (Santa Barbara: The Narrative Press, 2003).

Blunt, Wilfrid. *The Compleat Naturalist. A Life of Linnaeus* (London: Frances Lincoln, 2001).

Boettiger, Carl August. *Reise nach Wörlitz 1797* (Berlin: Deutscher Kunstverlag Berlin München, 1999).

Boyd, Julian P. et al. (ed.). *The Papers of Thomas Jefferson* (Princeton: Princeton University Press, 1950–).

Bradley, Richard. *General Treatise of Husbandry and Gardening* (London: T. Woodward and J. Peele, 1724).

———. *New Improvements of Planting and Gardening* (London: W. Mears, 1717).

Brazell, John Harold. *London Weather* (London: HMSO, 1968).

Brett-James, Norman G. *The Life of Peter Collinson* (London: E. G. Dunstan & Co., 1925).

Brewer, John. *The Pleasures of Imagination. English Culture in the Eighteenth Century* (London: HarperCollins, 1997).

Brigham, David R. "Mark Catesby and the Partronage of Natural History in the First Half of the Eighteenth Century," in *Empire's Nature. Mark Catesby's New World Vision*, eds. Amy R. W. Meyers and Margaret Pritchard (Chapel Hill and London: University of North Carolina, 1998).

Brimley Johnson, R. (ed.). *The Letters of Hannah More* (London: John Lane, 1925).

Bristow, Alex. *The Sex Life of Plants* (London: Barrie & Jenkins, 1979).

Browne, Janet. "Botany for Gentlemen: Erasmus Darwin and 'The Loves of the Plants,' " *Isis*, vol. 80:4 (December 1989).

Bryk, Felix (ed.). *Linnaeus im Auslande* (Stockholm: in eigenem Verlage, 1919).

Campbell, Susan. *Charleston Kedding. A History of Kitchen Gardening* (London: Ebury Press, 1996).

Campbell-Culver, Maggie. *The Origin of Plants. The People and Plants that Have Shaped Britain's Garden History since the Year 1000* (London: Headline, 2001).

Cannon, John F. M. "Botanical Collections," in *Sir Hans Sloane: Collector, Scientist, Antiquary. Founding Father of the British Museum*, ed. Arthur MacGregor (London: British Museum Press, 1994).

Carter, Harold B. "Sir Joseph Banks and the Plant Collection from Kew sent to the Empress Catherine II of Russia 1795," *Bulletin of the British Museum (Natural History)*, vol.4:5 (1974).

———. *Sir Joseph Banks, 1743–1820* (London: British Museum, 1988).

Cartwright, James Joel (ed.). *The Travels through England of Dr. Richard Pococke During 1750, 1751 and later Years* (London: Camden Society, 1889).

Catesby, Mark. *Natural History of Carolina, Florida, and the Bahama Islands* (London, 1731–48).

Cave, Kathryne, Kenneth Garlick and Angus MacIntyre (eds.). *The Diary of Joseph Farington* (New Haven and London: Yale University Press, 1978–84).

Chambers, Douglas. "Painting with Living Pencils: Lord Petre," *Garden History*, vol 19:1 (1991).

———. *The Planters of the English Landscape Garden* (New Haven and London: Yale University Press, 1993).

Chambers, Neil. *Joseph Banks and the British Museum: The World of Collecting, 1770–1830* (London: Pickering & Chatto, 2007a).

———. *The Scientific Correspondence of Sir Joseph Banks, 1765–1820* (London: Pickering & Chatto, 2007b).

———. *The Letters of Sir Joseph Banks. A Selection, 1768–1820* (London: Imperial College Press, 2000).

Charlesworth, M. (ed.). *The English Garden, Literary Sources and Documents* (Mountfield, East Sussex: Helm Publishing, 1993).

Clegg, John (ed.). *Gilbert White's Garden Kalendar, 1751–1771*, facs. from MS in the British Library (London: The Scolar Press, 1975).

Clutton, George and Colin Mackay. "Old Thorndon Hall, Essex," *Garden History Society, Occasional Papers*, no.2 (1970).

Coke, Lady Mary. *The Letters and Journals of Lady Mary Coke* (Bath: Kingsmead Reprints, 1970).

Colden, Cadwallader. *Letters and Papers of Cadwallader Colden* (New York: New York Historical Society, 1918–37).

Conway, Hazel. *People's Parks: the Design and Development of Victorian Parks in Britain* (Cambridge: Cambridge University Press, 1991).

Coombs, David. "The Garden at Carlton House," *Garden History*, vol. 25:2 (1997).

Cowell, John. *The Curious and Profitable Gardener* (London: Weaver Bickerton, 1730).

Crèvecoeur, J. Hector St. John de. *Letters from an American Farmer*, first published in 1782 (Oxford: Oxford University Press, 1998).

Cross, Anthony. "Russian Gardens, British Gardeners," *Garden History*, vol.19:1 (1991).

———. "The English Garden in Catherine the Great's Russia," *Journal of Garden History*, vol.13:3 (1993).

Curtis, William. *A Catalogue of the British, Medicinal, Culinary, and Agricultural Plants Cultivated in the London Botanic Garden* (London: B. White, 1783).

Dandy, James Edgar. *The Sloane Herbarium* (London: British Museum, 1958).

Darlington, William. *Memorials of John Bartram and Humphry Marshall* (Philadelphia: Lindsay & Blakiston, 1849).

Darwin, Erasmus. *The Botanic Garden. Part II: Containing Loves of the Plants. A Poem. With Philosophical Notes*, first published 1789 (London: J. Johnson, 1791).

Dawson, Warren R. *The Banks Letters: A Calendar of the Manuscript Correspondence of Sir Joseph Banks* (London: British Museum, 1958).

Defoe, Daniel. *A Tour Through the Whole Island of Great Britain*, first published 1724–6, (London: Penguin Books, 1971).

Denny, Margaret. "Naming the Gardenia," *Scientific Monthly*, vol.67:1 (1948).

Desmond, Ray. "The Transformation of the Royal Gardens at Kew," in *Sir*

Joseph Banks. A Global Perspective, ed. R. E. R. Banks (London: Royal Botanic Garden Kew, 1994).

————. *Great Natural History Books and Their Creators* (London: British Library, 2003).

————. *The History of the Royal Botanic Gardens Kew* (London: The Harvill Press, 1995).

Doren, Carl von. "The Beginnings of the American Philosophical Society," *Proceedings of the American Philosophical Society*, vol. 87:3 (1943).

du Roi, D. Johann Philipp. *Die Harbkesche Wilde Baumzucht* (Braunschweig: Verlage der Fürstl.Waisenhaus Buchhandlung, 1771).

Duyker, Edward, and Per Tingbrand (eds.). *Daniel Solander. Collected Correspondence, 1753–1782* (Oslo, Copenhagen, Stockholm: Scandinavian University Press, 1995).

Egerton, Frank N. "Richard Bradley's Relationship with Sir Hans Sloane," *Notes and Records of the Royal Society of London*, vol. 25:1 (1970).

Ehret, Georg Dionysius. "A Memoir, 1758," *Proceedings of the Linnean Society of London* (Nov. 1894–June 1895).

Elliott, Brent. *Flora. An Illustrated History of the Garden Flower* (London: Royal Horticultural Society, 2003).

————. *The Royal Horticultural Society. A History 1804–2004* (London: Phillimore, 2004).

Ellis, John. *A Description of the Mangostan and the Breadfruit, Directions to Voyagers, for bringing over these and other Vegetable Productions, which would be extremely beneficial to the Inhabitants of our West India Islands* (London, 1775).

Evelyn, John. *Sylva, or a Discourse of Forest-Trees, and the Propagation of Timber* (London: J. Martyn, 1664).

Fabricius, Johann Christian. "Einige Nähere Umstände aus dem Leben des Ritters von Linné, 1780," facsimile of the first edition, in *Linné und Fabricius, zu ihrem Leben und Werk*, ed. Julius Schuster (München: Verlag der Münchner Drucke, 1928).

Fairchild, Thomas. "An Account of Some New Experiments, Relating to the Different, and Sometimes Contrary Motion of Sap in Plants and Trees," *Philosophical Transactions*, vol. 33 (1724–5).

————. *The City Gardener* (London: T. Woodward, 1722).

Farber, Paul Lawrence. *Finding Order in Nature* (Baltimore and London: The Johns Hopkins University Press, 2000).

Festing, Sally. "Rare Flowers and Fantastic Breeds. The 2nd Duchess of Portland and her Circle," *Country Life*, vol.179ii (12 June 1986).

Field, Henry. *Memoirs, Historical and Illustrative, of the Botanick Garden at Chelsea* (London, 1878).

Fleming, Laurence, and Alan Gore. *The English Garden* (London: Michael Joseph, 1980).

Fornander, Andreas. *Herbationes Upsalienses* (Uppsala, 1753).

Fothergill, John. *Some Account of the Late Peter Collinson* (London, 1770).

Fox, Richard Hingston. *Dr. John Fothergill and His Friends* (London: Macmillan, 1919).

Freshfield, Douglas W. *Life of Horace Benedict de Saussure* (London: Edward Arnold, 1920).

Frick, George. *Mark Catesby, The Colonial Audubon* (Urbana: University of Illinois Press, 1961).

Fries, Th. M., J. M. Hulth and A. H. Uggla (eds.). *Bref och Skrifvelser af och till Carl von Linné* (Stockholm: Aktiebolaget Ljus, 1907–22).

Fries, Th. M. (ed.). *Johann Beckmann's Schwedische Reise in den Jahren 1765–1766* (Uppsala: Uppsala Universitets Årsskrift, 1911).

Frost, Alan. "Botany Bay: An Imperial Venture of the 1780s," *The English Historical Review*, vol.100:395 (April 1765).

Fry, Joel T. "Archaeological Research at Historic Bartram's Garden," *Bartram Broadside* (1998).

———. "An International Catalogue of North American Trees and Shrubs: The Bartram Broadside, 1783," *Journal of Garden History*, vol. 16:1 (1996).

Gascoigne, John. *Joseph Banks and the English Enlightenment: Useful Knowledge and Polite Culture* (Cambridge: Cambridge University Press, 1994).

———. *Science in the Service of Empire: Joseph Banks, the British State and the Uses of Science in the Age of Revolution* (Cambridge: Cambridge University Press, 1998).

Gerard, John. *The Herball, or Generall Historie of Plantes very much enlarged and amended by Thomas Johnson* (London: Adam Islip, Ioice Norton, Richard Whitaker,1636).

Gilbert, Samuel. *The Florists Vade-Mecum*, 2nd edition (London: J. Taylor and J. Wyat, 1693).

Goetzmann, William H. "John Bartram's Journey to Onondaga in Context," in *America's Curious Botanist. A Tercentennial Reappraisal of John Bartram 1699–1777*, eds. Nancy E. Hoffmann and John C. van Horne (Philadelphia: The American Philosophical Society, 2004).

Greene, Jack P. *The Intellectual Construction of America* (Chapel Hill: University of North Carolina Press, 1993).

Grohmann, J. G. (ed.). *Ideenmagazin für Liebhaber von Gärten, Englischen Anlagen und für Besitzer von Landgütern* (Leipzig, 1796–1811).

Hall, Elizabeth. "The Plant Collections of an Eighteenth-century Virtuoso," *Garden History*, vol. 14:1 (1986).

Hallock, Thomas. "Narrative, Nature and Cultural Contact in John Bartram's Observations," *America's Curious Botanist. A Tercentennial Reappraisal of John Bartram 1699–1777*, eds. Nancy E. Hoffmann and John C. van Horne (Philadelphia: The American Philosophical Society, 2004).

Hamilton, Jill Duchess of. *Napoleon, the Empress and the Artist: The Story of Napoleon, Josephine's Garden at Malmaison, Redouté and the Australian Plants* (East Roseville: Kangaroo Press, 1999).

Hankin, Christiana C. (ed.). *Life of Mary Anne Schimmelpenninck* (London, 1859).

Harper, Francis (ed.). "John Bartram. Diary of a Journey through the Carolinas, Georgia, and Florida from 1 July 1765, to April 10, 1766," *Transactions of the American Philosophical Society*, New Series, vol.33:1 (1942).

Harvey, John. "A Scottish Botanist in London in 1766," *Garden History*, vol.9:1 (1981).

———. "Lord Petre's Legacy: The Nurseries at Thorndon, pt. 2," *Garden History*, vol.24:2 (1996).

———. "The English Nursery Flora, 1677–1723," *Garden History*, vol 26:1 (1998).

———. *Early Gardening Catalogues* (London: Phillimore, 1972).

———. *Early Nurserymen* (London: Phillimore, 1974).

Hayden, Peter. *Russian Parks and Gardens* (London: Frances Lincoln, 2005).

————. "The Russian Stowe," *Garden History*, vol. 19:1 (1991).

Hayden, Ruth. *Mrs. Delany: Her Life and her Flowers* (London: British Museum, 1980).

Henrey, Blanche. *British Botanical and Horticultural Literature* (London: Oxford University Press, 1975).

————. *No Ordinary Gardener: Thomas Knowlton, 1691–1781* (London: British Museum, 1986).

Hindle, Brooke. *The Pursuit of Science in Revolutionary America 1735–1789* (Chapel Hill and London: University of North Carolina Press, 1956).

Hirschfeld, C. C. L. *Theory of Garden Art, 1779–85*, trans. and ed. Linda B. Parshall (Philadelphia: University of Pennsylvania Press, 2001).

Historical Manuscript Commission. *Fortescue*, Fourteenth Report, Appendix, pt. V, vol. 2 (London: HMSO, 1894).

Hobhouse, Penelope. *Plants in Garden History* (London: Pavilion, 2004).

Hodges, Alison. "Painshill Park, Cobham, Surrey (1700–1800): Notes for a History of the Landscape Garden of Charles Hamilton," *Garden History*, vol 2:2 (1973).

Hoffmann, Nancy E. and John C. van Horne (eds.). *America's Curious Botanist. A Tercentennial Reappraisal of John Bartram 1699–1777* (Philadelphia: The American Philosophical Society, 2004).

Holme, Thea. *Chelsea* (London: Hamish Hamilton, 1972).

Hoyles, Martin. *Gardeners Delight, Gardening Books from 1560 to 1960* (London: Pluto Press, 1994).

Hunt, John Dixon. *The Picturesque Garden in Europe* (London: Thames & Hudson, 2003).

————. *William Kent. An Assessment and Catalogue of His Designs* (London, A. Zwemmer Ltd., 1987).

Isaacson, Walter. *Benjamin Franklin, An American Life* (New York: Simon & Schuster, 2003).

Jackson, B. D. "The Visit of Carl Linnaeus to England in 1736," *Svenska Linné-Sällskapets Årsskrift*, vol. 9 (1926).

Jackson, Donald and Dorothy Twohig (eds.). *The Diaries of George Washington* (Charlottesville: University Press of Virginia, 1976–79).

Jacobs, James A. "Historic American Landscape Survey PA-1-A: The John Bartram House," *Bartram Broadside* (2005).

James, John. *Theory and Practice of Gardening* (London: Geo. James, 1712).

James, Lawrence. *The Rise and Fall of the British Empire* (London: Abacus, 2001).

Jardine, Lisa. *Ingenious Pursuits, Building the Scientific Revolution* (London: Little, Brown, 1999).

Jardine, N., J. A. Secord and E. C. Spary (eds.). *Cultures of Natural History* (Cambridge: Cambridge University Press, 1996).

Jarvis, Charlie. *Order Out of Chaos. Linnaean Plant Names and Their Types* (London: Linnean Society, 2007).

Jarvis, P. J. "Plant Introductions to England," in *Change in the Countryside*, ed. R. A. Butlin and H. S. A. Fox (London: Institute of British Geographers, 1979).

Jonsell, Bengt. "The Swedish Connection," in *Sir Joseph Banks. A Global Perspective*, ed. R. E. R. Banks (London: Royal Botanic Garden Kew, 1994).

Kalm, Pehr. *Kalm's Account of His Visit to England on His Way to America in 1748* (London: Macmillan & Co, 1892).

———. *Travels in North America*, first published 1772 (Barre, Mass.: Imprint Society, 1972).

Kielmansegge, Count Frederick. *Diary of a Journey to England in the Years 1761–1762*, trans. Countess Kielmansegg (London: Longmans, Green, and Co., 1902).

King-Hele, Desmond (ed.). *The Letters of Erasmus Darwin* (Cambridge: Cambridge University Press, 1981).

———. *Erasmus Darwin and the Romantic Poets* (London: Macmillan, 1986).

———. *Erasmus Darwin. A Life of Unequalled Achievement* (London: Giles de la Mare, 1999).

Koerner, Lisbet. "Carl Linnaeus in his Time and Place," in *Cultures of Natural History*, eds. N. Jardine, J. A. Secord and E. C. Spary (Cambridge: Cambridge University Press, 1996).

———. *Linnaeus, Nature and Nation* (Cambridge, Mass.: Harvard University Press, 1999).

Köhler, Marcus. "Friedrich Karl von Hardenberg's Journeys to England and His Contribution to the Introduction of the English Landscape Garden to Germany," *Garden History*, vol. 25:2 (1997).

———. *Frühe Landschaftsgarten in Russland und Deutschland. Johann Busch als Mentor eines neuen Stils* (Berlin: Aland Verlag, 2003).

Labaree, Leonard W. (ed.). *The Papers of Benjamin Franklin* (New Haven and London: Yale University Press, 1959–).

Laird, Mark. "From Callicarpa to Catalpa: The Impact of Mark Catesby's Plant Introductions on English Gardens of the Eighteenth Century," in *Empire's Nature. Mark Catesby's New World Vision*, eds. Amy R. W. Meyers and Margaret Beck Pritchard (Chapel Hill and London: University of North Carolina Press, 1998).

———. *The Flowering of the Landscape Garden* (Philadelphia: University of Pennsylvania Press, 1999).

Laird, Mark, and John Harvey. "The Garden Plan for 13 Upper Gower Street, London," *Garden History*, vol.25:2 (1997).

Lambert, Aylmer Bourke. "Notes Relating to Botany, Collected from the MS of the late Peter Collinson," *Transactions of the Linnean Society*, vol.10 (1811).

Langford, Paul. *A Polite and Commercial People, England 1727–1783* (Oxford: Oxford University Press, 1989).

Langley, Batty. *New Principles of Gardening* (London: A. Bettesworth and J. Battey, 1728).

Le Rougetel, Hazel. *The Chelsea Gardener, Philip Miller 1691–1771* (London: Natural History Museum, 1990).

Leapman, Michael. *The Ingenious Mr. Fairchild* (London: Headline, 2000).

Lee, James. *An Introduction to Botany* (London, 1794).

Leighton, Ann. *American Gardens in the Eighteenth Century. For Use or for Delight* (Boston: University of Massachusetts Press, 1988).

Leith-Ross, Prudence. *The John Tradescants. Gardeners to the Rose and Lily Queen* (London: Peter Owen, 1998).

Lemmon, Kenneth. *The Golden Age of Plant Hunters* (London: Phoenix House, 1968).

Lewis, W. S. (ed.). *Horace Walpole's Correspondence* (New Haven and London: Yale University Press, 1937–61).

Lincoln, Margarette (ed.). *Science and Exploration in the Pacific* (London: National Maritime Museum, 1998).

Lindeboom, G. A. (ed.). *Boerhaave's Correspondence* (Leiden: E.J. Brill, 1962–79).

Lindroth, Sten. "The Two Faces of Linnaeus," in *Linnaeus: the Man and his Work*, ed. Tore Frängsmyr (Berkeley: California University Press, 1983).

Lindsay, Ann. *Seeds of Blood and Beauty: Scottish Plant-Collectors* (Edinburgh: Birlinn, 2005).

Linnaeus, Carl. *Philosophica Botanica*, trans. Stephen Freer (Oxford: Oxford University Press, 2003).

———. *Species Plantarum*, facsimile of the first edition 1753 (London: Ray Society, 1957).

———. *Systema Naturae*, facsimile of the first edition 1735 (Nieuwkoop: B. de Graaf, 1964).

———. *The Critica Botanica of Linnaeus*, first published in 1737, trans. Sir Arthur Hort (London: Ray Society, 1938).

Llanover, Lady (ed.). *The Autobiography and Correspondence of Mary Granville, Mrs. Delany* (London: Richard Bentley, 1861).

Longstaffe-Gowan, Todd. *The London Town Garden, 1740–1840* (New Haven and London: Yale University Press, 2001).

Loudon, John Claudius. *Arboretum et Fruticetum Britannicum, or, the Trees and Shrubs of Britain* (London: Longman, 1844).

———. *The Green-House Companion* (London: Harding, Triphook, and Lepard, 1824).

Lyons, H. G. "The Fairchild Trust," *Notes and Records of the Royal Society*, vol. 3 (1940–41).

Maccubbin, Robert P. and Peter Martin. *British and American Gardens* (Charlottesville: University of Virginia, 1983).

MacGregor, Arthur. "The Life, Character and Career of Sir Hans Sloane," in *Sir Hans Sloane: Collector, Scientist, Antiquary. Founding Father of the British Museum*, ed. Arthur MacGregor (London, British Museum Press: 1994).

Mackay, David. "Agents of Empire: The Banksian Collectors and Evaluation of New Lands," in *Visions of Empire: Voyages, Botany and Representations of Nature*, eds. David Philip Miller and Peter Hanns Reill (Cambridge, Cambridge University Press: 1996).

———. *In the Wake of Cook. Exploration, Science and Empire, 1780–1801* (Wellington: Victoria University Press, 1985).

Malmeström, Elis. "Linnés Självkänsla," *Svenska Linné-Sällskapets Årsskrift*, vol. 10 (1927).

Manger, Heinrich Ludewig. *Baugeschichte von Potsdam* (Berlin and Stettin: Friedrich Nicolai, 1789).

March, Earl of. *A Duke and His Friends* (London: Hutchinson & Co, 1911).

Mare, Margaret, and W. H. Quarrell (trans. and eds.). *London in 1710, from the Travels of Zacharias Conrad von Uffenbach* (London: Faber & Faber, 1933).

———. *Lichtenberg's Visits to England as Described in his Letters and Diaries* (Oxford: Clarendon Press, 1938).

Markham, Gervase. *The Second Booke of the English Husbandman* (London: John Browne, 1615).

Marquardt, Karl Heinz. *Anatomy of the Ship. Captain Cook's Endeavour* (London: Conway Maritime Press, 2001).

Martin, Peter. *The Pleasure Gardens of Virginia* (Princeton: Princeton University Press, 1991).

Martyn, John. *The Gardener's and Botanist's Dictionary by the Late Philip Miller* (London: Rivington, 1807).

————. (ed. and trans.). *Tournefort's History of Plants Growing about Paris* (London: C. Rivington, 1732).

Martyn, Thomas. *The Language of Botany* (London: John White, 1807).

Masson, Francis. *Stapeliae Novae* (London: George Nicol, 1796).

Mavor, William. *The Lady's and Gentleman's Botanical Pocket Book* (London: J. Crowder, 1800).

McCann, Timothy J. " 'Much troubled with very rude company . . .' The 2nd Duke of Richmond's Menagerie at Goodwood," *Sussex Archaeological Collections*, vol. 132 (1994).

McClellan, James E. *Science reorganised. Scientific Societies in the Eighteenth Century* (New York: Columbia University Press, 1985).

McCracken, Donald P. *Gardens of the Empire. Botanical Institutions of the Victorian British Empire* (London and Washington: Leicester University Press, 1997).

McDougall, Walter A. *Freedom Just Around the Corner. A New American History, 1585–1828* (New York: HarperCollins, 2005).

McLean, Elizabeth. "A Preliminary Report on the 18th Century Herbarium of Robert James, Eighth Baron Petre," *Bartonia*, no.57 (1992).

————. "John and William Bartram: Their Importance to Botany and Horticulture," *Bartonia*, no. 50 (1984).

McLynn, Frank. *1759. The Year Britain Became Master of the World* (London: Jonathan Cape, 2004).

Meader, James. *The Planter's Guide: or, Pleasure Gardener's Companion* (London: G. Robinson, 1779).

Medicus, Friedrich Kasimir. *Über Nordamerikanische Bäume und Sträucher als Gegenstände der Deutschen Forstwirtschaft und der Schönen Gartenkunst* (Heidelberg, 1791).

Meynell, Guy. "Philip Miller's Resignation from the Chelsea Physic Garden," *Archives of Natural History*, vol. 14:1 (1987).

Michaelis-Jena, Ruth, and Willy Merson (trans. and eds.). *A Lady Travels. Journeys in England and Scotland from the Diaries of Johanna Schopenhauer* (London: Routledge, 1988).

Miller, David Philip, and Peter Hanns Reill (eds.). *Visions of Empire: Voyages, Botany and Representations of Nature* (Cambridge: Cambridge University Press, 1996).

Miller, Philip. "A Method of Raising Some Exotick Seeds, Which Have Been Judged Almost Impossible to Be Raised in England," *Philosophical Transactions*, vol. 35 (1728).

————. "An Account of Some Experiments, Relating to the Flowering of Tulips, Narcissus's, &c.," *Philosophical Transactions*, vol. 37 (1731–32).

————. "Remarks upon the Letter of Mr. John Ellis F.R.S. to Philip Carteret Webb," *Philosophical Transactions*, vol. 50 (1757–58).

————. *Abridgement of the Gardeners Dictionary* (London, 1735, and subsequent editions).

————. *Das Englische Gartenbuch, oder Philipp Millers Gärtner-Lexicon* (Nürnberg, 1750).

————. *Figures of the Most Beautiful, Useful, and Uncommon Plants Described in the Gardeners Dictionary* (London, 1755–60).

———. *Gardeners Kalendar* (London, 1733, and subsequent editions).

———. *Gardeners and Florists Dictionary* (London: Charles Rivington, 1724).

———. *Gardeners Dictionary* (London, 1731, and subsequent editions).

Minter, Sue. *The Apothecaries' Garden. A History of the Chelsea Physic Garden* (Stroud: Sutton Publishing, 2000).

Morris, Sandra. "Legacy of a Bishop: The Trees and Shrubs of Fulham Palace Gardens Introduced 1675–1713," *Garden History*, vol. 19:1 (1991).

Münchhausen, Otto von. *Der Hausvater* (Hanover, 1770).

Muyden, Madame van (trans. and ed.). *A Foreign View of England in the Reigns of George I & George II. The Letters of Monsieur César de Saussure to His Family* (London: John Murray, 1902).

Mylius, Christlob. "Christlob's Mylius Tagebuch seiner Reise von Berlin nach England," in *Johann Bernoulli's Archiv zur neuern Geschichte, Geographie, Natur—und Menschenkenntnis* (Leipzig: Georg Emanuel Beer, 1787).

Nelson, E. Charles. "Australian Plants Cultivated in England before 1788," *Telopea*, vol. 4:2 (1983).

Nettel, Reginald (trans. and ed.). *Carl Philip Moritz. Journeys of a German in England in 1782* (London: Jonathan Cape, 1965).

O'Neil, Jean. "Peter Collinson's Copies of Philip Miller's *Dictionary* in the National Library of Wales," *Archives of Natural History*, vol. 20:3 (1993).

Osborn, James M. (ed.). *Spence, Joseph, Observations, Anecdotes, and Characters of Books and Men* (Oxford: Clarendon Press, 1966).

Otis, Denise. *Grounds for Pleasure: Four Centuries of the American Garden* (New York: Harry N. Adams Publishers, 2002).

Parkinson, John. *Paradisi in Sole Paradisus Terrestris*, first published 1629 (Walter J. Johnson, Amsterdam, 1975).

———. *Theatrum Botanicum* (London: Tho. Cotes, 1640).

Parkinson, Sydney. *A Journal of a Voyage to the South Seas* (London: Charles Dilly, 1784).

Paterson, Daniel. *A New and Accurate Description of All the Direct and Principal Roads in Great Britain* (London: T. Carnan, 1771).

Pavord, Anna. *The Naming of Names. The Search for Order in the World of Plants* (London: Bloomsbury, 2005).

Peck, Robert McCracken. "Books from the Bartram Library," in *Contributions to the History of North American Natural History*, ed. Alwyne Wheeler (London, Society for the Bibliography of Natural History, 1983).

Penn, William. *Some Fruits of Solitude*, first published in 1693 (Pulborough: Praxis, 1994).

Platt, Hugo. *Floraes Paradise* (London: William Leake, 1608).

Porta, John Baptist. *Natural Magick in Twenty Bookes*, first published in 1558 as *Magia Naturalis* (London: R. Graywood, 1658).

Porter, Roy. *English Society in the Eighteenth Century* (London: Penguin, 1990).

———. *Enlightenment. Britain and the Creation of the Modern World* (London: Penguin, 2001).

———. *The Greatest Benefit to Mankind. A Medical History of Humanity from Antiquity to the Present* (London: Fontana Press, 1999).

Potter, Jennifer. *Strange Blooms. The Curious Lives and Adventures of the John Tradescants* (London: Atlantic Books, 2006).

Prévost, Abbé. *Adventures of a Man of Quality*, trans. Mysie E. I. Robertson (London: Routledge, 1930).

Quest-Ritson, Charles. *The English Garden Abroad* (London: Viking, 1992).

Rauschenberg, Roy Anthony. "Daniel Carl Solander: Naturalist on the 'Endeavour,' " *Transactions of the American Philosophical Society*, New Series, vol. 58:8 (1968).

Reinikka, Merle A. *A History of the Orchid* (Coral Gables, Fl.: University of Miami Press, 1972).

Repton, Humphry. "Fragments on the Theory and Practice of Landscape Gardening," 1816, *The Landscape Gardening and Landscape Architecture of the Late Humphry Repton*, ed. John Claudius Loudon (London: Longman & Co., 1840).

————. "Observations on the Theory and Practice of Landscape Gardening," 1803, *The Landscape Gardening and Landscape Architecture of the Late Humphry Repton*, ed. John Claudius Loudon (London: Longman & Co., 1840).

Richards, Sarah. "A Magazine for the Friends of Good Taste: Sensibility and Rationality in Garden Design in late eighteenth Century Germany," *Journal of Garden History*, vol. 20:3 (2000).

Ringenberg, Jörgen et al. *Dendrologischer Atlas der Wörlitzer Anlagen* (Hamburg and Wörlitz: Dölling and Galitz Verlag, 2001).

Rochefoucauld, Francois de la. *A Frenchman in England, 1784, being the Melanges sur l'Angleterre of Francois de la Rochefoucauld*, trans. S. C. Roberts (London, Caliban Books, 1995).

Rogers, John. *The Vegetable Cultivator* (London, 1839).

Rössig, K. *Handbuch für die Liebhaber Englischer Pflanzungen und für Gärtner* (Leipzig, 1796).

Rousseau, Jean Jacques. *Letters on the Elements of Botany addressed to a Lady*, trans. Thomas Martyn (London, 1785).

Rudé, George. *Hanoverian London* (London: Sutton Publishing, 2003).

Sambrock, James. "Painshill Park in the 1760s," *Garden History*, vol. 8:1 (1980).

Schuster, Julius (ed.). *Linné und Fabricius, zu ihrem Leben und Werk* (München: Verlag der Münchner Drucke, 1928).

Sckell, Friedrich Ludwig von. *Beitrage zur Bildenden Gartenkunst*, facsimile of 2nd edition 1825 (Worms: Werner'sche Verlagsgesellschaft Worms, 1982).

Scott, G. and F. A. Bottle (eds.). *The Journal of James Boswell* (Mount Vernon: W. E. Rudge, 1930).

Seward, Anna. *Memoirs of the Life of Dr. Darwin* (London: J. Johnson, 1804).

Sherburn, George (ed.). *The Correspondence of Alexander Pope* (Oxford: Clarendon Press, 1956).

Shteir, Ann B. *Cultivating Women, Cultivating Science. Flora's Daughters and Botany in England, 1760–1860* (Baltimore and London: Johns Hopkins University Press, 1996).

Siegesbeck, Johann Georg. *Botanosophiae verioris brevis sciagraphia in usum discentium adornata. Accedit ob argumenti analogiam, Epicrisis in clar. Linnaei nuperrime evulgatum systema plantarum sexuale, et huic superstructam methodum botanicam* (St. Petersburg, 1737).

Simond, Louis. *Journal of a Tour and Residence in Great Britain during the Years 1810 and 1811* (Edinburgh: Constable 1817).

Slaughter, Thomas P. *The Natures of John and William Bartram* (New York: Vintage Books, 1997).

Sloane, Hans. *A Voyage to the Islands of Madera, Barbadoes, Nieves, St. Christopher and Jamaica* (London: printed for the author, 1725).

Smith, Adam. *An Inquiry into the Nature And Causes of the Wealth of Nations,* first published 1776 (London: Penguin Books, 1982).

Smith, James Edward (ed.). *A Selection of the Correspondence of Linnaeus and Other Naturalists* (London: Longman, 1821).

Smith, Lady Pleasance (ed.). *Memoir and Correspondence of Sir James Edward Smith* (London: Longman, 1832).

Smith, Merril D. "The Bartram Women: Farm Wives, Artists, Botanists and Entrepreneurs," *Bartram Broadside* (2001).

St. Aubyn, Fiona. *A Portrait of Georgian London* (Churt: D. Leader, 1985).

Stearn, William T. "A Royal Society Appointment with Venus in 1769: The Voyage of Cook and Banks in the 'Endeavour' in 1768–1771 and Its Botanical Results," *Notes and Records of the Royal Society of London,* vol. 24:1 (1969).

Stroud, Dorothy. *Capability Brown* (London: Faber & Faber, 1975).

Stuart, David. *The Plants that Shaped our Gardens* (London: Frances Lincoln, 2002).

Summerson, John. *Georgian London* (London: Barrie & Jenkins, 1988).

Survey of London. Chelsea (London: London County Council, 1913).

———. *The Parish of St. Leonard, Shoreditch,* Forrest, G. Topham, James Bird and Phillip Norman. vol. 8 (London: Batsford, 1922).

Swem, Earl G. (ed.). "Brothers of the Spade. Correspondence of Peter Collinson, of London, and of John Custis, of Williamsburg, Virginia, 1734–1746," *Proceedings of the American Antiquarian Society,* vol. 58, pt.1 (1948).

Swinden, Nathaniel. *The Beauties of Flora Display'd* (London, 1778).

Switzer, Stephen. *Icnographia Rustica* (London, 1718).

Sydow, Carl Otto von. "Linnaeus and the Lapps," *Taxon,* vol. 25:1 (1976).

Symes, Michael. "A. B. Lambert and the Conifers at Painshill," *Garden History,* vol. 16:1 (1988).

———. "Charles Hamilton's Planting at Painshill," *Garden History,* vol. 11:2 (1983).

———. "Lord Petre's Legacy: The Nurseries at Thorndon," *Garden History,* vol. 24:2 (1996).

Symes, Michael, Alison Hodges and John Harvey. "The Plantings at Whitton," *Garden History,* vol. 14:2 (1986).

The Parliamentary History of England, 1765–71, vol. 16 (Hansard: London, 1813).

Thomas, Keith. *Man and the Natural World. Changing Attitudes in England 1500–1800* (London: Penguin Books, 1983).

Thornton, Robert. *The Temple of Flora* (London: Weidenfeld & Nicolson, 1981).

Tradescant, John. *Musaeum Tradescantianum* (London: John Grismond, 1656).

Trauzettel, Ludwig. "Wörlitz: England in Germany," *Garden History,* vol. 24:2 (1996).

Tromp, Heimerick. "A Dutchman's Visits to some English Gardens in 1791," *Journal of Garden History,* vol. 2:1 (1982).

Turner, D. (ed.). *Extracts from the Literary and Scientific Correspondence of Richard Richardson* (Yarmouth: Charles Sloman, 1835).

Tusser, Thomas. *Five hundreth points of good husbandry* (London: Richard Tottill, 1573).

Uggla, Arvid H. "Från Linné den Yngres Englandsresa," *Svenska Linné-Sällskapets Årsskrift*, vol. 36, 1953.

Uglow, Jenny. *A Little History of British Gardening* (London: Chatto & Windus, 2004).

———. *The Lunar Men. The Men Who Made the Future* (London: Faber & Faber, 2002).

Vertue, George. *Walpole Society* (Oxford, Oxford University Press), vol. 30, (1948–50).

Vines, Giles. "What's in a Name?," *Kew Magazine*, no. 55 (2006).

Volkmann, Johann Jacob. *Neueste Reisen durch England vorzüglich in Absicht auf die Kunstsammlungen, Naturgeschichte* (Leipzig: Caspar Fritsch, 1781).

Wall, Cecil. *A History of the Worshipful Society of Apothecaries of London* (London: Oxford University Press, 1963).

Walpole, Horace. "The History of Modern Taste in Gardening, 1780," in *The English Garden, Literary Sources and Documents*, Mountfield, ed. M. Charlesworth, vol. 2 (East Sussex: Helm Publishing, 1993).

Weber, Caroline. *Queen of Fashion. What Marie Antoinette Wore to the Revolution* (London: Aurum, 2006).

Weiss, Thomas (ed.). *Infinitely Beautiful. The Dessau-Wörlitz Garden Realm* (London: Frances Lincoln, 2007).

Wendeborn, Friedrich August. *A View of England*, first published 1785 (London: J. Robinson, 1791).

Weston, Richard. *Gardener's and Planter's Calendar* (London, 1773).

Whately, Thomas. *Observations on Modern Gardening*, first published 1770 (London: T. Payne, 1771).

Williams, Glyndwr. "The Endeavour Voyage: A Coincidence of Motives," in *Science and Exploration in the Pacific*, ed. Margarette Lincoln (London: National Maritime Museum, 1998).

Wolridge, John. *Systema Horti-Culturae or the Art of Gardening* (London: T. Burrell and W. Hensman, 1677).

Wright, Louis B., and Marion Tinling (eds.). *The Secret Diary of William Byrd of Westover, 1709–1712* (Richmond: The Dietz Press, 1941).

Wulf, Andrea, and Emma Gieben-Gamal. *This Other Eden. Seven Great Gardens and 300 Years of English History* (London: Little, Brown, 2005).

Zirkle, Conway. *The Beginnings of Plant Hybridization* (Philadelphia: Philadelphia University Press, 1935).

Notes

Prologue: The Fairchild Mule

6 Fairchild's hybrid: Bradley 1717, pt. 1, p. 23; RS JBC, vol. XI (1714–20), 4 February 1720, f. 437.

6 description of Fairchild: based on Richard van Bleeck's portrait of Thomas Fairchild, *c.*1723, held at the Plant Science Library, University of Oxford.

7 description of the hybrid: Collinson to Bartram, 22 July 1740, p. 136.

7 Brompton Park Nursery: there were too few plants that flowered from October to the end of January to be included in Henry Wise's list for the Brompton Nursery (*c.*1700), Harvey 1974, Appendix IV, p. 150.

7 "principally of extremely": de Saussure, 14 June 1726, van Muyden 1902, pp. 136–7.

8 William and Mary: Wulf and Gieben 2005, pp. 62ff.

8 Fairchild's plants from Holland: Fairchild received plants from Holland from Richard Bradley in 1714, Leapman 2000, pp. 61–9; Fairchild sent fellow gardener Thomas Knowlton to Holland in 1726 to procure rarities, Knowlton to Blair, 12 January 1728, Henrey 1986, p. 47.

10 "one in five hundred": *Castanea pumila*, for example, which had been introduced by 1699, remained a rarity, as "not one seed in five hundred sent over ever grew," Miller 1768, "Castanea pumila."

10 Tradescant in Virginia: Potter 2006, pp. 256–78; and for Tradescant's plant introductions see Leith-Ross 1998, pp. 181–95.

10 Fairchild's plants in Hoxton: Fairchild 1722, pp. 15–27, 51–6; and Fairchild, "List of those Plants which flower every Month in my Garden," 1722.

10 "scarlet flower'd horse chestnut": Aiton 1810–13; Fairchild, "List of those Plants which flower every Month in my Garden," 1722.

10 "the chief Gardens": Miller 1724, "Tulip Tree."

10 "purchas'd from abroad": Fairchild 1722, p. 22.

10 "are so rare": Miller 1724, "Hellebores."

10 Portugal laurel, etc.: Henrey 1986, p. 47; Harvey 1998, p. 67; Miller 1724, "Guernsey-Lily."

10 landmark for gardeners: Sherard to Richardson, 24 July 1722, Turner 1835, p. 184.

11 "[f]rom Pliny to": Wolridge 1677, p. 143.

11 "[w]hen you sow": Gilbert 1693, Appendix Monthly Directions, no page numbers.

11 traditional gardening advice: Gilbert 1693, Appendix Monthly Directions, no page numbers; Porta 1658, pp. 91–2; Markham 1615, pp. 33–6.

11 "when the wind": John Evelyn's "Directions to the Gardiner at Says Court," Notes for the Flower Garden, Fleming and Gore 1980, p. 53.

11 comets as signs: Jardine 1999, pp. 11–12.

12 plant superstitions: Gerard 1636, pp. 845–6; Thomas 1983, pp. 75, 78.
12 Fairchild's experiments: Fairchild 1724–5, p. 127; RS JBC, vol. XII
   (1720–26), 2 April 1724, f. 461; Bradley 1724, vol. 1, pt. 3, p. 130;
   vol. 2, pt. 4, p. 318; vol. 3, pt. 3, pp. 96–7.
12 "[T]here is noe thing": Locke to Sloane, 14 September 1694, de Beer
   1979, vol. 5, p. 128.
12 "for the improvement": Jardine 1999, p. 83.
13 Royal Society experiments: Jardine 1999, pp. 117–23.
13 Hans Sloane's elephant: Leapman 2000, p. 136.
13 Fairchild and the Royal Society: Richard Bradley, Patrick Blair and
   James Douglas, for example, asked Fairchild to experiment for them.
   Fairchild 1724–5, p. 127; RS JBC, vol. XII (1720–26), 2 April 1724,
   f. 461; Bradley 1724, vol. 1, pt. 3, p. 130; vol. 2 pt. 4, p. 318; vol. 3,
   pt. 3, pp. 96–7; Fairchild and the Royal Society publications, Fairchild
   1722, p. 5.
13 "the most rational": Bradley 1724, vol. 2, p. 313.
13 "true Bent of": Bradley 1724, vol. 1, p. 128.
14 Fairchild at the Royal Society: RS JBC, vol. XI (1714–20), 4 February
   1720, ff. 437–39.
14 description of the Royal Society: Christlob Mylius, 24 September 1753,
   Mylius 1787, p. 74.
14 Mule's conception: RS JBC, vol. XI (1714–20), 4 February 1720,
   ff. 437–9.
14 "the Power and": Fairchild 1722, p. 9.
14 "of a middle nature": RS JBC, vol. XI (1714–20), 4 February 1720,
   f. 437.
15 "[a] Curious Person": Bradley 1717, pt. 1, p. 23.
15 "[t]his art is very": Heimerick Tromp in 1791, Tromp 1982, p. 47.
15 "the true artists": Johanna Schopenhauer in 1803–5, Michaelis-Jena
   and Merson 1988, p. 5.
15 "planted with various": Sale by Auction, The Times, 24 October 1786,
   p. 4.
15 first gardening magazine: Curtis's Botanical Magazine.
16 "the wonderfull works": Thomas Fairchild's Will, 21 February 1729
   (13 October 1729), TNA PROB 11/632 and Lyons 1940–41, p. 81.
16 "never been able": Dr. Denne's sermon in 1733, Leapman 2000, p. 213.

Chapter 1: "Forget not Mee & My Garden"

19 Collinson to Custom House: Collinson to Bartram, 24 January 1735
   [1734], pp. 3–6 (Armstrong dates this 1734, Armstrong, 2002,
   pp. 11–15).
19 description of Collinson's London: unless otherwise referenced this is
   based on contemporary visitor accounts: Defoe 1971, p. 316; de Saussure, 16 December 1725 and 26 October 1726, van Muyden 1902,
   pp. 67, 80–81, 84, 95–6, 167; Lichtenberg to Ernst Gottfried Baldinger,
   10 January 1775, Mare and Quarrell 1938, pp.63–4; von Archenholz
   1789, vol. 1, pp.135–6.
19 "a forest of ships": Prévost 1930, p. 76.
19 "foster-mother": de Saussure, 16 December 1725, van Muyden 1902,
   p. 96.

20 exports to colonies: Langford 1989, pp. 168–9.
20 "our merchants are": Defoe in 1726, quoted in Rudé 2003, p. 52.
20 London's food consumption: Rudé 2003, p. 20.
20 "made entirely of": Lichtenberg to Ernst Gottfried Baldinger,
10 January 1775, Mare and Quarrell 1938, p. 63.
20 "the choicest merchandise": de Saussure, 16 December 1725, van
Muyden 1902, p. 81.
20 "little cottage": Collinson to Thomas Story, April 1731, Brett-James
1925, p. 29; Peckham: Defoe 1971, p. 176.
20 "the most Delightfull": Collinson to Colden, 7 March 1742, Armstrong
2002, p. 96.
20 "hurrys of town": Collinson to Thomas Story, April 1731, Brett-James
1925, p. 29.
20 "curious plants": and following description, Collinson's own notes on
his childhood, Lambert 1811, p. 271.
20 footnote: Collinson to Mary Collinson, 27 July 1728, Armstrong 2002,
p. 4.
21 "I must be peeping": Collinson to Michael and Mary Russell, August
1739, Armstrong 2002, p. 80.
21 "a Cockney who": Collinson to John Custis, 2 February 1741, Swem
1948, p. 86.
21 Pennsylvania's booming market: James 2001, p. 83.
21 "By your leave": de Saussure, 26 October 1726, van Muyden 1902,
p. 167.
21 "the wealthiest corner": de Saussure, 16 December 1725, van Muyden
1902, p. 80.
21 Pennsylvania Coffee House: Collinson to Bartram, 24 February 1739, this
letter was written from the Pennsylvania Coffee House, p. 112.
21 Custom House: de Saussure, 16 December 1725, van Muyden 1902,
p. 84.
22 "good friends among": Collinson to Bartram, 26 February 1737, p. 38.
22 plant cargo: and the following descriptions, Collinson to Bartram, 24
January 1735 [1734], pp. 3–6 (Armstrong dates this 1734, Armstrong,
2002, pp. 11–15).
22 Collinson and kalmias: Catesby 1731–48, vol.2, plate 98 (in which
Collinson's love for kalmias is mentioned).
22 "Forget not Mee": Collinson to George Robins, 6 October 1721, LS
MS Armstrong (transcriptions of original letters).
22 plant boxes jettisoned: Miller to Veltheim, 24 July 1756, LHASA, MD,
Rep. H Harbke 1856, f. 23.
22 "[o]ne may as soon": William Byrd III, quoted in Swem 1948, p. 167.
22–3 "it might be": Collinson to Custis, 12 November 1736, Swem 1948, pp.
51–2.
23 "[I] admire them": Collinson to Colden, 7 March 1742, Armstrong
2002, p. 96.
23 "Sublime Contemplation": Collinson to Bartram, 25 July 1762, p. 565.
23 "there we see": Penn (1693) 1994, Section Country Life; point 220.
23 "Man's work": Penn (1693) 1994, Section Country Life; point 222.
23 "nosegays": and Collinson on flowers in the house, Collinson to
Bartram, 20 January 1751, p. 313; Collinson to Linnaeus, 18 January
1744 and Collinson to Trew, 20 March 1746, Armstrong 2002,
pp. 112, 129; Fothergill 1770, p. 16.

24 classification of a fish: Collinson to Sloane, c.1740, Armstrong 2002, p. 86.
24 "rational pursuit": and gardening as virtuous pastime, Fothergill 1770, pp. 9–10.
24 "Wee Brothers of": Collinson to John Custis, 15 December 1735, Armstrong 2002, p. 37.
25 Franklin's subscription library: Armstrong, p. 9.
25 "greatly owing to": Benjamin Franklin to Michael Collinson, 8 February 1770, Labaree 1959–, vol. 17, p. 66.
25 Collinson selected for the library: Collinson to the Library Company, 22 July 1732, Armstrong 2002, p. 9.
25 "to get rid of my": Collinson mentioned Joseph Breintnall, the secretary of the Library Company, as the man who introduced him to Bartram as well as to Sam Chew; "An Account of the Introduction of American Seeds into Great Britain," 1766, Peter Collinson, NHM MSS Col. 1; Peter Collinson's Commonplace Book, LS MS 323b f. 82; Collinson to Bartram, 16 January 1744, p. 229.
25 transport times and climate: Miller 1731, "Transport of plants."
25 "darling study": Bartram to J. Slingsby Cressy, probably March or April 1740, and Bartram to Catcott, 26 May 1742, Berkeley and Smith Berkeley 1992, pp. 130, 193; Bartram to Collinson, 1 May 1764, p. 627.
25 Bartram's farm and family: Bartram to Jared Eliot, 24 January 1757, Berkeley and Smith Berkeley 1992, p. 415; Bell 2004, p. 4.
26 "God Save": Jacobs 2005, p. 5.
26 "plain Country Man": Collinson to Custis, 24 December 1737, Armstrong 2002, p. 59.
26 Bartram's grandfather: Berkeley and Smith Berkeley 1982, p. 2.
26 Bartram expelled from Darby Meeting: Letter of Disownment to John Bartram from the Darby Monthly Meeting, 1 February 1758, Berkeley and Smith Berkeley 1982, Appendix 9, p. 322.
26 "ye lattin pusels": Bartram to Colden, 25 January 1746, 1992, p. 271.
26 Catesby in the colonies: Catesby had been in Virginia from 1712 to 1719 and in Carolina and the Bahamas from 1722–1726; for Catesby in America: Brigham 1998; Frick 1961; for seeds he sent: Catesby 1731–1748, vol.1, p.V
26 Catesby's seeds for Fairchild: Bradley 1724, vol. 3, and Harvey 1974, pp.151–9.
27 Catesby's seeds for Collinson: Catesby to Sherard, 13 November 1723, and Catesby to Collinson, 5 January 1723, quoted in Brigham 1998, p. 105.
27 Catesby's specimens for English collectors: the twelve collectors included Hans Sloane, William Sherard and the Duke of Chandos among others, Catesby 1731–48, vol. 1, "List of Sponsors."
27 Collinson's financial help: Collinson's handwritten note in his copy of Catesby's *Natural History*, reprinted in Loudon 1844, p. 69; Catesby 1731–48, vol. 1, p. XI.
27 gardeners with a small salary: see Catesby's subscription list in *Natural History*, which included Philip Miller, Thomas Knowlton and other gardeners and nurserymen. Book prices: Philip Miller's *Figures of the most Beautiful, Useful, and Uncommon Plants described in the Gardener's Dictionary* (published 1755–60), which consisted of 300 coloured plates, cost twelve guineas; Johann Jacob Dillenius's *Historia*

Muscorum cost £1 10s.; and Hans Sloane's *History of Jamaica* was priced at £6 10s.

27 Catesby on American plants: Catesby 1731–48, vol. 1, p. IX; Catesby also assisted nurseryman Christopher Gray in compiling "A Catalogue of American Trees and Shrubs that will endure the Climate of England."

27 "what was common": "An Account of the Introduction of American Seeds into Great Britain," 1766, Peter Collinson, NHM MSS Col. 1.

28 "Pray send a root": Collinson to Bartram 20 January 1735, p. 14; Fairchild and hellebores, Miller 1724, "Hellebores."

28 "[p]lease to remember": Collinson to Bartram, 3 February 1735 [1736], p. 18 (this letter should be dated 1736 because Collinson refers here to his letter of 16 August 1735).

28 "thee knowest what": Collinson to Bartram, 31 March 1735 and 2 May 1738, pp. 9–10, 90.

28 packing instructions: Collinson to Bartram, 24 January 1735 [1734], pp. 3–6 (Armstrong dates this 1734, Armstrong, 2002, pp. 11–15).

28 "person or books": Bartram to Collinson, 1 May 1764, p. 627.

28 numbering of plant specimens: list of 208 specimens, Collinson to Bartram, 20 March 1737, pp. 50–57.

28–9 "I always write": Collinson to Bartram, 22 March 1751, p. 319. There are dozens of letters in which Collinson explained that he was "vastly hurried."

29 "I have much to do": Collinson to Bartram, 3 February 1742, pp. 180–81.

29 "I keep a regular": Collinson to Bartram, 3 February 1735 [1736], p. 17 (this letter should be dated 1736 because Collinson refers here to his letter of 16 August 1735).

29 "The seeds in No. 2": Collinson to Bartram, 12 February 1735, p. 20.

29 "I think thee has": Collinson to Bartram, 12 February 1735, p. 20; moisture of the roots: Collinson to Bartram, 1 March 1735, p. 8.

29 "oddly huddled together": Bartram to Collinson, 27 February 1737, p. 40; Collinson's response: Collinson to Bartram, 10 December 1737, p. 68.

30 special germinating bed: and seeds' acclimatization, Collinson to Bartram, 20 January 1735, p. 16.

30 "I never trust": Collinson to Bartram, 20 January 1756, p. 392.

30 "[t]his year Better": Collinson to Bartram, 19 June 1735, p. 10.

30 "very different from": Collinson to Bartram, 19 June 1735, p. 10.

30 "How often I survey'd": Collinson to Custis, 20 October 1734, Armstrong 2002, p. 19.

30 Bartram accused Collinson: Bartram's letter of 18 October 1735 has not survived, nor the one from 3 November 1735, but Collinson refers to them in his answers on 3 February 1736 and 20 March 1736, pp. 17, 26.

30 Bartram's compensation: Collinson to Bartram, 1 March 1735 and 20 February 1736, pp. 8, 21

31 "one Sixth part": Collinson quotes Bartram's accusations in his answer; Collinson to Bartram, 20 February 1736, p. 22.

31 "Pray give no body": Collinson to Bartram, 1 March 1735, p. 9.

31 "Thee may assure": Collinson to Bartram, 3 February 1735 [1736], p. 18 (this letter should be dated 1736 because Collinson refers here to his letter of 16 August 1735).

31 "Receive it in Love": Collinson to Bartram, 20 February 1736, pp. 21–3.
31 subscription system: Collinson to Bartram, 12 March 1736, p. 23.
31 "the gentlemen should": Collinson to Bartram, 21 April 1736, p. 27.
31 risk remained with Bartram: Dobbs to Bartram, 29 June 1751, Berkeley and Smith Berkeley 1992, p. 331.
32 Logan and Bartram's botanical books: Bartram to Sloane, 23 September 1743, Berkeley and Smith Berkeley 1992, p. 224.
32 Logan and Collinson: since Logan's visit to England they had corresponded regularly—Collinson, for example, read Logan's letters at the Royal Society; RS JBC, vol. XV (1734–6), 23 January 1735, f. 70; Hindle 1956, p. 22.
32 subscription system: Collinson to Bartram, 1 June 1736, p. 28.
32 "I shall make a present": James Logan to Collinson, 8 June 1736, Berkeley and Smith Berkeley 1982, p. 36.
32 "has shown a very": Collinson to Bartram, 28 August 1736, p. 34.

Chapter 2: "The bright beam of gardening"

34 Mercurial Thermometer: Miller 1731, "Thermometer."
35 "a Work of the greatest": Collinson to Custis, 21 March 1737, Armstrong 2002, p. 55; see also Collinson to Bartram, 20 January 1735, Armstrong 2002, p. 22.
35 *Dictionary* as a gift: Collinson to Bartram, 20 January, 1735, 20 May 1737, pp. 14, 48.
35 "Consult Millar [*sic*] on": Collinson to Bartram, 2 February 1738, 10 July 1737, pp. 83, 94.
35 "noble Present": Bartram to Collinson, 1 November 1737, 6 December 1739, pp. 65, 129.
35 "draw a Right or": James 1712, p. 87.
36 "rooted sufficiently": Miller 1731, "Geranium."
36 "in a natural Way": Bradley 1724, vol. 2, p. 176.
36 "in a methodical Order": and following quotes and information, Miller 1731, pp. x, xi.
36 Miller's education and life: Kalm 1892, pp.109–11; Rogers 1839, pp. 339–43.
36 "plain old-fashioned": Martyn 1807, p. vii.
36–7 "pure and uncontaminated": and the following quotes, Miller 1731, p. vi.
38 "wretched, with few plants": Uffenbach, 26 October 1710, Mare and Quarrell 1934, p. 161.
38 "no Wine or other": Garden Committee Minutes, 21 August 1722, WSA, MS 8228/1, f. 3
38 "ffor improving naturall": Sloane's Deed of Covenant to the Society of Apothecaries, 1722, CPG; Minter 2000, p. 11.
38 Miller's nursery in Southwark: Bradley 1724, vol. 2, p. 32.
38 Miller's avaricious landlord: Blair to Sloane, 30 December 1721, BL Sloane 4046, f. 168.
38 "superior to most": Blair to Sloane, 30 December 1721, BL Sloane 4046, f. 168.
38 "[P]lants as well as": Garden Committee Minutes, 25 February 1724, WSA, MS 8228/1, f. 4.

38 "Botanick art is": Clutton to Richardson, 13 April 1726, Turner 1835, p. 251.

38 list of rules: Garden Committee Minutes, 21 August 1722, WSA, MS 8228/1, ff. 1–3.

39 Fairchild's Mule: Miller, 1731, "Caryophyllus."

39 "working on [the] Tulips": Miller to Blair, 11 November 1721, Blair 1720–21, p. 216.

39 Miller's diary: Miller, *Gardeners Kalendar*, 1733, p. viii.

40 "no plants rootes": Garden Committee Minutes, 21 August 1722, WSA, MS 8228/1, f. 3.

40 "pray take no": Miller to Blair, quoted in Le Rougetel 1990, p. 38.

40 postage, packaging and travel: Garden Committee Minutes, 30 April 1750, WSA, MS 8228/1, ff. 42–3.

40 Miller and Bartram's boxes: Collinson to Bartram, 21 April 1736, p. 27.

40 Miller to the Netherlands: Knowlton to Blair, 12 January 1728, Henrey 1986, p. 90.

40 "should travel to": Miller 1724, vol. 1, "Gardener."

40 American torch-thistle: Miller 1731, "Cereus."

40 "Learned botanist": and the following quotes, Boerhaave to Miller, 16 August 1729, Lindeboom 1962–79, vol. 3, p. 139; Boerhaave to William Sherard, 17 February 1728, Lindeboom 1962–72, vol. 1, p. 161; Boerhaave to Miller, 16 August 1729, Lindeboom 1962–79, vol. 3, p. 137; Boerhaave to Miller, 2 March 1732, Lindeboom 1962–79, vol. 3, p. 141.

41 pineapples: Miller 1731, "Ananas"; Le Rougetel 1990, p. 50; Beauman 2005, pp. 60–63.

41 tanner's bark and hotbeds: Miller 1731, "Hotbeds" and "Tanners bark"; Henrey 1986, p. 41.

41 "in greater Strength": Bradley 1724, vol. 3, p. 134.

42 greenhouse at Chelsea: Miller 1731, "Greenhouse" and "Stoves."

42 stoves at Chelsea: Miller 1731, "Stoves."

42 hothouses at Hampton Court: Wulf and Gieben-Gamal 2005, pp. 62–4.

42 costs of greenhouse at Chelsea: 24 August 1732 Court of Assistants, Field 1878, p. 53.

42 footnote: Bale to Duchess of Beaufort,1 September 1692, BL Sloane 4062 f.246; and for Fairchild's hothouses, Bradley, 1724, vol. I, p. 178–180.

43 Miller's horticultural successes: Cowell 1730, p. 34; Miller 1728, pp. 485–8; Miller 1731–2, pp. 81–4.

43 "We have in flower": Miller to Blair, quoted in Le Rougetel 1990, p. 38.

44 "a single tree": Miller 1768, "Datura."

44 "blush rose colour": and following plant references, Aiton 1810–13; Miller 1731, Miller 1731 and 1768, "Ketmia"; Miller 1731, "Camphora."

44 cedar of Lebanon at Royal Society: RS JBC, vol. XIII (1726–31), 20 February 1729, f. 302.

44 "will in a short time": Society of Gardeners 1730, p. VIII.

44 "according to his own": Miller to Richardson, 19 August 1727, Turner 1835, p. 27.

44 "a Knave or a Blockhead": Society of Gardeners 1730, p. x.

45 "scarce one of twentie": Parkinson 1975, pp. 571, 336.

45 "the Mercenary Flower Catchers": Gilbert 1693, Epistle to the Reader, no page numbers.

45 "varies so much": Dillenius to Bartram, 26 April 1737, Berkeley and Smith Berkeley 1992, p. 48.

45 Society of Gardeners: Society of Gardeners 1730, pp. vi–xi.

45 "to compare such": Society of Gardeners 1730, p. ix.

45 "Mr. Catesbey's new": "A Catalogue of English and Foreign Trees, Sold by Robert Furber," 1727, p. 1 (the nurseryman Robert Furber was also a member of the Society of Gardeners); Society of Gardeners 1730.

45 Miller and the Society: Rogers 1839, pp. 336–7.

46 "Miller's early dwarf": "A Catalogue of Garden, Grass and Flower Seeds, Trees, Shrubs, Herbaceous, Green-House and Hot-House Plants, Sold by Russell, Russell, & Willmott, Nursery & Seedsmen, Lewisham, Kent," 1800.

46 "Miller 1st sort": Lady Skipwith's plant lists, 1785–1805, Leighton 1988, p. 288.

46 "Men of Letters": John Martyn at the Royal Society, RS JBC, vol. xiii (1726–31), 27 May 1731, f. 376.

46 "you draw all": Alexander Pope to Earl of Marchmont, July 1743, Sherburn 1956, vol. 4, p. 459.

46 "the chief book that": Ellis to Linnaeus, 1756 or 1757, Smith 1821, vol. 1, p. 84.

46 "as Miller directs": Gilbert White, 16 April 1752, 11 November 1754, Glegg 1975, f. 12, 20v.

46 "contributed more to the": *Curtis's Botanical Magazine*, vol. 1, 1787, plate 11.

46 subscribers of *Dictionary*: Miller 1731, List of Subscribers.

47 "the bright beam": Rogers 1839, p. 339.

Chapter 3: "My harmless sexual system"

48 Linnaeus in England: Jackson 1926, p. 3.

48 "sought him as": Afzelius 1826, p. 30.

48 "prince of botanists": and Linnaeus's opinion about himself, Afzelius 1826, pp. 39, 29.

48 "mother's milk": Afzelius 1826, pp. 104–5.

48–9 "were full of rare": Ehret Nov. 1894–June 1895, p. 50; Afzelius 1826, p. 27.

49 Linnaeus's appearance: Johann Beckmann in 1765, Fries 1911, p. 48.

49 "slow people": Afzelius 1826, p. 127.

49 "Linnaeus's ambition": Johann Christian Fabricius in 1780, Schuster 1928, no page numbers.

49 "God himself led": Afzelius 1826, p. 91.

49 "wild forest with untamed": and Linnaeus in Lapland, Afzelius 1826, p. 108; Black 1979, p. 15; Lindroth 1983, p. 59; Blunt 2001, pp. 56–70.

49 courting Sara Elisabeth: Linnaeus, January 1735, Translation of Linnaeus' Diary of 1735, Bryk 1919, vol. 1, p. 2.

50 Ehret drawings: Ehret Nov. 1894–June 1895, p. 51.

50 "We are all devoted": Linnaeus to Haller, 1 May 1737, Smith 1821, vol. 2, p. 242.

50 plants in Collinson's garden: Collinson to Bartram, 3 February 1735, 19 June 1735, 1 June 1736, 20 September 1736, pp. 10–18, 28–9, 35;

Catesby 1731–48, vol. 1, plate 39, 55, 57; Miller 1731, "Magnolia lauri folio subtus albicante" [*Magnolia virginiana*], "Cornus (Sassafras-Tree)"; Reinikka 1972, p. 18.

51 "ticklish plant": Collinson to Bartram, 1 June 1736, p. 28.

51 "understand anything that": Linnaeus to François Boissier de La Croix de Sauvages, 2 November 1737, Lindroth 1983, p. 23.

51 classification of botanists: Linnaeus (1751) 2003, pp. 13–29.

52 "Hot or sharpe": Parkinson 1640, Preface, no page numbers.

52 footnote: Pavord, 2005, p. 359.

53 Linnaeus's eyesight: Linnaeus (1735) 1964, p. 24.

55 DNA and classification: Vines 2006, p. 20.

55 *Systema Naturae* in London: copies were sent to the Duke of Bedford, William Cole, Richard Richardson, Hans Sloane, Richard Mead, John Martyn, Dillenius, Philip Miller, Thomas Shaw; Gronovius to Linnaeus, 19 October 1735, 18 January 1736, The Linnaean Correspondence, linnaeus.c18.net, letter L0047; Garden Committee Minutes, 24 March 1736, WSA, MS 8228/1, f. 14.

55 "so well made up": and following quote, Gronovius to Richardson, 30 August 1735, Turner 1835, p. 344.

56 "if you want to": and following quote, Gronovius to Linnaeus, 19 October 1735, The Linnaean Correspondence, linnaeus.c18.net, letter L0047.

56 Sloane spent £50,000: Thomas, Birch, *Memoirs relating to the Life of Sr Hans Sloane Bart formerly President of the Royal Society*, 1753, BL Add 4241, f. 24.

56 "on proper Notice": Thomas, Birch, *Memoirs relating to the Life of Sr Hans Sloane Bart formerly President of the Royal Society*, 1753, BL Add 4241, f. 21.

56 "to be worthy to": Boerhaave to Sloane, 18 July 1736, Lindeboom 1962–79, vol. 1, p. 199.

56 "FIRST OF THE SCHOLARS": Boerhaave to Sloane, 5 October 1731, Lindeboom 1962–79, vol. 1, p. 195.

57 note to Linnaeus: and Linnaeus's visit to Sloane, Cromwell Mortimer to Linneaus, 27 July 1736, The Linnaean Correspondence, linnaeus.c18.net, letter L0092 (Mortimer was Sloane's assistant); Jackson 1926, p. 6.

57 Sloane's museum: the description of Sloane's museum is based on several visitors' accounts, Zacharias Conrad von Uffenbach, 3 November 1710, Mare and Quarrell 1934, pp.185–8; Christlob Mylius, 18 October 1753, Mylius 1787, pp. 97–9; Sauveur Morand in 1729, reprinted in MacGregor 1994, Appendix 2, p. 31.

57 "all manner of curious": Uffenbach, 3 November 1710, Mare and Quarrell 1934, p. 187.

57 Händel's muffin: MacGregor 1994, p. 28.

57 230 volumes: this is the number Sauveur Morand saw in 1729. Today there are 337, which are bound together into 265 volumes, Cannon 1994, p. 137.

57 "true English fashion": Uffenbach, 19 July 1710, Mare and Quarrell 1934, p.127. One of the more chaotically arranged herbariums in Sloane's possession had belonged to James Petiver, an apothecary who had been demonstrator at Chelsea just before Philip Miller arrived.

57 "One must run over": Sherard to Richardson, 28 March 1721, Turner 1835, p. 166.

57–8 Linnaeus's classification cupboard: Afzelius, p. 231; Linnaeus (1751)
2003, pp. 330–31.

58 Sloane's lack of time: Sloane 1725, vol. 2, pt. 1, Preface.

58 "Many not named": and following quote, Cannon 1994, p. 139.

58 "a great Obstruction": Sloane 1725, vol. 2, pt. 1, p. xvii.

58 Ray's *Historia Plantarum*: Sloane's annotated copy is held at the
Natural History Museum in London; see also Cannon 1994, p. 138.

58 "What was he": Linnaeus to Haller, summer 1737, Smith 1821, vol. 2,
p. 281.

58 "Sloane's collection is": Linnaeus to Olof Celsius, November or
December 1736, Fries et al. 1907–43, vol. 1 pt. 5, p. 256.

58–9 "A foreign and unknown": Linnaeus, *Classes Plantarum* (1738), Danby
1958, p. 11.

59 Collinson to Bartram, 14 December 1737, p. 72.

59 "bridal bed": Linnaeus, *Præludia Sponsaliarum Plantarum* (1729),
Blunt 2001, p. 33.

59 "the lips of the": Linnaeus (1751) 2003, p. 105.

59 "Twenty males or more": and the following quotes, Linnaeus (1753)
1957, vol. 1, pp. 31–4.

59 Linnaeus's visit at Chelsea: the description is based on accounts by two
of his pupils to whom he told the story, Johann Beckmann in 1765, Fries
1911, p. 119, and Paul Giseke, quoted in Blunt 2001, p. 112.

59 "a difficult man": Johann Beckmann in 1765, Fries 1911, p. 119.

59 "did not want to contradict": Johann Beckmann in 1765, Fries 1911,
p. 119.

60 "the botanist of Clifford's": Paul Giseke, quoted in Blunt 2001, p. 112.

60 "scowled at me": Paul Giseke, quoted in Blunt 2001, p. 112.

60 "greatest perfection": Miller to Alston, 1 October 1737, EUL, Laing III
375, f. 22a.

60 "regaled him with": Johann Beckmann in 1765, Fries 1911, p. 119.

60 "a fine parcel for": Paul Giseke, quoted in Blunt 2001, p. 112; Afzelius
1826, p. 228.

60 "without the least reason": Miller to Alston, 1 October 1737, EUL,
Laing III 375, f. 22b.

60 "not having seen": Miller to Alston, 4 February 1756, EUL, Laing III
375, f. 24a*v*.

60 "will be of very": Miller to Alston, 1 October 1737, EUL, Laing III
375, f. 22b.

61 "the man who has": Paul Giseke, quoted in Blunt 2001, p. 113.

61 "under tears and": Afzelius 1826, p. 28; Linnaeus to Olof Celsius,
November or December 1736, Fries et al. 1907–43, vol. 1, pt. 5,
p. 256.

61 "a thorough insight": Dillenius to Richardson, 25 August 1736, Turner
1835, p. 347; Afzelius 1826, p. 28.

61 "I do not doubt": Dillenius to Linnaeus, 16 May 1737, Smith 1821,
vol. 2, p. 92.

61 "[A] whole bunch": Linnaeus to Olof Celsius, November or December
1736, Fries et al. 1907–43, vol. 1, pt. 5, p. 256; list of plants that
Linnaeus received, Linnaeus in *Hamburgische Berichte*, facs. 1737, Bryk
1919, vol. 1, pp. 140–41.

62 "altogether whimsicall and": Knowlton to Richardson, 31 October
1736, Turner 1835, p. 349.

62 "Science of Botany": Collinson to Colden, 4 September 1743, Armstrong 2002, p. 110.

62 "too smutty for": Charles Alston, quoted in Shteir 1996, p. 17.

62 "I doubt very much": Johann Amman to Sir Hans Sloane, 6 September 1736, BL Sloane 4054, f. 298v.

62 "the most obscene": *Encyclopaedia Britannica*, 1768, quoted in Thornton 1981, p. 9.

62 Americans and Linnaeus: Collinson to Linnaeus, 12 March 1745, Smith 1821, vol. 1, p. 12.

62 "The performance is": Logan to Bartram, 19 June 1736, Berkeley and Smith Berkeley 1992, p. 31.

62 "thou wilt . . . be fully": Logan to Bartram, 19 June 1736, Berkeley and Smith Berkeley 1992, p. 32.

62 "in awe": Logan to Linnaeus, 28 October 1738, The Linnaean Correspondence, linnaeus.c18.net, letter L0259.

62 "[N]o doubt but he": Collinson to Bartram, 14 December 1737, p. 72; and Bartram on dissecting plants: Bartram to Logan, 19 August 1737, Berkeley and Smith Berkeley 1992, pp. 61–3.

63 "a pretty amusement": Collinson to Bartram, 14 December 1737, p. 71.

63 "pray make this complaint": Collinson to Bartram, 20 December 1737, p. 72.

63 "very few like it": Collinson to Bartram, 14 December 1737, p. 72; Collinson to Colden, 26 April 1745, Armstrong 2002, p. 125.

63 "unrivalled science": Boerhaave to Linnaeus, 13 Jan 1737, Lindeboom 1962–79, vol. 3, p. 189.

63 "that one should be": Gronovius to Richardson, 22 July 1738, Turner 1835, p. 369.

63 "[W]ithout the system": Linnaeus, *Fundamenta Botanica* (1736), Lindroth 1983, p. 22.

63 "no one before": Afzelius 1826, p. 93.

63 "puffed up": Linnaeus to Haller, 3 April 1737, Smith 1821, vol. 2, p. 233.

64 "There is nobody": and following quote about Miller, Linnaeus to Haller, 1 May 1737, Smith 1821, vol. 2, p. 245.

64 Order of the Polar Star: Johann Beckmann in 1765, Fries 1911, p. 48.

64 "I perceive you take": Johann Amman to Linnaeus, 15 November 1737, Smith 1821, vol. 2, p. 193.

64 "my harmless sexual": and the following quotes, Linnaeus to Albrecht von Haller, 3 April 1737, Smith 1821, vol. 2, pp. 232, 234.

64 "mere gewgaws": Dillenius to Linnaeus, 16 May 1737, Smith 1821, vol. 2, p. 93.

64 "I feel . . . much displeased": Dillenius to Linnaeus, 18 August 1737, Smith 1821, vol. 2, p. 96.

64 "you are not very patient": Dillenius to Linnaeus, 28 November 1737, Smith 1821, vol. 2, p. 103.

65 calling himself "sensitive": Afzelius 1826, p. 127.

65 "lose his temper": Johann Christian Fabricius in 1780, 1928, no page numbers.

65 "The work is very": Johann Amman to Linnaeus, 15 November 1737, Smith 1821, vol. 2, p. 195.

65 "fictitious matrimony of": Siegesbeck 1737, p. 49; Johann Amman to Linnaeus, 23 January 1738, Smith 1821, vol. 2, p. 196.

65 "loathsome harlotry": and following quote, Linnaeus (1753) 1957, vol.
   1, p. 25; Siegesbeck, 1737, p. 49.
65 "my reputation must": Linnaeus to Haller, March 1738, Smith 1821,
   vol. 2, p. 320.
65 "all, with one voice": and following quote, Linnaeus to Haller, 12
   September 1739, Smith 1821, vol. 2, p. 335.

Chapter 4: "Pray go very Clean, neat & handsomely Dressed to Virginia"

66 Bartram's children: Isaac (1725, from Bartram's first marriage), James
   (1730), Moses (1732), Elizabeth (1734, died in infancy), Mary (1736),
   Elizabeth and William (1739), Ann (1741), John (1743), Benjamin
   (1748), Smith 2001, pp. 1–2.
66 "exact observations of": Collinson to Bartram, 12 March 1736, p. 23.
66 "Be sure [to] have": Collinson to Bartram, 12 March 1736, p. 24.
66–7 Bartram's travel accounts: Collinson to Bartram, 16 May 1741,
   p. 154.
67 "is there any account": Collinson to Bartram, 14 December 1737, p.
   70.
67 "Read & Travel": Collinson to Bartram, 14 December 1737, p. 70.
67 "[I] wish to bear": and Collinson's notes of distances, Collinson to
   Bartram, 3 February 1736, p. 18; Peter Collinson's Commonplace Book,
   LS MS 323b f. 175.
67 "Like the parson's": Collinson to Bartram, 7 June 1736, p. 31.
67 "not Look att the": Collinson to Custis, 24 December 1737, Armstrong
   2002, p. 59.
67 "the seeds in generall": Custis to Collinson, 29 July 1736, Swem 1948,
   p. 50.
67 "Pray go very Clean": Collinson to Bartram, 17 February 1738, p. 84.
67 "look phaps More": Collinson to Bartram, 17 February 1738, p. 84.
68 Catesby's engravings: Bartram to Collinson, probably 13 June 1738,
   p. 91.
68 Bartram's expedition: unless otherwise referenced, the description of the
   expedition is based on Bartram's *Journal of a Trip to Maryland and
   Virginia*, autumn 1738, Berkeley and Smith Berkeley 1992, pp. 101–3.
68 Custis's garden: Custis to Collinson, 1734, Swem 1948, p. 40.
68 horse chestnut: Collinson had sent horse chestnuts on 12 February
   1735, 26 January 1739, pp.19, 109; Bartram on horse chestnut:
   Bartram to Collinson, 1 April 1739, p. 116.
68 "I am told those": Custis to Collinson, 29 July 1736, Swem 1948, p. 49.
68 "Our poor country grows": Custis to Collinson, 18 July 1738, Swem
   1948, p. 71; and description of effect of winter and drought: Custis
   to Collinson, 28 August 1737, Swem 1948, p. 62 and Martin 1991,
   pp. 60–61.
68 "he is the most takeing": Custis to Collinson, probably 12 August
   1739, Swem 1948, p. 77.
68–9 "my dear friend Col": Bartram's *Journal of a Trip to Maryland and
   Virginia*, autumn 1738, Berkeley and Smith Berkeley 1992, p. 102.
69 Catesby and Byrd: William Byrd II, 5 June 1712, Wright and Tinling
   1941, p. 540; Martin 1991, p. 72.

69 "observing curious plants": Bartram's *Journal of a Trip to Maryland and Virginia*, autumn 1738, Berkeley and Smith Berkeley 1992, p. 102; and Bartram on Byrd's garden: Bartram to Collinson, 18 July 1739, p. 121.

69 "much esteemed in ye": and finding *Thuja occidentalis* and Eastern hemlock, Bartram to Byrd II, early November 1738, Berkeley and Smith Berkeley 1992, p. 99.

70 cost of Eastern hemlock: "List of Trees in the Thorndon Nurseries, 1740–42," TNA 30/8/74 ff. 150–51, reprinted in full in Harvey 1996, p. 281.

70 "for I cant afford": and following quote and descriptions, Bartram's *Journal of a Trip to Maryland and Virginia*, autumn 1738, Berkeley and Smith Berkeley 1992, p. 103.

70 1,100 miles: Bartram to Collinson, 10 December 1738, p. 104.

70–71 seeds in a jumbled mess: Bartram to Collinson, 10 December 1738, p. 104.

71 "I Long the arrival": Collinson to Bartram, 1 February 1739, p. 110; and captain's refusal to take boxes, Collinson to Bartram, 26 January 1739, p. 107.

71 "In pain for": Collinson to Bartram, 24 February 1739, p. 112.

71 "tasts in Life": Collinson to Bartram, March 1739, p. 113.

71 "as thee often": Bartram to Collinson, 18 July 1739, p. 121.

71 "rotten mouldy" hat: Bartram to Collinson, 18 July 1739, p. 122.

71 "My Cap it's True": Collinson to Bartram, March 1739, p. 115.

71 "I thought some sory": Bartram to Collinson, 18 July 1739, p. 122.

72 Bartram's response to Collinson's complaint: Bartram to Collinson, 12 April and 18 July 1739, pp. 118–19, 121–2.

72 "my kind & generous": and following quote, Bartram to Collinson, May 1738; Collinson to Bartram, March 1739, pp. 90, 115.

72 Bartram's garden: unless otherwise referenced, the description of Bartram's garden is based on Bartram to Collinson, 7 and 25 September 1740, pp. 141, 143; William Bartram, "A Draft of John Bartram's House and Garden as it appears from the River, 1758," watercolour on paper by William Bartram © The Right Hon. Earl of Derby, Knowsley.

72 Bartram and rhododendrons: Bartram to Collinson, 1 April 1739, p. 117.

72 "they seem to Die": Collinson to Bartram, 10 July 1738, 3 March 1743, pp. 93, 185.

74 "were lost in common": Garden to Colden, 4 November 1754, Colden 1918–37, vol. 4, p. 472.

74 "curious Trees, Shrups": George Washington, 10 June 1787, Jackson and Twohig 1976–9, vol. 5, p. 240.

74 "a perfect portraiture": Garden to Colden, 4 November 1754, Colden 1918–37, vol. 4, p. 472.

74 "Every den is": Garden to Colden, 4 November 1754, Colden 1918–37, vol. 4, p. 472.

74–5 Plants in Bartram's garden: Collinson to Bartram, 12 February and 3 September 1735, 22 March 1737, pp. 13, 20, 43.

75 "I shall show my": Bartram to Collinson, 1 November 1737, p. 65.

75 "in A few hours": Bartram to Collinson, May 1738, p. 90.

75 "plant them in": Collinson to Bartram, 28 June 1735, p. 12.

75 "patience & perseverance": Collinson to Bartram, 14 March 1737, p. 42.

75 "perished & disolved": and Bartram and Cedar of Lebanon, Bartram to Collinson, 29 July 1757, p. 424; Collinson to Bartram, 12 February 1735, p. 20.

76 "[Custis] Desired some": Collinson to Bartram, 20 December 1737, p. 73.

76 "no slothful forgetful": Collinson to Bartram, 12 August 1737, p. 60.

76 "to be better": Collinson to Bartram, 20 January 1738, p. 79.

76 "a Horse Load": Collinson to Bartram, 12 April 1739, p. 117.

76 "spruce firs": and Bartram's complaint about bad harvest, Bartram to Collinson, December 1744 and 10 December 1738, pp. 250, 104.

76 "for half A dozen": Bartram to Collinson, December 1744, p. 250.

76 American willow oak: Bartram to Collinson, May 1738, p. 89.

76 *Juniperus virginiana*: Bartram to Collinson, 7 September 1740, p. 141.

77 "[T]hee art not": Collinson to Bartram, 26 February 1737, p. 38.

77 Bartram in New Jersey 1739: Bartram to Collinson, November 1739, pp. 126–8.

77 Bartram cut his foot: Bartram to Collinson, November 1739, p. 127.

77 "a great affliction": Bartram to Collinson, spring 1745, p. 251.

77 Bartram's lonely journeys: see Bartram's letters, 31 July 1736, 13 June 1738, 7 September 1740, pp.32–3, 91–3, 139–42.

78 "[N]o mankind to be": Bartram to Collinson, 7 September 1740, p. 140.

78 "dismall travelling": Bartram to Collinson, 7 September 1740, p. 140.

78 "chewed it": Bartram to Collinson, 30 September 1763, p. 609.

78 white cedar: and prices, Collinson to Bartram, 26 February 1737, p. 38; "List of Trees in the Thorndon Nurseries, 1740–42," TNA 30/8/74, ff. 150–51, reprinted in full in Harvey 1996, p. 281.

78 "one of ye Silvan": Bartram to Collinson, probably 13 June 1738, p. 92.

78 "delighted most to dream": Bartram to Miller, 18 February 1759, Berkeley and Smith Berkeley 1992, p. 457.

79 "the path of wild": Bartram to Catcott, 26 May 1742, Berkeley and Smith Berkeley 1992, p. 194.

79 "perpendicular banks of": Bartram to Collinson, autumn 1753, p. 362.

79 "by runing & sliping": Bartram to Collinson, probably 22 July 1741, p. 163.

79 "thee must look": Collinson to Bartram, 25 February 1741, p. 149; Bartram to John Mitchell, 3 June 1743, Berkeley and Smith Berkeley 1992, p. 239.

79 Bartram's farm: Bartram bought 142 acres from Andrew Jonason on 4 September 1735 and 50 acres from Andrew Souplis Jr. on 3 April 1739 (I would like to thank Joel T. Fry, curator of Bartram's Garden, for providing this information).

79 "just in the very nick": Bartram to Collinson, July 1739, p. 120; as only copper coins were allowed to be sent from England to the North American colonies, Collinson dispatched £10 in halfpennies—4,800 coins. Bartram to Collinson, summer 1738 and Collinson to Bartram, 7 February 1739, pp. 94, 111.

79 "Rascally Spaniards": Collinson to Bartram, 20 December 1740, p. 147.

80 "[T]his untoward Warr": Collinson to Bartram, 25 February 1741, p. 150.

80 "winters amusements": Bartram to Slingsby Cressy, probably March or April 1740, Berkeley and Smith Berkeley 1992, p. 131.

80 botanical books at the Library Company: Bartram had to leave the value of the book as a deposit and pay sixpence for the hire, Bartram to Collinson, 6 December 1739, p. 29.

80 "In Reading of Books": Collinson to Bartram, 14 December 1737, p. 70.

80 "if solomon had loved": Bartram to Collinson, May 1738, p. 89.

80 Bartram's botanical books: Bartram to Collinson, December 1739; Catesby to Bartram, 20 May 1740; Sloane to Bartram, 16 January 1742; Bartram to Sloane, 23 September 1743; Bartram to Catcott, 24 November 1743, Berkeley and Smith Berkeley 1992, pp. 130, 132–3, 179, 224–5.

81 "I think he has": Bartram to Collinson, December 1739, p. 130.

Chapter 5: "All gardening is landscape-painting"

85 Collinson's summer tours: Fothergill 1770, p. 8; Collinson to Bartram, 20 January 1738, p. 79.

85 Collinson's character: Fothergill 1770, pp. 8, 16.

85 "most Valuable & Intimate": Collinson to Bartram, 1 March 1735, p. 8.

85 "chat by ye fire": Petre to Collinson, 22 January 1738, BL Add 28726, f. 38v, also ff. 24v, 74, 99.

85 "young people of fortune": Forthergill 1770, p. 9.

86 "[H]ow much I wish": Petre to Collinson, 10 April 1739, BL Add 28726, f. 71.

86 "ye oftener you come": and Petre being reluctant to make decisions, Petre to Collinson, 11 April 1742, BL Add 28726, f. 114v; Petre to Collinson, 31 January 1737, BL Add 28726, f. 19.

86 "often prevented young": Fothergill 1770, p. 9.

86 other aspects of Petre's life: Petre to Collinson, various letters from 1737 to 1741, BL Add 28726, ff. 21v, 23, 30v, 94.

86 Collinson shared boxes: Collinson to Bartram, 1 March 1735, p. 8.

86 "grown to great": Collinson to Bartram, 1 September 1741, p. 167.

86 "Lord Petre has raised": and following quotes, Collinson to Bartram, 20 January 1738, 1 June 1736, 21 July 1741, 16 June 1742, pp. 80, 28, 158, 196.

86 "Children of my own": Collinson to Trew, 20 March 1746, Armstrong 2002, p. 129.

87 "a Bushel of Red Cedar": and following quote and other additional seed orders, Collinson to Bartram, 1 February 1739, p. 110; Collinson to Bartram, 12 March 1736, 12 April 1739, pp. 24, 117 and most letters from the late 1730s.

87 900 tulip poplars: "List of Trees in the Thorndon Nurseries, 1740–42," TNA 30/8/74 ff. 150–51, in Harvey 1996, p. 281.

87 prices for tulip poplars and American sycamore: Robert Furber's bill for Carlton House, 14 October 1734, Coombs 1997, p. 169.

87 tulip poplars at Thorndon: Collinson to Bartram, 12 March and 1 June 1736; 1736, pp. 24, 28; Petre to Collinson, 9 May 1737, BL Add 28726, f. 26v; "List of Trees in the Thorndon Nurseries, 1740–42," TNA 30/8/74 ff. 150–51, in Harvey 1996, p. 281.

87 "20 Thousand Trees": Collinson to Bartram, 1 September 1741, p. 167.

87 200,000 exotics at Thorndon: Collinson to Linnaeus, 18 January 1744, Armstrong 2002, p. 112.

88 "ye finest Palm": Petre to Collinson, 9 September 1736, BL Add 28726, f. 17.

88 Collinson on *Camellia japonica*: Collinson to Sloane, *c.* 1740, Armstrong 2002, p. 87.

88 Bartram's clients: by the time Collinson visited Thorndon in summer 1741, Bartram had the following clients: 1734 Lord Petre, 1735 Philip Miller, 1736 the Duke of Richmond, 1738 William Hird, 1739 Alexander Colhoun, 1740 the Duke of Norfolk, 1741 Alexander Catcot.

88 description of Thorndon: unless otherwise referenced, the descriptions of Thorndon and the plants are based on: Collinson to Bartram, 1 September 1741, pp. 167–8; Lambert 1811, vol. 10, p. 273; Collinson's MS notes in his copy of John Evelyn's *Sylva* (1664) at the Royal Forestry Society, in Laird 1999; "List of Trees in the Thorndon Nurseries, 1740–42," TNA 30/8/74 ff. 150–51, in Harvey 1996, p. 280.

88 John Caryll on Petre: John Caryll to Dowager Lady Petre (Petre's mother), November 1734, Clutton and Mackay 1970, pp. 29–30.

88 "Herculean Undertaking": and references to tree transplantations, Collinson, quoted in Clutton and Mackay 1970, p. 35; Collinson's MS notes in Miller's *Dictionary* (1752), NLW.

88–9 mounts at Thorndon: Lambert 1811, p. 274; Collinson's MS notes in his copy of John Evelyn's *Sylva*, in Laird 1999, p. 70.

88 footnote: Lancelot Brown's tree-moving, Wulf and Gieben 2005, p. 127.

89 "greatest pfection": Collinson to Bartram, 1 September 1741, p. 168.

89 winter 1739/40: Brazell 1968, p. 9; Collinson to Custis, 31 January 1740, Armstrong 2002, p. 84.

89 Collinson's plants in Peckham: Collinson to Custis, 21 October 1741, Swem 1948, p. 93; Petre to Collinson, 8 January 1740, BL Add 28726, f. 79v.

89 callicarpa and *Magnolia grandiflora*: Miller 1752, "Magnolia"; Miller 1768, "Johnsonia."

89 night shifts in stoves: Hall 1986, p. 11.

89 "to keep Constant fires": Collinson to Custis, 31 January 1740, Armstrong 2002, p. 84.

89 "notwithstanding I have done": Petre to Collinson, 8 January 1740, BL Add 28726, f. 79v.

89 Thorndon's stoves: Miller 1731, "Greenhouse" and "Stoves"; Collinson to Linnaeus, 18 January 1744, and Collinson to Richard Richardson, 4 April 1746, Armstrong 2002, pp. 112–13, 133.

89 "The Great Stove is": Collinson to Richard Richardson, 4 April 1746, Armstrong 2002, p. 133.

89–90 Ananas Stove: Collinson to Linnaeus, 18 January 1744, Armstrong 2002, pp. 112–13; Miller 1731, "Ananas" and Plate II "Tan Stove for Young Anana Plants"; for pineapples in general see Beauman 2005, p. 87.

90 pineapple for Collinson: Petre to Collinson, 8 January 1740, BL Add 28726, f. 80.

90 Petre, the most innovative gardener: Philip Southcote on Petre, after 1751, Spence 1966, vol. 1, point 603.

90 Petre's character: Collinson to Bartram 3 July 1742, p. 199; Fothergill 1770, pp. 8–9

90 "my brother Gardeners": and references to the gardeners as a group, Petre to Richmond, 23 March 1740, West Sussex Record Office, Good-

wood MS 112 f. 297; Petre to Collinson, 13 May 1741, BL Add 28726, f. 97v; Petre to Collinson, 26 May 1736, BL Add 28726, f. 5; 9 September 1736, BL Add 28726, f. 17; 8 September 1738, f. 54v; Knowlton to Brewer, 18 or 20 November 1738, Henrey 1986, p. 182.

90 Petre employed Miller: Martyn 1732, preface (in the dedication Martyn praises the cultivation of plants at Thorndon "by the skilful hand of my friend Mr. Miller"); Miller's catalogue of plants at Thorndon is available on microfilm in the ECRO (T/A 671).

90 "to the care of": Petre to Collinson, 1 February 1739, BL Add 28726, f. 64v; 26 May 1736, f. 5.

90 plant swapping: Richmond to Miller, 26 January 1739, Museum of the History of Science, Oxford, MS Gunther 14/2, f. 24.

90 Miller at Goodwood: Laird 1999, pp. 178–84; Collinson to Richmond, 27 August 1744, March 1911, vol. 2, pp.440–42; Richmond to Collinson, 24 August 1744, BL Add 28726, f. 146.

91 "as to Miller": Knowlton to Brewer, 15 December 1741, Henrey 1986, p. 196.

91 "botanic wars": Rauthmell to Richardson, 31 October 1731, Turner 1835, p. 316.

91 "Lights and Shades": Miller 1739, "Wilderness."

91 "living pencils": Collinson to Southcote, 9 October 1752, Armstrong 2002, p. 159.

92 "all sorts of pines": Collinson to Bartram, 12 April 1739, p. 118.

92 "our people are": and following quote, Collinson to Bartram 25 February 1741, p. 151; Petre to Collinson, 3 February 1741, BL Add 28726, f. 93.

93 "they seem to like": Collinson to Bartram, 3 March 1743, p. 185.

93 "when I walk among": Collinson to Bartram, 1 September 1741, p. 167.

93 "Hints Borrowed from": Collinson to Southcote, 9 October 1752, Armstrong 2002, p. 159.

93 "appear accidental": Miller 1731, "Wilderness."

93 "Spontaneous acts of": Collinson to Southcote, 9 October 1752, Armstrong 2002, p. 159.

94 scented plants: Miller 1731; Miller 1739, "Wilderness."

94 "Prison Walls": Switzer 1718, vol. 3, p. 76.

94 "The whole . . . is planted": Collinson to Bartram, 1 September 1741, p. 167.

94 first footnote: Petre's plans for the Duke of Norfolk at Worksop, 1738, ECRO D/DP P150A.

94 second footnote: Joseph Addison, *The Spectator*, no. 414, 25 June 1712, and Alexander Pope, *Guardian*, no. 173, 29 September 1713.

95 "understood the colours": Philip Southcote, *c.*1753, Spence 1966, vol. 1, point 1124, 424.

95 "with his own hands": Collinson to Bartram, 14 March 1737, p. 41.

95 Petre's copies of the *Dictionary*: Miller 1731, List of Subscribers.

95 "cruel weather": and instructions to Goodwood steward, Richmond to Collinson, 16 February 1748, BL Add 28727, f. 5v; Richmond to Labbé, 2 December 1730, West Sussex Record Office, Goodwood MS 102 f. 116.

95 Prince of Wales works in shrubberies: Vertue 1948–50, p. 153.

95 deaths of Richmond and Prince of Wales: John Mitchell to Bartram, 30 March 1751, p. 320; Vertue 1948–50, p. 156.

96 catalogues of "Hardy-trees": Miller 1739, Addenda.

96 "some of those Rules": Miller 1739, "Wilderness."

96 Guglielmo: Hunt 1987, p. 13.

96 plants at Carlton House: and Robert Furber's bills between 1734–1738 for Carlton House, Coombs 1997, pp. 158, 169–77.

96 "leaped the fence": and following quote: Walpole (1780) 1993, pp. 402, 404.

96 "carried it farther": Philip Southcote on Petre, after 1751, Spence 1966, vol. 1, point 603; Philip Southcote was the owner of Wooburn Farm, the famous "ornamental farm" in Surrey which he had begun in 1734. The Duke of Richmond called him "my uncle Southcote," and he was a good friend of Collinson, as well as the son of Petre's guardian (Richmond to Collinson, 17 December 1742, BL Add MS 28726, f. 124v). England's first garden historian, Horace Walpole, also declared that Kent had failed to see the potential of the American trees and shrubs which were, as he explained "so peculiar to our modern landscape," Walpole (1780) 1993, p. 403.

96 "brown hills covered": Mary Delany to Mrs. Dewes, February 1741, Llanover 1861, vol. 2, p. 147.

Chapter 6: "Send no Seeds for him . . . all is att an End"

98 Purpose of Bartram's subscription: A Copy of the Subscription Paper, for the Encouragement of Mr. John Bartram, *Pennsylvania Gazette*, 17 March 1742, Labaree 1959–, vol. 2, pp. 356–7.

98 "[I]t is very hard": Collinson to Bartram, 25 February 1741, p. 149.

99 target to raise £50: Bartram to Collinson, 25 September 1740, p. 144.

99 "spend most of my": Bartram to Cressy, 29 March 1741, Berkeley and Smith Berkeley 1992, p. 153.

99 £20 had been raised: A Copy of the Subscription Paper, for the Encouragement of Mr. John Bartram, *Pennsylvania Gazette*, 17 March 1742, Labaree 1959–, vol. 2, p. 357.

99 "[T]he man I Loved": Collinson to Bartram, 3 July 1742, p. 198.

99 "a great cold": Petre to Collinson, 22 June 1742, BL Add 28726, f. 118v.

99 "All our Schemes": Collinson to Bartram, 3 July 1742, p. 198.

99 "Lost in Embrio": Collinson to Bartram, 3 July 1742, p. 198.

99 Catesby's illustrations as order catalogue: Collinson to John Blackburne, 20 October 1742, Armstrong 2002, p. 105.

99 Collinson's search for clients: and American subscription, Bartram to Collinson, 18 December 1742, p. 211.

100 "all round Him would": Collinson to Benjamin Smithurst, 9 September 1742, Armstrong 2002, p. 104.

100 "Oh what will become": Collinson to Richard Richardson junior, 12 August 1742, Armstrong 2002, p. 102.

100 "Young Trees are": Collinson to Mr. Leigh, September 1757, Armstrong 2002, p. 208.

100 "had a great of discourse": and reference to Duke of Bedford, Southcote to Collinson, 30 December (no year but certainly 1742), Peter Collinson's Commonplace Book, LS MS 323b ff. 16, 202.

100 garden at Worksop: and Norfolk as subscriber, Petre's plans for the Duke of Norfolk at Worksop, 1738, ECRO, D/DP P150A; like Petre, Norfolk subscribed to two boxes a year, Collinson to Bartram, 20 October 1740, p. 146.

100 footnote: Le Rougetel 1990, p. 59.

101 "some of the Curious": Richmond to Collinson, 5 December 1742, BL Add MS 28726, f. 122v.

101 "a vacant corner": Richmond to Collinson, 20 July 1742, BL Add MS 28726, f. 120.

101 "between the hours": Richmond to Collinson, 12 November 1742, BL Add MS 28726, f. 121v.

101 *Thuja occidentalis*: Richmond to Collinson, 5 December 1742, BL Add 28726, f. 123v.

101 "they are not to be gott": Richmond to Collinson, 17 December 1742, BL Add 28726, f. 124v.

101 "a mount Lebanon": Richmond to Collinson, 28 December 1742 and 5 December 1742, BL Add 28726, ff. 127v, 122v.

101 "I fear I am still": Richmond to Collinson, 28 December 1742, BL Add 28726, f. 127.

101 eighty tulip poplars: Richmond to Collinson, 29 January 1743, BL Add 28726, f. 129.

101 "the Dukes of Bedford": Richmond to Collinson, 28 December 1742, BL Add 28726, f. 128.

101-2 "an other Gentleman": Bartram to Colden, 26 June 1743, Berkeley and Smith Berkeley 1992, p. 219.

102 "[i]f ever I come": Bartram to Collinson, 21 June 1743, pp. 216–17.

102 Bartram's expedition to Iroquois Confederacy: Bartram (1751) 1973; Goetzmann 2004, pp. 97–105; Hallock 2004, pp. 107–25.

102 "to introduce A peaceable": Bartram to Colden, 26 June 1742, Berkeley and Smith Berkeley 1992, p. 219.

103 "looking for New Subscribers": and further instructions: Collinson to Bartram, 10 March 1744, p. 234.

103 "not to forget the pains": Collinson to Bartram, 10 March 1744, p. 235.

103 subscribers remained scant: Bartram to Fothergill, 24 July 1744, Berkeley and Smith Berkeley 1992, p. 240.

104 "Surprised att the": Collinson to Colden, 9 March 1744, Armstrong 2002, p. 113.

104 "secret pleasure of": Bartram to Colden, 29 April 1744, Berkeley and Smith Berkeley 1992, p. 238.

104 equivalent to the Royal Society: Bartram informed Collinson as early as 1737 about the idea of a scientific society, Bartram to Collinson, autumn 1737, p. 66.

104 "Infancy of your": Collinson to Bartram, 10 July 1738, p. 93.

104 "Useful Knowledge": A Proposal for Promoting Useful Knowledge, 14 May 1743, Labaree 1959–, vol. 2, p. 380.

104 "Curious persons": Bartram to Colden, 27 March 1744, in Berkeley and Smith Berkeley 1992, p. 237.

104 "first Drudgery of": Proposal for American Philosophical Society, 1743, Labaree 1959–, vol. 2, p. 380.

104 improved roads and postal services: and Franklin as Postmaster General, Greene 1993, pp. 78–9; Franklin to Collinson, 21 May 1751, Labree 1959– , vol. 4, p. 134; Doren 1943, p. 280.

104 "poor progress": Bartram to Colden, Berkeley and Smith Berkeley 1992, 7 April 1745, pp. 251–2.

104 "ye most curious": Bartram to Colden, Berkeley and Smith Berkeley 1992, 7 April 1745, p. 252.

104 "very idle Gentlemen": Franklin to Colden, 15 August 1745, Labaree 1959–, vol. 2, p. 379.

104 footnote: cost of letters, Doren 1943, p. 280.

105 "Most our thoughts": Bartram to Colden, Berkeley and Smith Berkeley 1992, 7 April 1745, p. 252.

105 Bartram's worries: and ship seized with his journal, Bartram to Fothergill, 7 December 1745, Berkeley and Smith Berkeley 1992, p. 267; Bartram to Collinson, 24 July 1744, p. 240.

105 boxes to brothers de Jussieu: Collinson to Colden, 30 March 1745, Armstrong 2002, p. 124.

105 "War keeps the Muses": Bernard de Jussieu to Linnaeus, 1 May 1745, Smith 1821, vol. 2, p. 215.

105 boxes forwarded to the botanic garden in Paris: René Antoine Ferchault de Réaumur to Abraham Trembley, 13 March 1746, Beer 1960, p. 212.

105 "half a Loaf is": Collinson to Bartram, 2 March 1747 and 20 July 1747, pp. 285, 289.

106 largest order: and good harvest, Collinson to Bartram, 2 March 1747, p. 285; Bartram to Collinson, 20 July 1747, p. 289.

106 "our people is daily": Bartram to Collinson, 30 January 1748, p. 292.

106 Petre's daughter's tooth: Lady Petre to Collinson, 24 May 1743, BL Add 28726, f. 134.

106 Miller supplied tanner's bark: Lady Petre to Collinson, 15 April 1744, BL Add 28726, f. 144.

106 "many plants droop": Lady Petre to Collinson, 12 July 1744, BL Add 28726, f. 142.

107 "the like opportunity": Collinson to Richmond, 24 October 1747, West Sussex Office, Goodwood MS 108, f. 794.

107 "2 Basketts & 5 Bundles": Collinson to Richmond, 29 March 1748, Armstrong 2002, p. 145.

107 "I must have them": Richmond to Collinson, 14 March 1748, BL Add 28727, f. 8v.

107 Thorndon sales and price list: Lady Petre to Collinson, 14 March 1748, BL Add MS 28727, f. 11; "List of Trees in the Thorndon Nurseries, 1740–42," TNA 30/8/74, ff. 150–51, in Harvey 1996, pp. 276–82.

107 "I hope you don't forget": Richmond to Collinson, 14 March 1748, BL Add 28727, f. 8v.

107 "Curious species for thee": Bartram to Collinson, 17 November 1742, p. 208.

107–8 "If I may boast": Collinson to Colden, 7 March 1742, Armstrong 2002, p. 96.

107 footnote: Lady Petre to Collinson, 14 March 1748, BL Add MS 28727 f. 11.

108 "scarcely a garden": Pehr Kalm, 10 June 1748, Kalm 1892, p. 67.

108 Collinson's garden: unless otherwise referenced, the description of Collinson's garden is based on Pehr Kalm's account in June 1748, Kalm 1892, pp. 66–8, 87.

108 Collinson hated the heat: Collinson to Linnaeus, 17 September 1765, Armstrong 2002, p. 265.

108 "I am no Stranger": Collinson to Colden, 7 March 1742, Armstrong 2002, p. 96.

108 flowerbeds and labels: Collinson to Peter Thompson, 27 October 1742, LS MS Armstrong (transcriptions of original letters); Collinson to Henrietta Maria Goldsborough, 1 February 1764, Armstrong 2002, p. 253; Collinson to Henrietta Maria Goldsborough, 26 June 1754, LS MS Armstrong (transcriptions of original letters).

108 *Monarda didyma*: Collinson to Christopher Jacob Trew, 18 January 1753, Armstrong 2002, p. 163.

108 *Dodecatheon meadia*: Collinson to Christopher Jacob Trew, 18 January 1753, Armstrong 2002, p.164; Catesby 1731–1748, vol. 2, Appendix, plate 1.

109 "Leaves are so much": Collinson to Linnaeus, 8 May 1746, Armstrong 2002, p. 135.

109 "came from my Garden": and first flowering of phlox in Collinson's garden, Collinson to Christopher Jacob Trew, 18 January 1753, Armstrong 2002, p. 164; Collinson to Colden, 23 August 1744, Armstrong 2002, p. 118.

109 *Yucca filamentosa*: and first flowering in Collinson's garden in summer 1747, Potter 2006, p. 270; Peter Collinson's Commonplace Book, LS MS 323a, f. 3.

109 "I received the first": and *Delphinium grandiflorum*: Collinson, 22 October 1746, Lambert 1811, vol. 10, p. 281.

109 "Every particle of Earth": Collinson to Bartram, 1 January 1764, p. 619.

109 description of Collinson's appearance: Fothergill 1770, p. 16; Collinson to Bartram, 16 June 1742, p. 196.

109 "Easily kill'd": Collinson to Custis, 31 January 1740, Armstrong 2002, p. 81.

109 "most unfortunately": Collinson to Bartram, 19 July 1753, p. 348.

109 "hand Engine": Collinson to Bartram, 12 April 1746, p. 276.

109 seagulls with clipped wings: Pehr Kalm, 19 April 1748, Kalm 1892, p. 163.

110 "how I Employe all": Collinson to Colden, 7 March 1742, Armstrong 2002, p. 96.

110 order of thirteen boxes: Collinson to Bartram, 6 May 1749, p. 299.

110 Bartram's new customers by 1750: the list of customers is based on Collinson's orders to Bartram during this time, for example in Berkeley and Smith Berkeley 1992, pp. 285, 299, 307; "An Account of the Introduction of American Seeds into Great Britain," 1766, Peter Collinson, NHM MSS, Col. 1.

110 "quite mad after planting": Richmond to Collinson, 16 February 1748, BL Add MS 28727, f. 5.

110 Thorndon's second plant auction: "List of Trees in the Thorndon Nurseries, 1740–42," TNA 30/8/74, ff. 150–51, in Harvey 1996, p. 276.

110 "the Laudable Spirit of": Collinson to Arthur Dobbs, 10 March 1750, Armstrong 2002, p. 151.

110 plant order from Prince of Wales: William Shirley to Bartram, 5 May 1749; Mitchell to Bartram 1 August 1750, Berkeley and Smith Berkeley 1992, pp. 305, 311.

110 "seatled a correspondance": Knowlton to Richardson junior, 13 November 1750, Turner 1835, pp. 406–7.

111 "cast a great damp": and following quotes, Collinson to Bartram, 24 April 1751, p. 324.

Chapter 7: "Commonwealth of Botany"

112 Linnaeus's botanizing excursions: description is based on Fornander 1753; Linnaeus's instructions in Linnaeus (1751) 2003, pp. 330–31; Afzelius 1826, p. 49; Linnaeus's pupils' accounts: Fabricius (1780) 1928, no page numbers; Beckmann (1765) 1911; and Koerner 1999, pp. 41–3.

112 Linnaeus's appearance: Fabricius (1780) 1928, no page numbers.

112 "army of botanists": Daniel Melanderhielm, quoted in Koerner 1999, p. 42.

113 "Vivat Linnaeus!": Blunt 2001, p. 170.

113 botanic garden and Linnaeus's family: Linnaeus to Haller, 18 July 1743, Smith 1821, vol. 2, p. 364.

113 augmented collection at Uppsala: Blunt 2001, p. 148.

113 "I brought the natural": Afzelius 1826, p. 49.

113 plants from Bartram to Uppsala: Collinson to Linnaeus, 10 April 1740, LS MS Armstrong (transcriptions of original letters); Catesby also sent Bartram's specimens and seeds to Linnaeus, Catesby to Linnaeus, 26 March 1745, Smith 1821, vol. 2, p. 440; Afzelius 1826, pp. 44, 115, 122.

113 "all fine, showey": Collinson to Bartram, 25 February 1741, p. 148.

114 Clifford's gardener and specimens: Linnaeus (1753) 1957, vol. 1, p. 154.

114 "I did not deserve": Clifford to Linnaeus, 20 April 1739, The Linnaean Correspondence, linnaeus.c18.net, letter L0280.

114 "What have I received": Dillenius to Linnaeus, 2 October 1746, Smith 1821, vol. 2, p. 129.

114 "The seeds & specimens": Collinson to Linnaeus, 27 March 1748, Armstrong 2002, p. 144.

115 Linnaeus and his seeds: Solander to Erik Gustaf Lidbeck, 25 March 1757, Duyker and Tingbrand 1995, p. 39.

115 "Commonwealth of Botany": Linnaeus (1751) 2003, p. 170.

115 "to send me plants": "To the Botanical Reader," 16 September 1750, Linnaeus (1751) 2003, p. 6.

115 Linnaeus's "apostles": Afzelius 1826, p. 48; Linnaeus (1753) 1957, vol. 1, p. 154.

116 "I long for him": Linnaeus to Abraham Bäck, 28 May 1751, Fries et al. 1907–22, vol. 1, pt. 4, p. 149.

116 vernacular plant names: Thomas 1983, pp. 81–5.

116 John Cowell's plant names: Cowell 1730, p. 48.

117 old names for *Kalmia angustifolia*: Miller 1768, "Kalmia angustifolia" and Catesby 1731–48, Appendix, plate 17.

117 "1 foot long": Linnaeus (1751) 2003, p. 246.

117 "[t]hese widely scattered": Linnaeus (1753) 1957, vol. 1, p. 154.

117 old names for *Pinus strobus*: Society of Gardeners 1730.

118 "every plant-name": Linnaeus (1751) 2003, p. 170.

118 "only genuine botanists": Linnaeus (1751) 2003, p. 169.

119 new generic names: in *Fundamenta Botanica* (1736) and *Critica Botanica* (1737) Linnaeus laid out the rules for naming plants.

119 "second Adam": Haller about Linnaeus, *Göttische Zeitungen von Gelehrten Sachen*, 1746, p. 670, quoted in Blunt 2001, p. 122.

119 "an Augean stable": Dillenius to Linnaeus, 18 August 1737, Smith 1821, vol. 2, p. 96.

119 "lead to worse": Amman to Linnaeus, 15 November 1737, Smith 1821, vol. 2, p. 195.

119 "unpardonable": Miller to Alston, 1 October 1737, EUL, Laing III 375, f. 22b.

119 "most part useless": Alston 1754, p. 71.

120 "so Vain as to": Earl of Bute's comment about *Species Plantarum*, Collinson to Linnaeus, 20 December 1754, Armstrong 2002, p. 178.

120 "Butt if you will": Collinson to Linnaeus, 10 April 1755, Armstrong 2002, p. 184.

120 "Thus Botany, which was": Collinson to Linnaeus, 20 April 1754, Armstrong 2002, p. 177.

120 "the Azaleas he": Collinson to Colden, 18 October 1753, LS MS Armstrong (transcriptions of original letters).

120 Vatican and Linnaeus: Hoyles 1994, p. 129.

120 "all answered with": Kalm 16 June 1748, Kalm 1892, p. 111.

120 "the Vanity of being": Miller to Alston, 4 February 1756, EUL Laing III 375, f. 24av.

120 "the most bitter abuse": Linnaeus (1753) 1957, vol. 1, p. 155.

120 Linnaeus's army of botanists: Afzelius 1826, pp. 95–6.

120 Linnaeus's self-confidence: Afzelius 1826, p. 93.

121 "be terrific": Linnaeus to Bäck, 28 June 1751, Fries et al. 1907–22, vol. 1, pt. 4, p. 154.

121 "masterpieces" and "jewels": and Linnaeus's own reviews, *Lärda Tidningar* and *Hamburgische Berichte* 1732–7, Bryk 1919; Blunt p. 180; Malmeström 1927, pp. 87–9.

121 "the greatest work": Afzelius 1826, p. 53.

121 "self-praise smells": Linnaeus to Archbishop Mennander, 1762, Malmeström 1927, p. 89.

121 "Such neatness!": Garden to Linnaeus, 15 March 1755, Smith 1821, vol. 1, pp. 284–5.

121 "attained to ye greatest": Bartram to Collinson, 11 June 1743, p. 216.

121 Jane Colden: Hindle 1956, pp. 42–3; Collinson to Bartram, 20 January 1756, p. 393.

121 "the favoured priest": Garden to Linnaeus, 12 April 1761, Smith, vol. 1, p. 304.

121 "papers of Some": Collinson to Franklin, 12 April 1747, Armstrong 2002, p. 138.

122 "there is no general": Bartram to Miller, 18 February 1759, Berkeley and Smith Berkeley 1992, p. 457.

122 "[P]ray run not": Collinson to Bartram, February 1755, p. 395.

122 "[w]hen the greatest lords": Kalm, 20 May 1748, Kalm 1892, p. 111.

122 old friends turned against Miller: Delany to Dewes, December 1754, Llanover 1861, p. 309; Miller to Veltheim, 24 July 1756, LHASA, MD, Rep. H Harbke 1856, f. 23; Knowlton to Brewer, 15 December 1741, Henrey 1985, p. 196.

122 "obliged to change": Ellis to Linnaeus, 1757, Smith 1821, vol. 1, p. 84.

122 classification of poison ivy: *Philosophical Transactions*, vol. 49, 1755–6, pp. 157–66 and vol. 50, 1757–8, pp. 430–40.

122 Miller embroiled in row with Ellis: Le Rougetel 1990, p. 46; Ellis to Linnaeus, 1756 or 1757, Smith 1821, vol. 1, p. 84.

122 "I shall be able": Ellis to Garden, 8 April 1761, Smith 1821, vol. 1, p. 507.

122 "those people that used": Ellis to Linnaeus, 31 May 1757, Smith 1821, vol. 1, p. 86.

123 "Now, my dear friend": Collinson to Linnaeus, 29 July 1755, Smith 1821, vol. 1, p. 37.

123 "after Linnaeus Method": Collinson to Bartram, 20 January 1756, p. 393.

123 "the only prize available": Linnaeus (1751) 2003, p. 182.

123 link between botanist and plant: Linnaeus (1737) 1938, p. 61.

124 "labour over acquiring": Linnaeus (1737) 1938, pp. 63–4; and *Milleria* in general, Le Rougetel 1990, p. 151.

124 "truly tenacious of his": Johann Beckmann in 1765, Beckmann 1911, p. 102.

124 "witness that you shone": Linnaeus to Rudbeck, 29 July 1731, quoted in Blunt 2001, p. 35.

124 "lowly, insignificant, disregarded": Linnaeus (1737) 1938, p. 64.

124 "natural abilityes": Johann Beckmann in 1765, Beckmann 1911, p. 102.

124 *"Ingratus cuculus"*: and row with Siegesbeck, Sten Carl Bielke to Linnaeus, 5 January 1745, Fries et al. 1907–22, vol. 1, pt. 3, p. 192.

124 "If someone said": and following quote, Linnaeus to Sten Carl Bielke, 24 April 1745, Fries et al. 1907–22, vol. 1, pt. 3, p. 194.

125 "a little mean-looking": and following quote about *Ellisia*, Ellis to Linnaeus, 21 September 1762, Smith 1821, vol. 1, p. 159.

125 *Bartramia*: Miller 1755–60, plate 298; Armstrong 2002, p. 113.

126 "giving Mee a species": Collinson to Linnaeus, 13 May 1739, Armstrong 2002, p. 72.

126 "for I had it In": Collinson to Bartram, 22 September 1739, p. 125.

126 "a pretty violent head-ach": Kalm, 15 October 1748, Kalm (1772) 1972, p. 106.

126 Collinson ordering *Collinsonia*: Collinson to Bartram, 16 September 1741, p. 169.

126 "retreated from the": Collinson to Gronovius, 24 March 1753, Armstrong 2002, pp. 166–7.

126 Collinson asked for Linnaeus's assistance: Gronovius to Linnaeus, 14 March 1737, The Linnaean Correspondence, linnaeus.c18.net, letter L0154.

126 "unless I had one of your": Collinson to Linnaeus, 12 May 1756, Armstrong 2002, p. 18.

126 Solander in Uppsala: Duyker and Tingbrand 1995, p. 1.

126 "greater progress in Natural": Testament for Daniel Solander from Linnaeus, 5 March 1759, Duyker and Tingbrand 1995, p. 43.

126 "I almost spend more": Solander to Erik Gustaf Lidbeck, 25 March 1757, Duyker and Tingbrand 1995, p. 40.

126 "that Herr Linnaeus has": Pehr Löfling to Solander, 14 May 1753, Duyker and Tingbrand 1995, p. 27.

128 Solander's appearance and character: Ellis to Linnaeus, 21 December 1762; 19 August 1768, Smith 1821, vol. 1, pp. 160, 232; Rauschenberg 1968, p. 13.

128 "a philosophical gossip": Frances Burney's father on Solander, Fanny Burney to Mr. Crisp, 22 January 1780, Barrett 1904, vol. 1, p. 318.

128 "Pray desire him to": Ellis to Linnaeus, 24 October 1758, Smith 1821, vol. 1, p. 108.

128 Lisa Stina and Solander: Rauschenberg 1968, p. 15; Fabricius (1780) 1928, no page numbers.

128 "as I would my own": Linnaeus to Ellis, 30 May 1759, Smith 1821, vol. 1, p. 124.

128 "found tranquil asylum": Linnaeus to Ellis, 30 May 1759, Smith 1821, vol. 1, p. 123.

128 "Is no certain Advice": Collinson to Linnaeus, 25 July 1759, Armstrong 2002, p. 218.

129 "not to let them hear": Solander to Linnaeus, 15 October 1759, Duyker and Tingbrand 1995, p. 44; see also other letters written from Skåne, pp. 60–93.

129 "my dear Solander": Linnaeus to Ellis, 29 April 1760, Smith 1821, vol. 1, p. 127.

129 Solander's arrival in London: Solander to Linnaeus, 1 July 1760, Duyker and Tingbrand 1995, p. 105.

129 a pilgrimage of botanists to Woodford: Ellis to Linnaeus, 21 July 1758, Smith 1821, vol. 1, p. 99.

129 James Gordon made £500: Ellis to Garden, 8 April 1761, Smith 1821, vol. 1, p. 507; Solander to Linnaeus, 31 October 1760, Duyker and Tingbrand 1995, p. 143.

129 "Every body is in love": Ellis to Garden, 8 April 1761, Smith 1821, vol. 1, p. 507.

129 specimen to Linnaeus: Linnaeus to Ellis, 29 September 1758, Smith 1821, vol. 1, p. 103.

129 called new plant *Warneria*: Ellis to Linnaeus, 21 July 1758, Smith 1821, vol. 1, pp. 99–100.

130 Warner refused name *Warneria*: Ellis to Linnaeus, 1 August 1758, Smith 1821, vol. 1, p. 101.

130 "may be a most useful": Ellis to Linnaeus, 31 May 1757, Smith 1821, vol. 1, p. 86 (this refers to another plant which Ellis had originally singled out as "*Gardenia*"); and Ellis suggested *Gardenia* in 1760: Ellis to Linnaeus, 13 June 1760, Smith 1821, vol. 1, p. 130.

130 "far more insight than": and following quote, Solander to Linnaeus, 1 July 1760, Duyker and Tingbrand 1995, p. 106.

130 "Let Mr. Ellis": Linnaeus to Solander, 23 July 1760, Duyker and Tingbrand 1995, p. 128.

130 Solander promises to send a living plant: Solander to Linnaeus, 21 July 1760, Duyker and Tingbrand 1995, p. 117.

130 Ellis begged, threatened and flattered: Ellis to Solander, 24 August 1760, Duyker and Tingbrand 1995, p. 140.

130 "I wish to guard": Linnaeus to Ellis, 11 August 1760, Smith 1821, vol. 1, p. 134.

130 Ellis's "botanical reputation": Solander to Linnaeus, 31 October 1760, Duyker and Tingbrand 1995, pp. 141–2.

131 "taken by some hasty": Miller 1768, "Jasminum" and "Gardenia."

131 "the light that will lead": Linnaeus's speech to the Swedish Royal
Society, 1759, quoted in Koerner 1999, p. 23.

Chapter 8: "The English are all, more or less, gardeners"

132 description of London: based on visitor accounts Pehr Kalm, 10–11 May
1748, Kalm 1892, pp. 33–5; Carl Philip Moritz , 2 June 1782, Nettel
1965, p. 29; de Saussure, 14 June 1726, van Muyden 1902, p. 138.
132 "the pleasant enjoyments": Pehr Kalm, 25 June 1748, Kalm 1892,
p. 85.
133 "I am in quite great": Linnaeus to Solander, 23 July 1760, Duyker
and Tingbrand 1995, p. 127.
133 "[m]ake yourself informed": Linnaeus to Solander, 1 January, 1762,
Duyker and Tingbrand 1995, p. 204.
133 "it will never be": Linnaeus to Ellis, 16 September 1761, Smith 1821,
vol. 1, p. 149.
133 1,000 cedars of Lebanon from a butcher: Collinson 8 June 1761,
Lambert 1811, vol. 10, p. 274.
133 "this extraordinary Multiplication": Miller 1759, preface, no page
numbers.
133 "Your country seems": Linnaeus to Ellis, 6 November 1759, Smith
1821, vol. 1, p. 126.
133 "fagg End of the World": Collinson to Linnaeus, 9 March 1744,
Armstrong 2002, p. 114.
134 species for Uppsala: Linnaeus to Solander, 1 January, 1762, Duyker
and Tingbrand 1995, p. 204.
134 James Gordon's nursery: and following description, Solander to
Linnaeus, 1 July 1760, Duyker and Tingbrand 1995, p. 106.
134 *Calycanthus floridus*: Miller 1759, "Basteria."
134 "Imagine if one could": Linnaeus to Solander, 1 January 1762, Duyker
and Tingbrand 1995, p. 204.
134 "[W]here in Sweden": Solander to Linnaeus, 1 July 1760, Duyker and
Tingbrand 1995, p. 106.
134 "gain the most": Solander to Linnaeus, 1 July 1760, Duyker and
Tingbrand 1995, p. 107.
134 Gordon's cultivation methods: John Hope in September 1766, Harvey
1981, p. 58.
135 "the dusty Seeds": Collinson's MS notes in Miller's *Dictionary*
(1756–9), NLW.
135 seed shop in Fenchurch Street: Collinson to Bartram, 10 June 1760,
p. 485.
135 nurseries and seedsmen: Harvey 1974, pp. 5–6.
135 Bartram's customers: "An Account of the Introduction of American
Seeds into Great Britain, 1766," Peter Collinson, NHM, MSS Col. 1.
135 selection rose from 200 to 600 seeds: this was the nurseryman John
Webb, who remained one of Bartram's most faithful customers, Laird
1999, p. 218.
135–6 Bartram's plants in the English garden: Laird 1999, pp. 221, 142–4,
158, 164.

136 "the English are all": D. Georg Leonhart Huth's preface to Miller 1750, vol. 1, no page numbers.

136 "der Englische Wald": Medicus 1791, pp. 3, 48; Veltheim to Miller, 24 March 1755, Köhler 2003, p. 16.

136 "a large Nursery of Trees": Miller to Veltheim, 4 May 1757, LHASA, MD, Rep. H Harbke 1856, f. 52.

136 "I have been so much": Miller to Alston, 24 November 1758, EUL, Laing III 375, f. 25a.

137 "Mr. Miller at Chelsea": Veltheim to Grafen Podewils, October 1759, LHASA, MD, Rep. H Harbke 1857, ff. 75–8.

137 "maybe too old": Veltheim to Hofsekretär von Florencourt in Braunschweig, 22 March 1764, LHASA, MD, Rep. H Harbke 1859, f. 157.

137 standardisation of seed boxes: Collinson to Bartram, 24 April 1751; Nathaniel Powell to Bartram, 14 August 1753; Bartram to Collinson 8 August 1763, Berkeley and Smith Berkeley 1992, pp. 323, 352, 603–4.

137 Bush attracted Miller's clients: Veltheim to Bush, 4 December and 12 December 1759, LHASA, MD, Rep. H Harbke 1857, ff.118–24, 132–5; Veltheim to Otto von Münchhausen, 26 February 1760, LHASA, MD, Rep. H Harbke 1857, ff. 170–72. Between 1760 and 1767 Bush ordered two to six boxes from Bartram every year, "An Account of the Introduction of American Seeds into Great Britain, 1766," Peter Collinson, NHM, MSS Col. 1.

137 Bush's three-guinea profit: Veltheim to Podewils, October 1759, LHASA, MD, Rep. H Harbke 1857, f. 75–8.

137 "whole forests from England": Veltheim to Ober-Forstmeister Hoym, 4 December, LHASA, MD, Rep. H Harbke 1857, f. 123.

137 Bartram's plant list: *Gentleman's Magazine*, vol. 24, February 1754, p. 65.

138 "for I Love to Ballance": Collinson to Bartram, 1 December 1751, p. 336.

138 "Careless person": Collinson to Bartram, 10 May 1763, p. 591.

138 "[T]his was not ye": Bartram to Collinson, 8 August 1763, p. 603.

138 "the fire of friendship": Collinson to Bartram, 6 December 1763, p. 615.

138 "Gordon & I keep": Collinson to Bartram, 13 February 1754, p. 371.

138 "for thy Amusement": Collinson to Bartram, 5 October 1762, p. 572.

138 "[M]y Garden ... is Enriched": Collinson to Linnaeus, 20 July 1759, Armstrong 2002, p. 217.

138–9 Collinson's move to Mill Hill: Lambert 1811, vol. 10, p. 282.

139 "[T]he Robin Red Breasts": Collinson to Bartram, 20 January 1756, p. 392.

139 daily coach service: Collinson and Michael Collinson to Franklin, 14 July 1767, Labaree 1759–, vol. 14, p. 214.

139 Collinson's garden at Mill Hill: Fox 1919, p. 172.

139 Tree of Heaven: Aiton 1810–13.

139 "a most beautiful": and *Arbutus andrachne*, John Hope, 13 September 1766, Harvey 1981, p. 59; Hall 1986, p. 22.

139 ff. Collinson's plants in Mill Hill: all plant descriptions are based on Collinson's own words. Sometimes I have used his descriptions from later letters, but only for plants which actually flowered in the summer of 1760, when Solander visited for the first time. Collinson to Gronovius,

5 June 1755; Collinson to Linnaeus, 12 May 1756 and 1 May 1765; Collinson to Bartram, 20 July 1756, 10 June 1760, 15 September 1760, 1 August 1761, 25 July 1762, Armstrong 2002, pp. 188, 197–8, 262; Berkeley and Smith Berkeley 1992, pp. 409, 485, 491–3, 531, 566; John Hope on 13 September 1766, Harvey 1981, p. 59; Ellis to Garden, 16 June 1756, Smith 1821, vol. 1, p. 381; Miller 1768, "Robinia hispida."

139 "I am Charm'd": Collinson to Bartram, 10 June 1760, p. 485.

139 "its fragrance is smelt": Collinson to Bartram, 25 July 1762, p. 566.

139 "a glorious show": Collinson to Bartram, 11 June 1762, p. 562.

140 "so different from all": Miller 1752, "Sarracenia."

140 "the Ravishing Sight": Collinson to Bartram, 11 June 1762, p. 562.

140 "Absent Friends": and following descriptions, Collinson to Colden, 25 February 1764, Armstrong 2002, p. 256.

141 seeds and cuttings for Linnaeus: Solander to Linnaeus, 23 July 1760, Duyker and Tingbrand 1995, p. 119.

141 "fast and indistinctly": Solander to Linnaeus, 1 July 1760, Duyker and Tingbrand 1995, p. 106.

141 "a pair of shoes": Ellis to Solander, 8 August 1760, Duyker and Tingbrand 1995, p. 136.

141 "[I]t is terribly expensive": Solander to Peter Jonas Bergius, 21 July 1760, Duyker and Tingbrand 1995, p. 124.

141 "fondness for me": and following quotes, Solander to Linnaeus, 31 October 1760, Duyker and Tingbrand 1995, pp. 144–5.

142 "the multitude of foreign": Solander to Linnaeus, 11 August 1761, Duyker and Tingbrand 1995, p. 169.

142 Goodwood: Richard Pococke, 23 September 1754, Cartrwright 1889, vol. 2, p. 111.

142 "I have a good": Solander to Linnaeus, 14 August 1761, Duyker and Tingbrand 1995, p. 169.

142 "become the absolute": Stroud 1975, p. 39.

142 "shrubberies planted of all": Walpole to Mann, 20 June 1760, Lewis 1937–61, vol. 21, p. 418.

142 Duke of Argyll and his guests: *The World*, 13 June 1754, no. 76, pp. 456–8.

142 footnote: Pehr Kalm, 29 May 1748, Kalm 1892, p. 59.

143 height of trees at Whitton: Joseph Spence's visit in 1760, Laird 1999, p. 86.

143 "another world": and Argyll promises plants for Linnaeus: Christlob Mylius on 28 October 1753, Mylius1787, p. 107; Solander to Linnaeus, 31 October 1760, Duyker and Tingbrand 1995, p. 143.

143 "peculiar manner": Solander to Linnaeus, 11 August 1761, Duyker and Tingbrand 1995, p. 170.

143 Solander's visits to Painshill: his last visit was in 1781 together with Linnaeus's son Carl.

143 dark "hanging" woods: Whately (1770) 1771, p. 186.

143 description of Painshill: based on following visitor accounts, Count Kielsmansegge in 1761, Kielsmansegge 1902, pp. 55–7; William Gilpin, "Mr. Hamilton's Gardens at Cobham in Surrey," 20 May 1765, Hodges 1973, pp. 48–9; Whately (1770) 1771, pp. 184–94; Sir John Parnell's Journals of 1763 and 1769, Sambrock 1980, pp. 91–106; Symes 1983, pp. 117–24 (plant lists).

143 Hamilton as Bartram's customer: Charles Hamilton had ordered the first box from Bartram in 1748, and had subscribed to the 1739 edition of Miller's *Dictionary*.

143 "the finest Persian": Sir John Parnell, 9 June 1769, Sambrock 1980, p. 99.

143 "contributed essentially to": Walpole (1780) 1993, vol. 2, p. 403.

144 "to adorn Pleasure-gardens": Miller 1752, "Bignonia."

144 "more common": Miller 1752, "Chionanthus."

144 "raised in great Plenty": Miller 1752, "Gleditsia triacanthos" and "Acer glauca."

144 "great Demand": Miller 1759, "Fraxinus americana."

144 flowering dogwood: Miller 1768, "Cornus florida."

144 "curious Persons in": Miller 1737, "Phaseoloides Carolinianum"; Miller 1752, "Glycine frutescens"; Miller 1759, "Glycine Carolina Kidney Bean Tree."

144 Capability Brown and American plants: Brown ordered from Bartram's best nurseryman customer John Williamson, Laird 1999, pp. 136, 277-8.

144 "a heavy-timbered American": Louis Simond in 1810, Laird 1998, p. 213.

145 "I met ... the famous": Ellis to Solander, 12 September 1762, Duyker and Tingbrand 1995, p. 242.

145 "people have taste in": Solander to Peter Jonas Bergius, 1 July 1760, Duyker and Tingbrand 1995, p. 124.

145 "same fickleness of": Solander to Linnaeus, 5 February 1761, Duyker and Tingbrand 1995, p. 160.

145 "Throw him where": Scott and Bottle 1930, vol. 14, p. 182.

145 "[T]ell them that everything": Solander to Linnaeus, 5 February 1761, Duyker and Tingbrand 1995, p. 160.

145 Solander decided to stay in London: Ellis to Garden, 8 April 1761, Smith 1821, vol. 1, p. 508.

145 "[O]nce one begins to": Solander to Linnaeus, 16 October 1759, Duyker and Tingbrand 1995, p. 47.

145 "I wish I had had": Linnaeus to Solander, 14 September 1761, Duyker and Tingbrand 1995, p. 178.

145 "Pray let me know": Linnaeus to Ellis, 16 September 1761, Smith 1821, vol. 1, p. 152.

145 "What sirens or": and following quotes: Linnaeus to Solander, 11 December 1761, Duyker and Tingbrand 1995, p. 197.

146 "advised to subjoin a short": Miller's *Gardeners Kalendar,* 1760, preface. James Lee's *Introduction to the Science of Botany* was published in the same year. With Solander in the country the Linnaean system continued to attract botanists and gardeners. The professor of botany at Cambridge, Thomas Martyn, for example, was the first to lecture about the system in 1762, and William Hudson's *Flora Anglica* (1762) was the first book to use both the classification and the binominal nomenclature.

146 "look at the future": and following quote, Linnaeus to Solander, 11 December 1761 and 5 March 1762, Duyker and Tingbrand 1995, pp. 197-8, 212.

146 "I have been longing": Linnaeus to Solander, 5 March 1762, Duyker and Tingbrand 1995, p. 212.

146 "[T]he more he is": and Solander's coloured waistcoats, Ellis to
Linnaeus, 21 December 1762, Smith 1821, vol. 1, p. 160; Rauschenberg
1968, p. 12.

146 "without letting it known": Ellis to Solander, 14 April 1762, Duyker
and Tingbrand 1995, p. 226.

147 Solander accepts St. Petersburg offer: Solander to Linnaeus,
13 May 1762, Duyker and Tingbrand 1995, p. 228; Linnaeus to
Ellis, 23 November 1762, Smith 1821, vol. 1, p. 158.

147 "[B]y your regard": Linnaeus to Ellis, 23 November 1762, Smith
1821, vol. 1, p. 158.

147 "Lost & Sunk into": and following quotes, Collinson to Linnaeus,
16 November 1762, Armstrong 2002, p. 243.

147 Solander promised to Linnaeus's daughter: Lisa Stina was promised to
marry Solander, and in January 1762 Linnaeus had signed off a letter
to Solander, "My wife, Yours and Carl [his son] . . . send their greetings." Linnaeus to Solander, 1 January 1762, Duyker and Tingbrand
1995, p. 204.

147–8 Collinson and British Museum: many of Collinson's acquaintances
were surprised that he was not made trustee of the Museum as he was
"one of the founder's most antient and intimate Friends," Fothergill
1770, p. 15.

148 "the Philosphick World": and following quote, Collinson to William
Watson, 5 October 1762, Armstrong 2002, p. 242.

148 "That is not Kind": Collinson to Linnaeus, 2 September 1762,
Armstrong 2002, p. 240.

148 "Must I not have": Collinson to Linnaeus, 17 September 1765,
Armstrong 2002, p. 266.

148 "our beloved Solander": Peter Collinson Commonplace Book, 16 June
1762, LS MS 323b, f. 37.

148 "a Chearful Glass": Collinson to Linnaeus, 2 September 1762,
Armstrong 2002, p. 239.

148 "most knowing naturalist": Peter Collinson Commonplace Book,
16 June 1762, LS MS 323b. f. 37.

148 "[A]fter consulting many": Ellis to Linnaeus, 21 December 1762,
Smith 1821, vol. 1, p. 159.

Chapter 9: "See what a complete empire we have now got within ourselves"

149 "[Y]e barbarous inhuman": Bartram to Collinson, 21 February 1756,
p. 400.

149 "the War which is": Miller to Veltheim, 24 July 1756, LHASA, MD,
Rep. H Harbke 1856, f. 23.

149 "Oh stupid obstinate": Bartram to Collinson, July 1755, p. 386.

150 "The Preservation and Enlargement": and following quote, Collinson
to Newcastle, 25 February 1756, Armstrong 2002, pp. 195–6.

150 "This is the old": Collinson to Colden, 19 May 1756, Armstrong
2002, p. 200.

150 "[T]his Cruel Warr": Collinson to Gronovius, 29 April 1757,
Armstrong 2002, p. 205.

150 ships captured: Miller to Veltheim, 4 May 1757, LHASA, MD, Rep.
   H Harbke 1856, f. 52; Garden to Ellis, 6 June, Smith 1821, vol. 1,
   p. 412.
151 "learned [rather] then": Bartram to Collinson, 27 April 1755, p. 385.
151 "Our affairs from": Bartram to Collinson, 11 November 1757,
   p. 431.
151 "nothing done to ye": Bartram to Franklin, 29 July 1757, Berkeley
   and Smith Berkeley 1992, p. 425.
151 "Wee have great": Collinson to Bartram, 10 March 1759, p. 459.
151 "Our bells are worn": Walpole to George Montagu, 21 October 1759,
   Lewis 1937-61, vol. 9, p. 251.
151 "unnatural boasting of": Solander to Linnaeus, 5 February 1761,
   Duyker and Tingbrand 1995, p. 160.
152 "Thou will be able": Collinson to Bartram, 10 March 1759, p. 459.
152 "heated by ye Botanick": Bartram to Templeman, 6 July 1761,
   Berkeley and Smith Berkeley 1992, p. 525.
152 "most barbarous creatures": Bartram to Collinson, 21 February 1756,
   p. 401.
152 "thy Penetrating Eye": Collinson to Bartram, 1 August 1761, p. 531.
152 double boxes: and following quote, Collinson to Bartram, 7 May
   1761, p. 513.
152 "seeds of allmost every": and following quote, Bartram to Collinson,
   14 August 1761, p. 534.
152 Bartram's injuries: Bartram to Collinson, 22 May and 14 August
   1761, pp. 517, 533.
152 "[C]limbing trees is over": and following quote, Bartram to Collinson,
   14 August 1761, p. 534.
153 Bartram's competition: Garden to Ellis, 25 July 1761, Smith 1821, vol.
   1, p. 512.
153 "Dear John don't be": Collinson to Bartram, 7 May, 12 June and
   1 August 1761, pp. 513, 521, 530.
153 Bartram's expedition: Bartram to William Bartram, 5 October 1761,
   Berkeley and Smith Berkeley 1992, p. 536.
153 "disrobed" Carolina evergreens: Bartram to Collinson, 10 May 1762,
   p. 558.
153 "flowering Shrubs that": Thomas Ord to Bartram, 10 October 1763,
   Berkeley and Smith Berkeley 1992, p. 610.
153 "Oh Carolina Carolina": Bartram to Collinson, 10 May 1762,
   p. 558.
153 Bartram's Carolina plants: see letters in particular in the later 1750s
   and early 1760s for Bartram's southern correspondents and his plant
   lists, also Bartram to Bouquet, 3 May 1762, Berkeley and Smith
   Berkeley 1992, pp. 555-7; Peter Collinson's Commonplace Book, LS
   323a; "An Account of the Introduction of American Seeds into Great
   Britain," 1766, Peter Collinson, NHM MSS Col. 1.
153 "happy Alacrity": Garden to Colden, 4 November 1754, Colden
   1920, vol. 4, p. 472.
153-4 "pass thro fire": Bartram to Collinson, 30 May 1763, p. 594;
   Martha Logan was the daughter of Robert Daniels, the deputy
   Governor. She regularly advertised her plants and seeds in the *South
   Carolina Gazette*.

154 "I would have all: Bartram to Collinson, 29 August 1762, p. 570.

154 "life or limb": Bartram to Collinson, 8 November 1761, p. 538.

154 William Bartram: in 1761 William Bartram left Philadelphia and moved to Carolina to set up as a merchant, but he returned home in June 1762. William Bartram to John Bartram, 20 May 1761, Berkeley and Smith Berkeley 1992, p. 515.

154 "tipitiwitchet sensitive": Bartram to Collinson, late summer 1762 and 29 August 1762, pp. 565, 570.

154 "burst with laughing": and following quote, Bartram to Collinson, 29 August 1762, p. 570.

155 "I had ye most prosperous": and description of expedition 1762, Bartram to Moses and William Bartram, 9 November 1762; Bartram to Eliot, 1 December 1762, Berkeley and Smith Berkeley 1992, pp. 574, 577.

155 "While the French": Collinson to Bartram, 10 December 1762, p. 580.

155 "Terrestrial Paradise": Collinson to Bartram, 23 February 1763, p. 586.

156 vessel was returned to the British: Collinson to Bartram, 11 March 1763, p. 587.

156 "Jewel" [Venus flytrap]: and following quotes, Collinson to Bartram, 30 June 1763, p. 600.

156 "for his skills Exceeds": Collinson to Bartram, 6 December 1763, p. 616.

156 "It is now in vain": Collinson to Bartram, 30 June 1764, p. 633.

156 "Now my dear John": Collinson to Bartram, 10 December 1762, Armstrong 2002, p. 245.

157 "[D]oes not the ardour": Collinson to Bartram, 7 April 1763, p. 588.

157 "very palace garden": Bartram to Collinson, 11 November 1763, p. 614.

157 "when peace is proclaimed": Bartram to Solander, 26 April 1763, Berkeley and Smith Berkeley 1992, p. 589.

157 "find more curiosities": Bartram to Collinson, 11 November 1763, p. 614.

157 *Magnolia grandiflora*: Collinson to Bartram, 15 September 1760, p. 492.

157 "the most beautiful": Miller 1731, "Larger Laurel leave'd Tulip Tree" [*Magnolia grandiflora*].

157 "it stares mee in": Collinson to Bartram, 12 June 1761, p. 521.

157 "My Dear John": and following quote, Collinson to Bartram, 5 October 1762, pp. 571–2.

158 Bartram's success: Bartram to Garden, 14 March 1756, and William Bartram to John Bartram, 18 May 1761, Berkeley and Smith Berkeley 1992, pp. 402, 515; Pehr Kalm, 26 September 1748, Kalm (1772) 1972, p. 67.

158–9 Bartram's house: the original four-room house had been completed by 1731, but Bartram had continued to alter it as his family and wealth grew. During the 1740s he added the kitchen to the house and in the mid-1750s he began to work on the major extensions, including the columns, his study and a new roof (Jacobs 2005, pp.1–12); the Library Company held, for example, James Gibbs's *A Book of Architecture* (1728) and Isaac Ware's translation of Andrea Palladio's *The Four Books of Palladio's Architecture* (1738).

159 his study, or "Chapel": description of study and treatment of visitors, Bartram to Collinson, 27 April 1755; Collinson to Bartram, 6 April 1759, Bartram to Franklin, 24 November 1770, Berkeley and Smith Berkeley 1992, pp. 381, 462, 735; Crèvecoeur (1782) 1998, p. 177.

159 "so that when my friends": Bartram to Sloane, 16 November 1743, Berkeley and Smith Berkeley 1992, p. 225.

159 "My garden now makes": and *Bignonia capreolata*, Bartram to Collinson, 30 May 1763, p. 594.

159 Portugal laurel: Joseph Paxton, for example, used it at Chatsworth, *Gardener's Magazine*, vol. 15, 1839, p. 451.

159 "take ten guineas": Bartram to Collinson, 20 February 1760, p. 481.

159 "one of the finest": Collinson to Bartram, 1 January 1764, p. 618.

159 "iron colored": Bartram to Collinson, 10 May 1762, p. 559.

159 European plants in Bartram's garden: Bartram to Collinson, 24 June 1760; Collinson to Bartram, 2 October 1760, pp. 486, 496.

159–60 "ye brightest colors": Bartram to Collinson, 19 July 1761, p. 529.

160 linaria becomes weed: Bartram to Collinson, early 1759, pp. 451–2.

160 "was much admired": Bartram to Collinson, 30 May 1763, p. 594.

160 "A gigantick monster": Bartram to Collinson, 20 July 1760, p. 488; Bartram had already experimented with hybrids in 1739, Bartram to Boyd, summer 1739, Berkeley and Smith Berkeley 1992, p. 120.

160 "Pennsylvania fireplace": during the archaeological excavations at Bartram's garden in 1980 the ornamented front plate of the stove which Bartram had purchased in 1760 was found (Fry 1998, p. 6).

160 "[T]hese I am proud": Bartram to Collinson, 8 August 1763, p. 604.

160 "I can chalenge any garden": and for plants in Bartram's garden, Bartram to Collinson, 19 July 1761, p. 529; Bartram to Collinson, 23 September 1764, 8 November 1761, 22 May 1761; Miller to Bartram, 10 November 1759; Gordon to Bartram, 28 August 1763, Berkeley and Smith Berkeley 1992, pp. 638, 538, 517, 477, 607.

161 Bartram's American customers: Martha Logan to Bartram, 20 February 1761, 20 December 1760; Sir John St. Clair to Bartram, 27 February 1761, Berkeley and Smith Berkeley 1992, pp. 500, 509, 506.

161 *Callistephus chinensis*: Collinson to Bartram, 12 February 1735, p. 20.

161 "have my mony": Bartram to Collinson, late summer 1762, p. 565.

161 "emerging out of their": Collinson to Templeman, 24 June 1762, Armstrong 2002, p. 236.

161 "I think powel": Bartram to Collinson, 27 April 1755, p. 383.

161 "that you send us": and following quote, Bartram to Collinson, 22 May 1761, p. 517.

161 Plants from England: Gordon to Bartram, 28 August 1763, Berkeley and Smith Berkeley 1992, p. 607.

161 will not keep to: and following quote, Bartram to Collinson, 23 October 1763, p. 612.

162 "Wee are the Agressors": and Collinson on Indians, Collinson to Bartram, 6 December 1763; Collinson to Bartram, 1 January 1764, pp. 615, 618; Collinson was passionate about the unfair treatment of the Native Americans and published some articles and a proposal for a solution of the conflicts in the *Gentlemen's Magazine* in October 1763.

162 "we have more reason": Bartram to Collinson, 4 March 1764, p. 622.
162 "as a conquer'd People": Franklin to Governor William Shirley, 4 December 1754, Labaree 1959–, vol. 5, p. 445.
162 "give us Members": Franklin to Richard Jackson, 1 May 1764, Labaree 1959–, vol. 11, p. 186.
162 "We are in your Hands": Franklin to Collinson, 30 April 1764, Labaree 1959–, vol. 11, p. 181.
163 "Admitting we are": John Adams (anonymously as Humphry Plough-flogger), Boston Gazette, August 1765.
163 "the importance of": Benjamin Franklin, "On the Propriety of Taxing America," The London Chronicle, 9–11 April 1767, Labaree 1959–, vol. 14, p. 116.
163 "sudden preferment": and following quote Bartram to Collinson, 23 September 1764, pp. 638–9.
163 "Neglect" and "I wish some Notice may": Franklin to Collinson, 24 September 1764, Labaree 1959–, vol. 11, pp. 352–3.
163 "I hope thee will": Bartram to Collinson, 15 October 1764, p. 641.
163 "Our Friend Peter": Franklin to Bartram, 14 February 1765, Berkeley and Smith Berkeley 1992, pp. 642–3.
163 Bartram as King's Botanist: Collinson to Bartram, 9 April 1765, p. 644.
164 "Old-England": Examination of Dr. Benjamin Franklin in the House of Commons, January 1766, The Parliamentary History of England (1765–71) 1813, p. 141.
164 "Sons of Liberties": McDougall 2005, pp. 217–18.
164 "die is thrown for": Garden to Ellis, 16 December 1765, Smith 1821, vol. 1, pp. 543–4.
164 "riotous Mobs": and following quote: Collinson to Bartram, 28 December 1765, p. 656.
164 "all commerce that": examination of Dr. Benjamin Franklin in the House of Commons, January 1766, The Parliamentary History of England (1765–71) 1813, p. 147.
165 "all Animosities will": Collinson to Colden, 20 March 1766, Armstrong 2002, p. 268.
165 Bartram left St. Augustine: Harper 1942, p. 49.
165 "I have brought home": Bartram to Collinson, June 1766, pp. 668–9.
165 Bartram's plants from Florida expedition: Bartram on 1, 3, 6 October 1765, 21 November 1765, Harper 1942, pp. 32, 31, 35, 83.
165 thieves in Collinson's garden: Collinson to Bartram, 5 October 1762, 28 December 1765, 28 May 1766, pp. 571, 657, 667.
165 "my most Curious": and following quote, Collinson to Bartram, 28 December 1765, p. 657.
165 "next in Beauty": Collinson's MS notes in Miller's Dictionary (1756–9), NLW.
166 Parliamentary Act: Collinson to Bartram, 21 August 1766, p. 674; "Act for Encouraging the Cultivation, and for the better Preservation of Trees, Roots, Plants and Shrubs," PA, HL/PO/PU/1/1766/6 G3n178.
166 "I am Ruined": Collinson to Bartram, 21 August 1766, p. 674.
166 Collinson's will: 9 December 1766, TNA PROB 11/941.
166 "I cut the other shin": and Bartram's other illness, Bartram to Collinson, 18 December 1766, p. 678; Bartram to Collinson, June 1766, p. 669.

166 "Boxes of pills": and Collinson's ailments: Collinson to Bartram, 21 August 1766; Bartram to Collinson, 18 December 1766; Collinson to Bartram, 19 September 1765, p. 655; Collinson to Bartram, 3 February 1767, pp. 674, 678, 655, 679.

166 "Our good frd B": Collinson to Bartram, May 1765, p. 648.

166 "struts along ye": Bartram to Collinson, 5 December 1766, p. 677.

166 "scrawl[ed] on": Bartram to Collinson, 10 April 1767, p. 684.

166 "Think how happy": and following quote, Collinson to Bartram, 31 July 1767, pp. 684, 686.

167 "I should be ashamed": Collinson to Bartram, 29 February 1768, p. 699.

167 "extremely beautiful": Miller 1768, "Dianthus."

167 British Museum: Solander to the Trustees of the British Museum, 13 February 1765, 22 February 1765, 29 June 1765, Duyker and Tingbrand 1995, pp. 262, 264, 265.

167 "O, Botany Delightfullest": and out-of-season fruit and vegetables, Collinson to Bartram, 30 June 1763; Collinson to Bartram, 25 December 1767, pp. 600, 694.

167 "so great is the": and *Gardenia*, Collinson to Bartram, 25 December 1767, pp. 693–4.

168 "Double Box": and following quote, Collinson to Bartram, 4 August 1763, pp. 602–3.

168 Collinson's thoughts on Thorndon: based on his notes after a visit in August 1762, Peter Collinson's Commonplace Book, LS MS 323b, f. 208.

168 "seized with a total": Fothergill 1770, p. 17.

168 "I have not any so": Bartram to Franklin, 5 November 1768, Berkeley and Smith Berkeley 1992, p. 708.

168 "I am at A loss": Bartram to Franklin, 10 April 1769, Berkeley and Smith Berkeley 1992, p. 709.

168 "without stooping to": Bartram to Fothergill, 26 November 1769, Berkeley and Smith Berkeley 1992, p. 725.

168 "[M]y travails as": Bartram to Franklin, 10 April 1769, Berkeley and Smith Berkeley 1992, p. 709.

169 "utterly in ye dark": and Bartram's worries, Bartram to Fothergill, 30 September 1770, 11 October 1770, Berkeley and Smith Berkeley 1992, p. 734.

169 "I know not whither": and Bartram's worries, Bartram to Fothergill, 30 September 1770, Bartram to Franklin, 24 November 1770, Berkeley and Smith Berkeley 1992, pp. 734–5.

169 Bartram's retirement: Bartram to Franklin, 29 April 1771, Berkeley and Smith Berkeley 1992, p. 739.

169 William Bartram's return: Slaughter 1997, p. 223.

169 Bartram's letter to his children: "To his Children," John Bartram, 1777, Berkeley and Smith Berkeley 1992, pp.772–5.

Chapter 10: "Ye who o'er Southern Oceans wander"

173 trustees grant leave: Solander to Trustees of British Museum, 24 June 1768, Duyker and Tingbrand 1995, pp. 268–70.

174 "in a most violent": John Thomas Stanley, 20 February 1790, Adeane 1899, p. 94; Banks to Hasted, February 1782, DTC, vol. 2, f. 99.

174 Banks at Chelsea: Carter 1988, p. 25; Banks also gave Philip Miller seeds and specimens after his voyage to Newfoundland, Garden Committee Minutes, 21 May 1767, WSA, MS 8228/1, f. 102.

174 Banks at the British Museum: Banks received his first reader ticket in August 1764, "Admissions to the British Museum Reading Room, 1762–1781," BL Add 45869, f. 14.

174 "esteemed as a savant": Fabricius on Solander in 1768, Fabricius (1780) 1928, p. 97.

174 Banks wanting to visit Linnaeus: Thomas Pennant to Banks, 10 June 1768, DTC, vol. 1, f. 8.

175 "temper": and "Secret Instructions issued to Cook 30 July 1768," TNA ADM 2/1132; for instructions in general: Williams 1998, pp. 3–18.

175 "Every blockhead does": Banks quoted in Chambers 2007a, p. 6.

175 Banks's £10,000 bill: Banks to Perrin, 16 August 1768, Chambers 2000, p. 2; Ellis to Linnaeus, 19 August 1768, Smith 1821, vol. 1, p. 231.

175 Banks's annual income: Carter 1988, p. 536.

175 "I am a man of": Banks to Perrin, 1 December 1768, Chambers 2000, p. 8.

175 "excellent opportunity": and following quote, Banks to Johan Alströmer, 16 November 1784, Chambers 2000, p. 78.

176 Endeavour: details of the refitting and construction of the Endeavour: Ship Plans Endeavour, 16 April and July 1768, NMM, ZAZ6588, ZAZ6590 and ZAZ6593; Marquardt 2001.

177 Endeavour at Plymouth: the dates and chronology of events throughout the Endeavour voyage are, unless otherwise stated, based on Cook's and Banks's daily journals.

177 Cook's express letter: Cook, 14 August 1768, Beaglehole 1955–74, vol. 1, p. 2.

177 "to expect he would": Daines Barrington to Thomas Pennant, 24 August 1771, Beaglehole 1962, p. 55; Carter 1988, p. 67.

177 Banks on matrimony: Banks to Perrin, 28 February 1768, Derbyshire Record Office, reference at the Sir Joseph Banks Archive Project, NHM.

177 "desperately in love": Horace Benedict de Saussure, 15 August 1768, and the description of Harriet and Banks on 15 August, Freshfield 1920, pp. 105–6.

177 "[t]he Ceremony of taking": Banks to George Leonard Staunton, 18 August 1792, Chambers 2000, p. 145.

177–8 provisions of the Endeavour: Aughton 2003, p. 10.

178 "almost frighten[s] me": Banks to Perrin, 16 August 1768, Chambers 2000, p. 1; for Banks's provisions, Ellis to Linnaeus, 19 August 1768, Smith 1821, vol. 1, p. 231.

178 "gentlemens Cabbins": Cook, 17 August 1768, Beaglehole 1955–74, vol. 1, p. 2.

178 "museum people": Solander to Ellis, 25 August 1768, Duyker and Tingbrand 1995, p. 272.

179 "Gentlemen of Fortune": Lloyd's Evening Post, 20 June and 3–5 August 1768; The Public Advertiser, 20 June 1768.

179 transit of Venus: and international collaborations, McClellan 1985, pp. 215–16.

179 "Solander . . . assured me": Ellis to Linnaeus, 19 August 1768, Smith 1821, vol. 1, p. 231.

179 "the laziest of mortals": Thomas Pennant to Banks, 3 July 1768, Chambers 2007b, vol. 1, p. 41.

179 "[n]o people ever": Ellis to Linnaeus, 19 August 1768, Smith 1821, vol. 1, p. 231.

179 "begs his kind": Ellis to Linnaeus, 1 November 1768, Smith 1821, vol. 1, p. 236.

180 Banks kept healthy: Banks, 27 September, 24, 25 October 1768, 1 April 1769, Beaglehole 1962, vol. 1, pp. 167, 176–8, 176, 243–4.

180 "I promisd myself": Banks to Morton, 1 December 1768, Chambers 2000, p. 10.

180 "impossible that the King": Banks to Perrin, 1 December 1768, Chambers 2000, p. 7; Viceroy of Brazil to Banks, 20 November 1768, BL Add 34744, f. 44.

180 "I am a Gentleman": Banks to Viceroy of Brazil, 17 November 1768, BL Add 34744, f. 41

180 "muster master of": Solander to Linnaeus, 1 December 1768, Duyker and Tingbrand 1995, p. 281.

180 "cursd, swore, ravd": Banks to Perrin, 1 December 1768, Chambers 2000, p. 7.

181 "French man laying": Banks to Perrin, 1 December 1768, Chambers 2000, p. 7.

181 "we hardly cou[l]d": Solander to Lord Morton, 1 December 1768, Duyker and Tingbrand 1995, p. 278.

181 "as greens and sallading": Solander to Ellis, 1 December 1768, Duyker and Tingbrand 1995, p. 274.

181 abseiled from a cabin: and plants collected by Banks and Solander, Parkinson 1784, p. 4; Stearn 1969, p. 69.

181 woollen clothes for sailors: and Banks's own clothes, Cook, 6 January 1769, Beaglehole 1955–74, p. 39; Banks, 25 October 17686, January 1769, Beaglehole 1962, vol.1, pp. 178–212.

181 "a very disagreeable": and following quote, Banks, 6 January 1769, Beaglehole 1962, vol. 1, p. 212.

181 "so lively and at": Banks, 7 January 1769, Beaglehole 1962, vol. 1, p. 213.

182 "all hands g[o]t abominably": Banks, 25 December 1768, Beaglehole 1962, vol. 1, p. 207.

182 disaster at Tierra del Fuego: Banks, 16 January 1769, Beaglehole 1962, vol. 1, pp. 218–22.

182 "No botanist has ever": Banks, 20 January 1769, Beaglehole 1962, vol. 1, p. 226.

182 ff. Tahiti: description of their time there is based on Banks, Beaglehole 1962, vol. 1, pp. 251–386.

183 "the truest picture": Banks, 13 April 1769, Beaglehole 1962, vol. 1, p. 252.

183 "keep him as a curiosity": Banks, 12 July 1769, Beaglehole 1962, vol. 1, pp. 312–13.

183 "a very pretty girl": Banks, 14 April 1769, Beaglehole 1962, vol. 1, p. 255.

183 "of putting their politeness": Banks, 14 April 1769, Beaglehole 1962, vol. 1, p. 254.

183 "of entertaining my friends": and following quote, Banks, 17 April 1769, Beaglehole 1962, vol. 1, p. 258.

183 Banks and Solander as traders: Banks, 3 May 1769, Beaglehole 1962, vol. 1, p. 270.

183 "Tåpáne and "Toråno": Banks, 10 May 1769, Beaglehole 1962, vol. 1, p. 275.

183 "unveiling all her charms": Banks, 12 May 1769, Beaglehole 1962, vol. 1, p. 275.

183 "superior to any thing": and description of Tahitian women, "Banks, Manners & Customs of S. Sea Islands," Beaglehole 1962, vol. 1, p. 334.

184 "a turban of": and Banks's clothes in Tahiti, Banks, 3 June 1769, Beaglehole 1962, vol. 1, p. 284; Parkinson 1784, p. 31.

184 "exposing their nakedness": and description of customs, Parkinson 1784, p. 33; Banks, 22 April, 5 July 1769, Beaglehole 1962, vol. 1, pp. 261, 309.

184 "a most excellent dish": Banks, 20 June 1769, Beaglehole 1962, vol. 1, p. 293.

184 "strange diversion": Banks, 29 May 1769, Beaglehole 1962, vol. 1, p. 283.

184 "no pretensions to be": Banks, 10 June 1769, Beaglehole 1962, vol. 1, p. 289.

185 "Vernerial distemper": and following quote, Cook, 6 June 1769, Beaglehole 1955–74, pp. 98–9.

185 "had disposd of themselves": Banks, 28 May 1769, Beaglehole 1962, vol. 1, p. 282.

185 300 species in herbarium: Carter 1988, p. 82.

185 first footnote: Porter 1999, p. 301.

186 "magical spying glass": and the following description, Banks, 3 October 1769, Beaglehole 1962, vol. 1, p. 396.

186 "is by much the best": and following quote, and Banks on food on the *Endeavour*, Banks, 23 September 1769, Beaglehole 1962, vol. 1, pp. 393–4.

187 "great diligence in": Garden Committee Minutes, 30 April 1750, WSA, MS 8228/1, ff. 42–3.

187 new committee's demands: Garden Committee Minutes, 17 August 1769, WSA MS 8228/1, f. 119; 25 August 1769, MS 8228/2, ff. 1–2.

187 "several Donors had": and Miller's complaints, Garden Committee Minutes, 26 January 1770, 29 August, 22 September, 6 October 1769, WSA, MS 8228/2, ff. 12, 3, 5.

187 numerical botanical catalogue: Alchorne, Stanesby, Index Horti Chelseiani, 1772–3, CPG MS. The ordering of index sticks began in October 1769, Garden Committee Minutes, 6 October 1769, 16 February 1770, WSA, MS 8228/2 ff. 5, 14.

188 "future disappointment": Garden Committee Minutes, 27 April, 16 March 1770, WSA, MS 8228/2, ff. 43, 35.

188 landing at Botany Bay: Banks, 28 April 1770, Beaglehole 1962, vol. 2, pp. 54ff.

188 "rank cowards": Banks, 4 May, 28 April 1770, Beaglehole 1962, vol. 2, pp. 59, 54.

188–9 plants at Botany Bay: Nelson 1983, p. 349; Stearn 1969, pp. 69, 79–84; Aiton 1810–13.

189 "of an animal clawd": and expeditions at Botany Bay, Banks, 1 May 1770, Beaglehole 1962, vol. 2, p. 57, Parkinson 1784, "New Holland," p. 136.

189 "Our collection of Plants": Banks, 3 May 1770, Beaglehole 1962, vol. 2, p. 58.

189 plants collected in New Zealand: Stearn 1969, p. 77.

190 reef accident: Banks, 10–12 June 1770, Beaglehole 1962, vol. 2, pp. 77–81; Cook, 11–12 June 1770, Beaglehole 1955–74, vol. 1, pp. 343–7.

190 "[F]ear of Death": Banks, 11 June 1770, Beaglehole 1962, vol. 2, p. 79.

191 committee's problems with Miller: Garden Committee Minutes, 15 June 1770, WSA, MS 8228/2, ff. 51–2.

191 "he would give": and Miller's excuses, Garden Committee Minutes, 8 June 1770, WSA, MS 8228/2, ff. 48–9.

191 "if he values his": Court Book, 15 June 1770, WSA, quoted in Meynell 1987, p. 80.

191 "the usual rudeness": Garden Committee Minutes, 28 September 1770, WSA, MS 8228/2, f. 57.

191 "all the plants from": Garden Committee Minutes, 12 October 1770, WSA, MS 8228/1, f. 118.

191 "refractory" behaviour: and taking case to the Court, Garden Committee Minutes, 15 June and 31 August 1770, WSA, MS 8228/2, ff. 51, 56.

191 "without any direction": and following quote, Garden Committee Minutes, 26 October 1770, WSA, MS 8228/2, ff. 76–7.

191 "according to his": and felling of plane tree, Garden Committee Minutes, 30 November, 24 November 1770, WSA, MS 8228/2, ff. 86, 85.

191 footnote: Garden Committee Minutes, 28 August 1770, WSA, MS 8228/1, f. 116.

192 ground down planks: Banks, 14 November 1770, Beaglehole 1962, vol. 2, p. 192.

192 malaria grips the *Endeavour* crew: Banks, 28 October 1770 and entries for November 1770, Beaglehole 1962, vol. 2, pp. 189ff.

192 "obstinacy and impertinence": and Miller's resignation, Ellis to Linnaeus, 28 December 1770, Smith 1821, vol. 1, p. 255; Garden Committee Minutes, 30 November 1770, WSA, MS 8228/2, f. 86.

192 *Endeavour* seeds at Chelsea: Garden Committee Minutes, 20 July 1771, WSA, MS 8228/2, f. 109.

192 "I may flatter myself": Banks to Hasted, February 1782, DTC, vol. 2, f. 97.

Chapter 11: "An Academy of Natural History"

193 "It is his discoveries": and other references, *London Chronicle*, 20 July 1771; *London Evening Post*, 20 July 1771.

193 "BANKSIA, from its": Linnaeus to Ellis, 20 December 1771, Smith 1821, vol. 1, p. 275.

193 "I am Mad, Mad": Banks to Thomas Pennant, 13 July 1771, Chambers 2000, p. 14.

193 Banks and Solander back in London: Lady Mary Coke, 9 August 1771, Coke 1970, vol. 3, p. 435; Franklin to Jonathan Shipley, 19 August 1771, Labaree 1959–, vol. 18, p. 209; *Gentleman's Magazine*, 10 August 1771, vol. 41, p. 375.

193 "the many narrow": Ellis to Linnaeus, 16 July 1771, Smith 1821, vol. 1, p. 263.

194 "they lik'd it": Franklin to Jonathan Shipley, 19 August 1771, Labaree 1959–, vol. 18, p. 209.

194 "too volatile a temper": Daines Barrington to Thomas Pennant, 24 August 1771, Beaglehole 1962, vol. 1, p. 55.

194 Banks as scientific libertine: "Histories of the Tête-à-Tête annexed; or, Memoirs of The Circumnavigator and Miss B–n," in *Town and Country Magazine*, September 1773, p. 458; Lady Mary Coke, 11 August 1771, Coke 1970, vol. 3, p. 437.

194 "most distressing": Daines Barrington to Thomas Pennant, 24 August 1771, Beaglehole 1962, vol. 1, p. 55.

194 Banks and Solander met King: Carter 1988, p. 96; *Gentlemen's Magazine*, 10 August 1771, vol. 41, p. 375.

194 "[v]ast numbers of": and information on Banks's and Solander's return, *Gentleman's Magazine*, 10 August 1771, vol. 41, p. 375; *Middlesex Journal, or Chronicle of Liberty*, 16 July 1771; *General Evening Post*, 13–16 July 1771.

194 "see fruits of his": and invites to nurserymen and gardeners, Mary Delaney, 1771, quoted in Hayden 1980, p. 114; Solander to William Forsyth, 23 December 1771, Duyker and Tingbrand 1995, p. 292.

194 *Endeavour* seeds and bulbs germinated: Lady Mary Coke, 26 August 1771, Coke 1970, vol. 3, p. 443.

195 30,000 dried specimens: Carter 1988, p. 95.

195 incident at Endeavour River: Banks, 26 June 1770, Beaglehole 1962, vol. 2, p. 85.

195 "laden with the greatest": Ellis to Linnaeus, 16 July 1771, Smith 1821, vol. 1, p. 263.

195 "to see this great": Linnaeus to Ellis, 8 August 1771, Smith 1821, vol. 1, p. 264.

195 "Surely none but": Linnaeus to Ellis, 8 August 1771, Smith 1821, vol. 1, p. 266.

195 "forty dozen of islands": and following quote, Walpole to Lady Mary Coke, 22 August 1771; Walpole to Horace Mann, 20 September 1772, Lewis 1937–61, vol. 31, p. 157, vol. 23, p. 436.

196 "the learned world": and following quote, Linnaeus to Ellis, 22 October 1771, Smith 1821, vol. 1, pp. 267–8.

196 "[T]hey have been": Ellis to Linnaeus, 19 November 1771, Smith 1821, vol. 1, p. 272.

196 "to persuade Solander": Linnaeus to Ellis, 20 December, 1771, Smith 1821, vol. 1, p. 273.

196 "If your intercession": Linnaeus to Ellis, 20 January 1772, Smith 1821, vol. 1, p. 280.

196 Banks and the *Resolution*: Beaglehole 1962, vol. 2, Appendix V, "Banks" and Cook's Second Voyage, Correspondence Banks and Sandwich and miscellaneous Memoranda," pp. 335–55; Cook to Banks, 2 June 1772, DTC, vol. 1, f. 35; Carter 1988, p. 101; Banks to Lord Sandwich, 30 May 1772, Chambers 2000, pp. 25–9.

196 "He swore & stamp'd": Elliott, John, "Memoirs of the Early Life of John Elliott, of Elliott House, Esq. And Lieut of the Royal Navy," 1772, BL Add 42714, f. 10v.

197 "of the whole world": Porter 2001, p. 297.

197 "an elephant, quite": James Boswell, quoted in Carter 1988, p. 124.

197 "a kind of superintendence": and Banks at Kew, Banks to the Spanish Ambassador, 1796, quoted in Desmond 1994, p. 106; Desmond 1995, p. 92.

198 Cook's plants in Banks's collection: Solander to Banks, 22 August and 5 September 1775, Duyker and Tingbrand 1995, pp. 356, 359.

198 plants crossing the globe: David van Royen to Banks, 4 January 1779, BL Add 8096, f. 162; Jan Deutz to Banks, 4 February 1776, BL Add 8094, ff. 60–61 and 14 December 1777, BL Add 8094, ff. 144–5; Henry de Ponthieu from 1779 to 1791, Dawson 1958, pp. 679–80; and for Antipodean requests also Peter Simon Pallas to Banks, March 1779, BL Add 8094, ff. 237–8.

198 Banks's contacts: Robert Brooke to Banks, 17 June 1787, Kew JBK/1/4: Banks's Letters, vol. 1, f. 275; Julius Philipp Benjamin von Rohr, Dawson 1958, p. 709; Peter Simon Pallas, Dawson 1958, pp. 644–6.

198 "the first Patron of": Lord Bute to Banks, 1 January 1774, DTC, vol. 1, f. 67.

198 "living Toad of about": and mermaids, Thomas Walker to Banks, 19 November 1780, DTC, vol. 1, f. 313; John Sinclair to Banks, 8 July 1809, DTC, vol. 17, ff. 318–24.

199 "under one of the": Banks and Solander to Blagden, 21 August 1773, Duyker and Tingbrand, 1995, p. 319.

199 "heat chamber": Carter 1988, p. 127.

199 Banks advised fellow thinkers: Rev. Daniel Lysons to Banks, 7 September 1772, DTC, vol. 8, ff. 75–6; James Bruce to Banks, 11 January 1774, DTC, vol. 1, f. 67.

199 Banks, a field botanist: George Colman, July 1775, quoted in Carter 1988, p. 131.

199 "a Newfoundland dog": Sarah Martha Holroyd to Lady Maria Josepha Stanley, 16 July 1800, Adeane 1899, p. 200.

199 "carried before a justice": Banks to Ferryman, 2 February 1788, DTC, vol. 6, f. 6.

199 "highway robber": "Histories of the Tête-à-Tête annexed; or, Memoirs of The Circumnavigator and Miss B–n," in *Town and Country Magazine*, September 1773, p. 458.

199 "the Purple Emperor": Banks to Ferryman, 2 February 1788, DTC, vol. 6, f. 6.

199 "Most think its": Banks to Ferryman, 2 February 1788, DTC, vol. 6, f. 6.

200 "turn'd science into": George Colman, July 1775, quoted in Carter 1988, p. 131.

200 Banks and Royal Society: when John Pringle, the previous president, had resigned, Banks had been in the country but Solander had immediately dashed off a letter, insisting "all look to you," Solander to Banks, January 1777, DTC, vol. 1, f. 140.

200 meetings and breakfasts at Soho Square: John Thomas Stanley to Dr. Scot, 21 February 1790, Adeane 1899, p. 93; Joseph Farington, 5 December 1794 and 23 November 1807, Cave et al. 1978–84, vol. 1, p. 189, vol. 8, p. 3147; Carter 1988, p. 379.

201 "perfectly free from": and description of library: Simond 1817, vol. 1, p. 39; Carter 1988, pp. 279–80; Banks had commissioned Sir Nathaniel Dance-Holland to paint Cook's portrait in 1776.

201 "cheerful and talkative": John Cullum to George Ashby, 22 May
1779, quoted in Carter 1988, p. 154.
201 "his fondness for company": Banks to Alströmer, 16 November 1784,
Chambers 2000, p. 79.
201 Sarah Wells: Sarah Wells was Banks's mistress from around 1776 until
his marriage in 1779, Carter 1988, pp. 152–3; Solander to John Lloyd,
5 June 1779, Duyker and Tingbrand 1995, p. 386.
201 "Mrs. Banks, 20 years": and the following quotes: Solander to John
Lloyd, 5 June 1779, Duyker and Tingbrand 1995, p. 386.
202 "an Academy of": Carl Thunberg in late 1777, quoted in Carter
1988, p. 173.
202 herbarium at Soho Square: Robert Jamieson's description in August
1793, quoted in Carter 1988, p. 336.
202 Philip Miller's herbarium: Solander to Ellis, 21 December 1774,
Duyker and Tingbrand 1995, p. 346.
202 George Clifford's herbarium: Carter 1988, Appendix VI "General
Basis of the Herbarium of Sir Joseph Banks," p. 555.
202 "if the Herbarium": Banks to Dryander, 22 October 1791; Dryander to
Banks, 28 October 1793, Dawson 1958, p. 278.
202 "livd almost wholly": Banks to Jacques Julien de La Billardière,
15 July 1797, Chambers 2000, p. 194.
202 "[f]or godsake when": Banks to Dryander, 6 October 1786, Chambers
2007b, p. 216.
202 "free from the Shackles": Banks's Speech to the Royal Society, 1778,
quoted in Carter 1988, p. 147.
203 "I have escapd a Million": Banks to Franklin, 9 August 1782, Labaree
1959–, vol. 37, p. 716.
203 "[Y]ou are welcome": Banks to Samuel Purkis, 2 March 1794, Chambers 2007b, vol. 4, p. 276.
203 "for the Benifit of": Franklin's passport for Cook "To All Captains
and Commanders of American Armed Ships," 10 March 1779, Labaree
1959–, vol. 29, pp. 86–7. The news that Cook had been murdered in
February 1779 reached Europe only in January 1780.
203 "I respect you as": Banks to Franklin, 29 March 1780, Chambers
2000, p. 54. Banks insisted that Franklin received the medal as an
individual rather than the American Congress—which would have
probably been too much of a political statement in times of war.
203 "a bird of peace": Banks to John Hunter, 30 March 1797, Chambers
2000, p. 190.
203 passport for Bonpland: Alexander von Humboldt to Banks, 15 August
1798, BL Add 8099, ff. 71–2.
203 "the science of two": Banks to Jacques-Julien de La Billardière,
9 June 1796, Chambers 2000, p. 171.
203 "at no small Personal": Banks to Thomas Coutts, 24 December 1803,
Chambers 2000, p. 251.
203 "other scientific persons": Banks to William Pitt the Younger,
17 March 1797; Banks to Jean Charretié (French Commissary in
London), 18 March 1797, Chambers 2000, pp. 185, 188.
204 La Billardière's collection: Banks to Jacques-Julien de La Billardière,
9 June 1796, Chambers 2000, p. 171.
204 "I shall not retain": and Banks on La Billardière's collection, Banks to
Antoine Laurent de Jussieu, 10 August 1796, Chambers 2007b, vol. 4,

pp. 434–5; Banks to Jacques-Julien de La Billardière, 15 July 1797, Chambers 2000, p. 194.

204 "ungrateful Solander": and Solander's and Banks's behaviour towards Linnaeus. Linnaeus on Solander, 1772, Afzelius 1826, p. 68; Ellis to Linnaeus, 14 January 1772, Smith 1821, vol. 1, p. 276.

204 Carl wanted duplicate specimens: Banks to Linnaeus the Younger, 5 December 1778, Chambers 2000, p. 51.

204–5 "the friendliness that Banks": Carl Linnaeus the Younger to Claes Alströmer, 28 April 1781, quoted in full Uggla 1953, p. 73.

205 "Nobody will yet receive": Dryander to Thunberg, 3 July 1781, quoted in Jonsell 1994, p. 25.

205 Banks's surprise: Carl Linnaeus the Younger to Claes Alströmer, 28 April 1781, quoted in full Uggla 1953, p. 75.

205 "klen Karl": Solander about Carl Linnaeus the Younger, Uggla 1953, p. 77.

205 "the greatest collection": Linnaeus to Sara Linnaeus, 2 March 1776, Blunt 2001, p. 239.

205 Solander in Tierra del Fuego: Banks, 16 January 1769, Beaglehole, vol. 1, pp. 218–21.

205–6 Solander's symptoms and paralysis: Bladgen to Banks, 8 May 1782, Chambers 2007b, vol. 1, p. 324; Carter 1988, pp. 180–81.

206 "extreme danger": Bladgen to Banks, 8 May 1782, Chambers 2007b, vol. 1, p. 324.

206 Hunter's autopsy report: Carter, 1988, p. 181.

Chapter 12: "As good-humoured a nondescript Otatheitan as ever!"

207 "[T]o write about the Loss": Banks to Lloyd, 16 June 1782, see also Banks to the Duchess of Portland, 10 June 1782, Chambers 2007b, vol. 1, pp. 331, 329.

207 "forces me to draw": and following quote, Banks to Alströmer, 16 November 1784, Chambers 2000, pp. 79–80.

207–8 "book of information": Banks to Yonge, 15 May 1787, Chambers 2000, p. 89.

208 "a great botanical exchange": Banks to Henry Dundas, 15 June 1787, quoted in Desmond 1995, p. 126.

208 Miller sent cotton to Georgia: Minter 2003, p. 24.

208 Linnaeus suggested growing tea in Sweden: Solander to Linnaeus, 19 December 1760, Duyker and Tingbrand 1995, p. 156.

208 Banks and King George at Kew: Banks to James Lind, 23 February 1789, Chambers 2000, p. 120; Desmond 1994, p. 106.

208 "Farmer George": Black 2006, p. 137.

208 "by nature intended": Banks to William Devaynes, 27 December 1788, Chambers 2000, p. 117.

209 "To exchange between": Banks to Sir George Yonge, 15 May 1787, Chambers 2000, p. 90.

209 state trade policies: Gascoigne 1998, pp. 4–34.

209 Banks as expert on botany: Matthew Wallen to Banks, Memorandum, 1785, Chambers 2007b, vol. 3, p. 2.

209 East India Company and Banks: Thomas Morton to Banks, 13 April 1785, BL Add 33978, f. 9.

209 "Knowledge, Philanthropy": Matthew Wallen, Memorandum, 1785, Chambers 2007b, vol. 3, p. 2.

209 "For godsake be active": Banks to Aiton, 29 August 1785, Hyde Collection, copy of the letter at the Sir Joseph Banks Archive Project, NHM.

210 "beauty or curiosity": Banks's Instructions for Hove, April 1787, DTC, vol. 5, f. 126.

210 shortage of cotton: Mackay 1985, pp. 160, 144–6.

210 botanic garden in Calcutta: Robert Kyd to Banks, 1 June 1786, DTC, vol. 7, f. 37; Banks to Sir George Yonge, 15 May 1787, Chambers 2000, p. 89.

210 "solely for the promotion": Banks to East India Company, 17 January 1790, DTC, vol. 7, f. 3.

210 "With such example": Banks to Sir George Yonge, 15 May 1787, Chambers 2000, p. 90.

211 "my favourite Colony": Banks to Philip Gidley King, 3 March 1797, DTC 10:2, f. 76.

211 "extremely cowardly": Banks in the House of Commons, 1 April 1779, quoted in Carter 1988, p. 164; for later recommendations, Frost 1765, pp. 310–11.

211 Captain Phillip at Soho Square: Carter 1988, p. 234.

211 "like a Small Green": Masson to Banks, 13 November 1787, quoted in Carter 1988, p. 232.

211 breadfruit: Lieutenant Vancouver, "Some Reflections upon the Bread fruit of the South Sea Islands, and upon a method for bringing it to Jamaica," February 1787, DTC, vol. 5, f. 227.

212 "a wholesome pleasant Food": Hinton East to Banks, 19 July 1784, Kew JBK/1/3: Banks's Letters, vol. 1, f. 169.

212 Banks and the breadfruit expedition: Hinton East to Banks, 19 July 1784, Kew BC 1, f. 168; Matthew Wallen to Banks, 23 September 1784, Chambers 2007b, vol. 2, p. 310; Hinton East visited Banks in August 1786; Banks to Charles Jenkinson, 30 March 1787, DTC, vol. 5, ff. 143–6; Thomas Townshend to Banks, 15 August 1787, DTC, vol. 5, f. 208; George Yonge to Banks, 7 September 1787, DTC, vol. 5, ff. 245–6. Both Solander and Banks had advised John Ellis for his publication *A Description of the Mangostan and the Breadfruit, Directions to Voyagers, for bringing over these and other Vegetable Productions, which would be extremely beneficial to the Inhabitants of our West India Islands* (1775).

212 "I wish such dablers": Banks to Sir Evan Nepean, 9 September 1787, DTC, vol. 5, f. 294.

212 David Nelson: Banks to Charles Jenkinson, 30 March 1787, DTC, vol. 5, f. 146.

212 "One day, or even": and following quote: Banks's Instructions for David Nelson, 1787, DTC, vol. 5, f. 220.

212 "never interfere in": Banks to Sir George Yonge, 9 September 1787, Chambers 2000, p. 100.

212 "[T]he Master & Crew": Banks to an unknown correspondent, *c.* February 1787, DTC, vol. 5, f. 210.

213 cabins on the *Bounty*: Plans of the Decks etc of the *Bounty*, 20 November 1787, NMM, ZAZ 6668; Banks to an unknown

correspondent, *c.* February 1787, DTC, vol. 5, f. 210; Bligh 2003, p. 2.

213 holes for plant pots: Nelson to Banks, 18 December 1787, DTC, vol. 5, f. 293.

213 Great Cabin on the *Bounty*: Banks to an unknown correspondent, *c.* February 1787, DTC, vol. 5, ff. 210–11.

213 *Bounty* voyage: Alexander 2004; Bligh 2003.

213 "no cause to inflict": Bligh to Banks, 17 February 1788, SLNSW Banks Series 46.21, Frame Number CY 3004/106 (Banks received the letter 20 August 1788).

213 last letter from the *Bounty*: Bligh to Banks, 20 June 1788, SLNSW Banks, Series 46.24, Frame Number CY 3004/124 (Banks received the letter 22 October 1788).

214 "an article of the": Banks to William Devaynes, 27 December 1788, Chambers 2000, p. 118.

214 Chinese tea planters to India: Banks to William Devaynes, 27 December 1788, Chambers 2000, p. 116.

214 "[t]heir patient industry": James Rennell to Banks, 22 December 1788, DTC, vol. 6, f. 101.

214 Australia and cinnamon tree in Barbados: Banks's "Instructions for James Austin and George Austin, the two Gardeners," July 1789, DTC, vol. 6, ff. 196–203; Carter pp. 253–4; Joshua Steele to Banks, 16 May 1789, BL Add 33978, f. 242.

214 "the numberless plants": *Monthly Review*, January 1790, p. 50

215 sample of cotton for waistcoat: Mackay 1985, p. 160; Samuel Felton to Banks, 17 January 1790, Kew JBK/1/5: Banks's Letters, vol. 2, f. 1.

215 "I am now so ill": Bligh to Banks, 13 October 1789, SLNSW Banks, Series 46.27 Frame Number CY 3004/130.

215 breadfruit on the *Bounty*: Bligh 2003, p. 86.

216 "a Fever": Bligh to Banks, 13 October 1789, SLNSW Banks, Series 46.27 Frame Number CY 3004/153.

216 Australia plant transport hit iceberg: Carter 1988, p. 254.

216 "I am tired of": Banks to William Roxburgh, 9 August 1798, BL Add 33980, f. 160.

216 "cursd the Chairman": and Banks's problems with the East India Company, Banks's annotation in William Maclean (Inspector of the East India Company at Custom House) to Banks, 23 June 1798, BL Add 33980, f. 156v; Banks to William Maclean, 22 June 1798, Dawson 1958, pp. 294–5; Banks to the Secretary of the East India Company, 30 June 1798.

216 "our Kew Hospital": Banks to William Roxburgh, 9 August 1798, BL Add 33980, f. 159.

216 Bligh and the *Providence*: Bligh to Banks, 24 October 1790, DTC, vol. 7, f. 160; Carter 1988, p. 257.

216 Babul—*Acacia Arabica*: Robert Wissett to Banks, 8 March 1790, RS Misc MSS 6, f. 38.

217 *Phormium tenax*: Banks, 11 October 1769, Beaglehole 1962, vol. 1, p. 407; Carter 1988, pp. 213–14.

217 "a quarterly statement": Banks to East India Company, 17 January 1791, DTC, vol. 7, f. 304.

217 botanist to Africa: Wilberforce to Banks, 21 December 1791, DTC, vol. 7, f. 293; Banks to Wilberforce, 8 April 1793, DTC, vol. 8, f. 196.

217 "Sir Joseph Banks": Bligh to Banks, 26 July–1 August 1793, SLNSW Banks, Series 50.28, Frame Number CY 3004/317.

217 breadfruit arrived in West Indies: James Wiles and Christopher Smith (gardeners on the *Providence*) to Banks, 17 December 1792, SLNSW Banks, Series 52.09, Frame Number CY 3004/379.

217 "there can be no excuse": Bligh to Banks, 26 July–1 August 1793, SLNSW Banks, Series 50.28, Frame Number CY 3004/314.

217 plants brought to England on the *Providence*: Christopher Smith, "Plants on Board of the *H.M.S. Providence*," August 1793, SLNSW Banks, Series 52.16, Frame Number CY 3004/401–6.

217 "never before seen plants": Banks to William Pitt, 1 September 1793, SLNSW Banks, Series 54.01, Frame Number CY 3004/486.

218 largest living collection of plants: Banks to William Pitt, 1 September 1793, SLNSW Banks, Series 54.01, Frame Number CY 3004/486; Banks to William Roxburgh, 29 May 1796, BL Add 33980, f. 65.

218 Tahitian to West Indies: Bligh to Banks, 16 December 1792, DTC, vol. 8, f. 131.

219 breadfruit trees flourished: Wiles to Banks, 16 October 1793, Kew JBK/1/5: Banks's Letters, vol. 2, f. 103.

219 "thriving with greatest": and following quote, Henry Shirley to Banks, 20 December 1794, DTC, vol. 9, ff. 141, 143.

219 "thrives (if possible)": Alexander Anderson to Banks, 30 March 1798, DTC, vol. 10, pt. 1, f. 25.

219 "perfectly naturalised": James Wiles to Banks, 16 May 1801, Kew JBK/1/7: Banks's Letters, vol. 2, f. 248.

220 sheep in Australia: Banks to John Maitland, 31 March 1804, Chambers 2000, pp. 253–5; Carter pp. 174, 228–9, 238–40, 427ff.

220 "for the Consideration of": Banks was sworn in on 29 March 1797, Carter 1988, p. 313.

220 "His Majesty's Ministre": Lord Auckland to Lord Grenville, 6 November 1791, HMC, Fortescue 1894, Fourteenth Report, Appendix, pt. v, vol. 2, p. 225.

220 "sprawls upon the Grass": Maria Josepha Holroyd about Banks, 2 August 1795, Adeane 1899, p. 94.

220 "attention is very much": Joseph Farington, 20–22 January 1796, Cave et al. 1978–84, vol. 2, p. 478.

220 Sunday soirées: Joseph Farington, 23 November 1807 and 1 February 1795, Cave et al. 1978–84, vol. 8, p. 3147 and vol. 2, 1978, p. 300.

220 Banks's collectors: Mackay 1996, p. 39.

220 first footnote: James 2001, p. 150.

220 second footnote: Banks to George Harrison, 1 September 1814, Chambers 2007b, vol. 6, p. 141–2; Banks to Sigismund Bacstrom, 20 August 1791, DTC 7 f. 245.

221 "favourite project": Banks to Captain Henry Essex Bond, 1 August 1794, DTC, vol. 9, f. 75.

221 "honor to the Science": Banks to George Harrison, 1 September 1814, Chambers 2007b, vol. 6, p. 144.

221 "commerce and manufactures": Smith 1982, p. 508.

Chapter 13: "Loves of the Plants"

222 Banks's finances: Carter 1988, p. 191.

222 interest in Linnaeus's collection: James Edward Smith to his father, 24 December 1783, 12 January 1784, Smith 1832, vol. 1, pp. 93, 97.

223 give Smith "honour": Smith 1832, vol. 1, p. 92.

223 Smith's father released funds: James Edward Smith to his father, 24 December 1783, 12 January, 10 April 1784, Smith 1832, vol. 1, pp. 93, 97, 103. The final price for the collection was £900 plus transport.

223 unpacking Linnaeus's collection: Smith 1832, vol. 1, pp. 126–8.

223 "most decidedly sets": Bishop of Carlisle to Smith, January 1785, Smith 1832, vol. 1, p. 134.

223 Botany and the middle classes: Seward 1804, pp. 167–8; Shteir 1996, p. 50.

223 "Botany Flourishes here": Banks to Smith, 15 August 1787, Chambers 2007b, vol. 3, p. 309.

223 "Yes, even for Women": Linnaeus, *Deliciae Naturae*, 1773, quoted in Koerner 1996, p. 148.

224 "fair daughters of": Abbot 1798, p. vi.

224 "It belongs, I believe": "Botanical Conversation," *The New Lady's Magazine, or Polite and Entertaining Companion for the Fair Sex*, May 1786, p. 177.

224 "prevents the tumults": Rousseau to Madame Delessert, 22 August 1771, Rousseau 1785, p. 21.

224 "the most delicious": Walpole to Mary Berry, 28 April 1789, Lewis 1937–61, vol. 11, p. 10.

224 "The Linnaean system is": Darwin quoted in King-Hele 1999, p. 110.

224 carriage as bespoke study: Hankin 1859, pp. 126–7; Seward 1804, p. 167.

224 "INCHANTED GARDEN": Darwin (1789) 1791, Address to the Reader.

225 "Beaux and Beauties": Darwin (1789) 1791, line 9.

225 "moss-embroider'd beds": Darwin (1789) 1791, line 232.

225 "little pictures suspended": Darwin (1789) 1791, Address to the Reader.

225 "First the tall CANNA": Darwin (1789) 1791, lines 39–44.

225 "folds her infant": Darwin (1789) 1791, line 206.

225 "chaste MIMOSA": Darwin (1789) 1791, line 301.

225 "ten fond brothers": Darwin (1789) 1791, line 58.

225 "blushing captives of": Darwin (1789) 1791, 120.

225 "six gay Youths": Darwin (1789) 1791, line 218.

225 footnote: Uglow 2004, p. 424.

227 Darwin "sublime" and "divine": Walpole to Lady Ossory, 8 September 1791, Lewis 1937–61, vol. 34, p. 123.

227 "shed lustre over": Seward 1804, p. 379.

227 "the dazzling manner": and following quotes, in King-Hele 1986: Wordsworth on Darwin, pp. 67–8, Cowper on Darwin, p. 149; Crabbe on Darwin, pp. 163ff.; Walter Scott on Darwin, p. 155; Coleridge on Darwin, p. 1.

227 "by poring through": Walpole to Mary Berry, 28 April 1789, Lewis 1937–61, vol. 11, p. 11.

227 Darwin's translations: the *System of Vegetables* was published in 1783/4 and the *Families of Plants* in 1787. See also Darwin to Banks, 29 September 1781, King-Hele 1981, pp. 112–14.

227 "to cast your eye": and Darwin and Banks, Darwin to Banks, 24 October 1781; 29 September 1781; 24 October 1781; 1 November 1781, 23 February 1782, King-Hele 1981, pp. 112–23.

228 "much encourage or retard": Darwin to Banks, 13 September 1781, King-Hele 1981, p. 109.

228 "immense price": Seward 1804, p. 167.

228 "as every line puts": Hankin 1859, p. 207.

228 "wish to become": *The Botanical Magazine*, vol. 1, 1787, title page; for botanical publications in general: "Table 1: The Rate of Publication on Botanical, Horticultural and Related Subjects, 1700–1800," Gascoigne 1994, p. 109.

228 "or the labours of": *The Botanical Magazine*, vol. 1, 1787, title page and preface.

228 "keep pace with": James Edward Smith, quoted in Desmond 2003, p. 107.

228 "herbarizing excursions": "Proposals for a course of herbarizing excursions by Mr. Curtis," Curtis 1783, no page number.

228 *Botanical Pocket Book*: Mavor 1800.

228–9 "*Natural History* and *Botany*": Benjamin Smith Barton to William Bartram, 19 February 1788, Hindle 1956, p. 310.

229 "Linnean [*sic*] science": Seward 1804, p. 127.

229 "I have specimens of": Richard Twiss to Francis Douce, 13 June 1791, Longstaffe-Gowan 2001, p. 141.

229 "a botanical arrangement": Repton (1816) 1840, p. 552.

229 "Botanical Walk": Richard Pococke, 29 November 1754, Cartwright 1889, vol. 2, p. 166.

229 "great wages": and admiration of English gardens, Archenholz 1789, vol. 2, p. 221; Wendeborn 1791, vol. 2, p. 231.

229 "foreign plants in": Volkmann 1781, vol. 1, p. 440.

229–30 affordable American species: "A Catalogue of forest-Trees, Fruit-Trees, Ever-Green and Flowering-Shrubs, sold by John and George Telford, Nursery-Men and Seeds-Men," York, A. Ward, 1775, reprinted in full in Harvey 1972, Appendix IV; "A Catalogue of Forest Trees, Fruit Trees, Evergreen and Flowering Shrubs, Hot-House, Green-House and Herbaceous Plants, Kitchen-Garden and Flower Seeds, sold by John Mackie, Nursery and Seeds-Man," Norwich, 1790 (for more plant prices see nursery catalogues in the Bibliography and Glossary).

230 manuals for shrubberies: Meader 1779; Weston 1773; Abercrombie and Mawe 1778.

230 garden in Upper Gower Street: Richard Twiss to Francis Douce, 1791, Laird and Harvey 1997, Appendix I, pp. 206–7.

230 "*rus in urbe*": Carl Philip Moritz, 17 June 1782 about Grosvenor Square, Nettel 1965, p. 59.

230 plants in Soho Square: Longstaffe-Gowan 2001, p. 220.

230 "Shall I bring you": Walpole to Mason, 7 August and 10 July 1775, Lewis 1937–61, vol. 28, pp. 218, 221.

230 "a specimen of English": and English garden in Caserta: Hamilton to Walpole, 19 February 1788, Lewis 1937–61, vol. 35, p. 439; Hamilton

to Banks, 20 February 1785, DTC, vol. 4, f. 128; Quest-Ritson 1992, pp. 102-3.

231 "Botanic garden": and following quote, Hamilton to Banks, 26 September 1786, DTC, vol. 5, f. 75; 20 October 1789, DTC, vol. 6, f. 264; 19 August 1788, DTC, vol. 6, f. 69.

231 "beautiful Botany Bay": Hamilton to Banks, 31 May 1797, DTC, vol. 10, pt. 1, f. 143.

231 "British custom of": Count della Torre di Rezzonico in 1790, Quest-Ritson 1992, p. 103.

231 Catherine's "anglomania": Catherine to Voltaire, 25 June 1772, Hayden 2005, p. 85.

231 "visit all the notable": Catherine to Vasily and Pyotr Neyelov, December 1770, Hayden 2005, p. 84.

231 Catherine's anglophile lovers: for example Stanislaus Augustus Poniatowski, Hayden 2005, p. 77; for her British gardeners see Cross 1991, pp. 12-20.

231 Bush and Bartram: Busch continued to order seed boxes from John Bartram during his time in Russia, James Freeman to Bartram, 13 July 1771, Berkeley and Smith Berkeley 1992, p. 743; Köhler 2003, p. 100.

231 Banks dispatched plants from Kew: and Tsarskoe Selo, Hayden 2005, p. 87; Cross 1991, p. 14; Banks, 29 December 1795, MS diary notes, reprinted in Carter 1974, p. 356.

231 "gardening in that": Jefferson to John Page, 4 May 1786, Boyd et al. 1950-, vol. 9, p. 445.

231 "to cut out the superabundant": Jefferson, "Objects of Attention for an American," 1788, Boyd et al. 1750-, vol. 13, p. 269.

232 "Any of the Nobility": James Meader, 22 October 1780, Cross 1993, p. 177.

232 foreign gardeners in England: for example, Friedrich Kuckuck went in 1762; J. C. Schlüter in 1761; gardener Daniel August Schwarzkopf went with his employer Earl Daniel Christoph von der Schulenburg for ten months in 1762; Johann Andreas Graefer went in 1763 and stayed in England; Johann Jonas Tatter went in 1765; Friedrich Ludwig von Sckell in 1773, Köhler 2003, pp. 89, 17, 25, 82, 16.

232 "still had an extremely": and foreign gardeners in Kew, Duchess de la Tremoille, 27 October 1774, Coke 1970, vol. 4, p. 419; Köhler 2003, pp. 70-77; Banks to George Caley, 7 March 1795, Chambers 2007b, vol. 4, p. 348.

232 continental books on English gardens: for example, C.C.L Hirschfeld's *Theorie der Gartenkunst*, 1779-85; René-Louis Girardin's *De la Composition des paysages*, 1777; Ercole Silva's *Dell'arte de' giardini inglesi*, 1801; Grohmann 1796-1811; see also Richards 2000.

233 "every distant part": "A Catalogue of Hardy Trees and Shrubs, Greenhouse and Stove Plants, herbaceous Plants, and Fruit Trees ... sold by Luker and Smith, Nurserymen and Seedsmen," London, 1783, p. v.

233 prominent flowerbeds: the first of these flower gardens was at Lord Harcourt's Nuneham Courtenay in Oxfordshire and Lady Elizabeth Lee's Hartwell in Buckinghamshire. Humphry Repton also brought flowers back into the garden, for example, at Sheringham in Norfolk and Endsleigh in Devon. Laird 1999, pp. 331-60; Wulf and Gieben-Gamal 2005, pp. 200-203.

233 "seats in a Theatre": Swinden 1778, p. 1.

233 tiered flowerbeds: the description is based on Swinden 1778, the planting plans for Hartwell House (1799), which are reproduced in Laird 1999, pp.366-70, and Tromp (1791) 1982, pp. 55-6.

233 plants in Kew: unless otherwise referenced, the following plant lists are based on Aiton 1789 and Aiton 1810-13.

233 "one of the most desirable": The Botanical Magazine, vol. 12, 1798, plate 406.

234 "acquainted with the plants": Banks to William Townsend Aiton, 17 July 1802, BL Add 33981, f. 40v.

234 "not choosing to trust": Sir James Edward Home, quoted in Desmond 1995, p. 94.

234 "[W]e have this summer": and Australian plants, Banks to Sir Thomas Gery Cullum, 15 November 1789, Chambers 2007b, vol. 3, p. 517; Richard Molesworth to Banks, 12 June 1789, Kew JBK/1/4: Banks's Letters, vol. 1, f. 350; "A Catalogue of Garden, Grass and Flower Seeds, Trees, Shrubs, Herbaceous, Green-House and Hot-House Plants, Sold by Russell, Russell, & Willmott, Nursery & Seedsmen, Lewisham, Kent," London, 1800; Campbell-Culver 2001.

234 hardenbergia and coral dusky pea: The Botanical Magazine, vol. 8, 1794, plate 263, 268; vol. 22, plate 865.

234 "one of the most ornamental": The Botanical Magazine , vol. 8, 1794, plate 263.

234 grevilleas were "common": The Botanical Magazine, vol. 22, 1805, plate 862.

234 footnote: Nelson 1983, pp. 352, 347.

235 "in most of the Nurseries": The Botanical Magazine, vol. 8, 1794, plate 260.

235 "an infinity of Plants": Banks to Abbé Pierre André Pourret, October 1791, Chambers 2007b, vol. 4, p. 71.

235 number of plants at Kew: Hill 1768 (c. 3,400 species); Aiton 1789 (c. 5,600 species); Aiton 1810-13 (c. 11,000 species); also Desmond 1994, p. 112.

235 "our King at Kew": Banks to George Staunton, 23 January 1796, Sutro Library, California, copy of letter at the Sir Joseph Banks Archive Project, NHM.

235 "vast variety of exotics": Banks to Georges Louis Marie Dumont de Courset, 20 December 1790, BL Add 8097, f. 401v.

235 "a most elegant addition": Banks to John Hunter, 30 March 1797, Chambers 2000, p. 189.

235 bird-of-paradise: The Botanical Magazine, vol. 4, 1790, plate 120.

235 Masson's stapelias: Masson 1796; Aiton 1810-13; Carter 1988, p. 304.

236 "every garret and cottage": Smith, 1819, quoted in Lindsay 2005, p. 53.

236 "beyond the power": The Botanical Magazine, vol. 12, 1798, plate 402.

236 systematic hybridisation: this was William Rollisson at Tooting, Elliott 2003, p. 171.

236 "found in most": The Botanical Magazine, vol. 11, 1797, plate 366; vol. 13, 1799, plates 440, 447.

236 "scarcely any period": The Botanical Magazine, vol. 10, 1796, plate 343.

237 "immediately set out": Banks's Instructions for James Cunningham and James Bowie, 18 September 1814, Kew, Kew Collectors, vol. 1,

f. 15 ; George Annesley to Banks, 13 December 1802, DTC, vol. 13, f. 325; Banks to Archibald Menzies, 22 February 1791, Chambers 2000, p. 129.

237 "till they have as": Walpole to Mason, 7 August 1775, Lewis 1937–61, vol. 28, p. 223.

237 "never come to perfection": Wendeborn 1791, vol. 2, p. 232.

237 "extensive shrubbery": Sale by Auction, *The Times*, 28 May 1791.

237 "Rare Shrubs, and": Sale by Auction, *The Times*, 9 May 1792.

237 "capital collection": Sale by Auction, *The Times*, 24 April 1801.

237 "a London garden": Richard Twiss to Francis Douce, 1791, Laird and Harvey 1997, Appendix I, p. 206.

237 "main criterion": Johanna Schopenhauer in 1803–5, Michaelis-Jena and Merson 1988, p. 3.

237–8 hothouse a sought-after accessory: like exotics and shrubberies, hothouses and greenhouses were also highlighted in the sales particulars, see for example: Sales by Auction, *The Times*, 24 October 1786; 20 April 1789; 28 May 1791; 9 May 1792; 24 April 1801.

238 "in most of the gardens": François de la Rochefoucauld in 1784, Rochefoucauld 1995, p. 45.

238 "the numerous tribes": Repton (1803) 1840, p. 217.

238 "an appendage to": Loudon 1824, p. iii.

238–40 botanical playing cards: Botanical Pastimes . . . Calculated to Facilitate the Study of the Elements of Botany, *c.*1810, Shteir 1996, p. 12.

240 "bunch of flowers": Carl Philip Moritz on 2 June 1782, Nettel 1965, p. 24.

240 "metamorphosed . . . into gardens": Archenholz 1789, vol. 2, p. 179.

240 rhododendrons and kalmias for hire: Longstaffe-Gowan 2001, p. 162.

240 "Fool that She Likes": Banks to Perrin, 19 November 1798, Chambers 2000, p. 199.

240 "a whole *jardin*": François Métra about Marie Antoinette's *pouf*, Weber 2006, p. 14.

240 "Eleven damsels": Hannah More to one of her relatives, 1777, Brimley Johnson 1925, p. 57.

240–41 "[S]carce a person": Richard Weston's introduction to his *Gardeners pocket-calendar* 1779, Henrey 1975, vol. 2, p. 469.

Epilogue

243 planting at Wörlitz: see also Ringenberg 2001.

243 "[i]t is now a fashion": Münchhausen 1770, vol. 5, p. 224.

243–4 Bartram's plants in Europe: Rössig 1796 (plant lists).

244 painting with plants: Hirschfeld 2001, p. 218; Sckell 1982, pp. 112ff.

244 Miller's advice: Miller 1739, "Wilderness."

244 Prinz Franz in England: Franz von Anhalt-Dessau visited England in 1763–4, 1766–7, 1775 and 1785.

244 foreigners visit Britain: Weiss 2007, p. 18.

245 "the secret birthplace": Weiss 2007, p. 14.

245 "Garten Bibliothek": Boettiger 1999, p. 26.

245 South Sea artefacts: Boettiger 1999, p. 24.

245 "Otahaitisches Kabinett": Weiss 2007, p. 188.

Picture Credits

I would like to thank the Wellcome Library, London, for their (almost free) image database and for digitalising so many illustrations to use in this book and the Linnean Society, London, for generously waiving the reproduction fees for illustrations from their collection.

First Colour Plate Section

Callicarpa americana, Mark Catesby's *Natural History of Carolina, Florida, and the Bahama Islands* (1731–1748), vol. 2, plate 47, reproduced with permission of the Linnean Society, London

Cypripedium calceolus var. parviflorum, Mark Catesby's *Natural History of Carolina, Florida, and the Bahama Islands* (1731–1748), vol. 2, plate 73, reproduced with permission of the Wellcome Library, London

Catalpa bignonioides, Mark Catesby's *Natural History of Carolina, Florida, and the Bahama Islands* (1731–1748), vol. 1, plate 49, reproduced with permission of the Wellcome Library, London

Kalmia angustifolia, Mark Catesby's *Natural History of Carolina, Florida, and the Bahama Islands* (1731–1748), vol. 2, plate 98, reproduced with permission of the Wellcome Library, London

Pinus strobus, Society of Gardeners' *Catalogus Plantarum* (1730), plate 17, reproduced with permission of the Linnean Society, London

Liriodendron tulipifera, The Botanical Magazine, 1794, vol. 8, plate 275, reproduced with permission of the Linnean Society, London

Monarda didyma, The Botanical Magazine, 1802, vol. 15, plate 546, reproduced with permission of the Wellcome Library, London

Liquidambar styraciflua, Mark Catesby's *Natural History of Carolina, Florida, and the Bahama Islands* (1731–1748), vol. 2, plate 65, reproduced with permission of the Wellcome Library, London

Platanus occidentalis, Mark Catesby's *Natural History of Carolina, Florida, and the Bahama Islands* (1731–1748), vol. 1, plate 56, reproduced with permission of the Wellcome Library, London

Carl Linnaeus in Lapland dress, oil painting after Martin Hoffman's original portrait of 1737, reproduced with permission of the Wellcome Library, London

Linnaea borealis, Carl Linnaeus's *Flora Lapponica*, 1737, reproduced with permission of the Wellcome Library, London

Frontispiece, Philip Miller's *Gardeners Dictionary*, 1764, vol. 1, reproduced with permission of the Wellcome Library, London

Plan of proposed design of Thorndon Park, by Bourgignon d'Anville (based on Lord Petre's original plans), 1733 © reproduced by courtesy of Essex Record Office

Second Colour Plate Section

Robinia hispida, The Botanical Magazine, 1795, vol. 9, plate 311, reproduced with permission of the Linnean Society, London

Joseph Banks, by Joshua Reynolds, 1771–3 © National Portrait Gallery, London

Transplanting of the Breadfruit Trees from Otaheite, by Thomas Gosse, 1796 © Captain Cook Memorial Museum, Whitby

Erica cerinthoides, The Botanical Magazine, 1793, vol. 7, plate 220, reproduced with permission of the Wellcome Library, London

Banksia serrata, Henry Andrews's *Botanists Repository*, vol. 2. plate 82, reproduced with permission of the Linnean Society, London

Strelitizia reginae, watercolour, no artist, reproduced with permission of the Wellcome Library, London

Pursuit of the Ship Containing the Linnaean Collection by Order of the King of Sweden, *Robert John Thornton's New Illustration of the Sexual System of Carolus von Linnaeus*, pt. 3 "Picturesque Botanical Plates, Illustrative of the sexual system of Carolus von Linnaeus," 1799–1810, no plate number, reproduced with permission of the Wellcome Library, London

Gardiners, Henry William Pybe's *Microcosm: or, a Picturesque Delineation*, 1808, vol. 2, reproduced with permission of the Wellcome Library, London

Callistemon citrinus, The Botanical Magazine, 1794, vol. 8, plate 260, reproduced with permission of the Wellcome Library, London

Pandorea pandorana, The Botanical Magazine, 1805, vol. 22, plate 865, reproduced with permission of the Wellcome Library, London

Kennedia rubicunda, The Botanical Magazine, 1794, vol. 8, plate 268, reproduced with permission of the Wellcome Library, London

Stapelia pedunculata, The Botanical Magazine, 1804, vol. 21, plate 793, reproduced with permission of the Wellcome Library, London

Three Men Botanizing, Frontispiece to Mr. Curtis' *Flora Londinenses*, Robert John Thornton's *New Illustration of the Sexual System of Carolus von Linnaeus*, pt. 3 "Picturesque Botanical Plates, Illustrative of the sexual system of Carolus von Linnaeus," 1799–1810, no plate number, reproduced with permission of the Wellcome Library, London

Battlesden Park, possibly by George Shepherd, *circa* 1818 © Bedfordshire & Luton Archives Service

Images in Text

Prologue: Formal Garden, frontispiece, in Society of Gardeners' *Catalogus Plantarum*, 1730, reproduced with permission of the Linnean Society, London

Chapter 1: Peter Collinson, engraving by Joseph Miller, 1770, reproduced with permission of the Wellcome Library, London

Chapter 1: John Bartram, as illustrated by Howard Pyle, "Bartram and His Garden" in *Harper's New Monthly Magazine*, February 1880, vol. 60

Chapter 2: Chelsea Physic Garden, engraving published by T. Cadell and W. Davies, 1795, reproduced with permission of the Wellcome Library, London

Chapter 2: Tan Stove, Philip Miller's *Gardeners Dictionary*, 1731, plate 1, reproduced with permission of the Wellcome Library, London

Chapter 3: Carl Linnaeus's Sexual System, in James Lee's *Introduction to Botany*, 1794, reproduced by courtesy of the London Library

Chapter 4: A Draft of John Bartram's House and Garden as it appears from the River, 1758, watercolour on paper by William Bartram © The Right Hon. Earl of Derby/The Bridgeman Art Library

Chapter 5: Detail of Lord Petre's drawings for Worksop, Lord Petre's Architectural Elements for Worksop, 1738, reproduced by courtesy of Essex Record Office

Chapter 6: Detail of Plant Transport Boxes, frontispiece, in John Ellis's *Directions for bringing over seeds and plants from the East-Indies and other Distant Countries in a State of Vegetation*, 1770, reproduced with permission of the Wellcome Library, London

Chapter 7: Linnaeus's Botanic Garden at Uppsala, 1745, in Linnaeus's *Horti Upsaliensis*, reproduced with permission of the Linnean Society, London

Chapter 7: Detail of *Siegesbeckia*, in Linnaeus's *Hortus Cliffortianus*, 1738, plate XXIII (detail), reproduced with permission of the Wellcome Library, London

Chapter 7: *Collinsonia canadensis*, in Linnaeus's *Hortus Cliffortianus*, 1738, plate V, reproduced with permission of the Wellcome Library, London

Chapter 8: Daniel Solander, drawn by James Sowerby, engraved by James Newton, 1784, reproduced with permission of the Wellcome Library, London

Chapter 9: Venus Flytrap, 1769, John Ellis, reproduced with permission of the Linnean Society, London

Chapter 10: Plan of His Majesty Bark *Endeavour* as fitted at Deptford in July 1768, ZAZ6590 © National Maritime Museum, London

Chapter 10: Heads of divers Natives of Othaheite, Huaheine, & Oheiteroah, in Sydney Parkinson's *A Journal of a Voyage to the South Seas, in His Majesty's ship, the Endeavour*, 1784, plate viii, reproduced with permission of the Wellcome Library, London

Chapter 11: Joseph Banks's Library at Soho Square, sepia wash drawing by Francis Boott, 1820 © Natural History Museum, London

Chapter 12: Detail of Plans of *Bounty* Lower Deck, 13 May 1790, ZAZ6668, © National Maritime Museum, London

Chapter 12: Eastern and Western Hemisphere, in Sydney Parkinson's *A Journal*

Acknowledgements

"I promisd myself three years uninterrupted enjoyment of my Favourite pursuit," Joseph Banks wrote from the *Endeavour* in 1768—which is exactly what I have done for the past three years. I'm grateful to the many people and institutions who have helped and supported me, so that I could bury myself in the lives and letters of "my" six men. It is with the greatest pleasure that I say thank you to:

the staff at Bartram's Garden in Philadelphia, in particular Joel T. Fry for a delightful day in the garden, plant lists and continuous assistance—thank you for your generosity; Eva Björn at the Linnaeus Museum in Uppsala for giving me a tour of the house and sharing her knowledge; the staff at the Chelsea Physic Garden, in particular Rosie Atkins whose enthusiasm both for the garden and Philip Miller is always infectious and David Frodin; the staff at the Rare Books and Manuscript Departments at the British Library, particularly Arnold Hunt for solving a palaeographical problem; Andrew Choong at the Historic Photographs and Ship Plans Section of the National Maritime Museum; Dee Cook at The Worshipful Society of Apothecaries of London; Ann Henderson at Special Collections at Edinburgh University Library; Catrin Holland at the Parliamentary Archives, London; the staff at the Sir Joseph Banks Archive Project at the Natural History Museum in London, in particular its Executive Director Neil Chambers who in the midst of his own deadlines gave me his time, unparalleled knowledge, books, articles and many transcripts of Banks's letters as well as pointers to letters which I would have failed to find otherwise—thank you so very much; Monica Knutson for her tour of Linnaeus's house at Hammarby; the staff at the Lindley Library; the Linnean Society for so generously waiving the reproduction fees for their images, in particular Gina Douglas for her kind help and for putting me in contact with the Swedish curators and gardeners, as well as Lynda Brooks and Ben Sherwood—thank you also for cups of tea and scones; the charming and wonderful staff at the London Library; Magnus Lidén at the Botanic Garden in Uppsala for a wonderful, enlightening day at Linnaeus's garden at Uppsala and Hammarby and for insightful comments after reading the Linnaeus chapters; Karin Martinsson at Botanic Garden in Uppsala; Jennifer Ramkalawon at the British Museum's Department of Prints and Drawings; Joe Rowntree from the Sheffield Botanical Garden for his help on the bones in Peter Collinson's garden; the staff at the Royal Society; Tony Simcock at the Museum of the History of Science in Oxford; Ernie Schuyler, Curator Emeritus of Botany, the Academy of Natural Science of Philadelphia; John Styles for his continuous support; Bob Tomlins for a transcript of one of Bartram's plant lists; and Liz Thornton.

I'm indebted to the following archives and libraries for their permission to quote from their manuscripts: Chelsea Physic Garden; Edinburgh University Library; Landeshauptarchiv Sachsen-Anhalt, Abteilung Magdeburg, Aussenstelle Wernigerode; Linnean Society, London; Museum of the History of Science, Oxford; The National Archives; National Library of Wales; the President and Council of the Royal Society; Sir Joseph Banks Archive Project at Natural History Museum in London; Trustees of the Goodwood Collection; the Trustees of the Royal Botanic Gardens Kew; British Library; The Worshipful Society of Apothecaries.

I am also grateful for the generous scholarship from the Wingate Foundation without which it would have been much more difficult to write this book and to travel abroad to visit John Bartram's garden in Philadelphia and to follow his footsteps in the Shenandoah Valley and the Catskills, to see Linnaeus's life and work in Sweden and "der Englische Garten" in Germany.

At Alfred A. Knopf I would like to thank Edward Kastenmeier, Tim O'Connell, Sara Eagle, Gabrielle Brooks, Iris Weinstein, Roméo Enriquez and Victoria Pearson for their hard work.

At William Heinemann (past and present) I would like to thank: Joy de Menil and Ravi Mirchandani for believing in this book, as well as the wonderful Jason Arthur and his dream team: the indefatigable Laurie Ip Fung Chun and Emma Finnigan. Most of all I'm ever so grateful to Rebecca Carter, who is the most perfect editor an author could wish for and who has made the edit of *The Brother Gardeners* a real pleasure. Thank you, for your sharp eye and brilliance—and thank you for picking my book.

I owe a great deal to all my friends and family: Kate Colquhoun for her insights and for making me laugh (always); Charlotte Desai for her clever comments on the original proposal; Anna Helleday and Hans Oskarson for their generous hospitality in Stockholm; Tom Holland for being the most fabulous Latin Dictionary in the world (I still think that Linnaeus's sexual system is more titillating than the Holy Lance) and for enjoying the madness of the "Vegetable Sermon" on the devilish date 06–06–06; Leo Hollis for your ingenious editorial eye, honesty and making me change the first third of the book; Mark Kubale for his comments on the *Endeavour* chapter and assistance on the Australian plants; Nick Leech, as always, I bow to your horticultural knowledge—thank you for letting me ask so many stupid questions and bearing my ignorance, looking through all these endless plant lists and thank you for giving me a beautiful garden (all horticultural mistakes are entirely mine—in the book and in the garden); Shefali Malhoutra for reading the entire manuscript despite not being interested in gardens and your clever comments; your ladyship Saskia for the usual stuff; Lisa O'Sullivan for sending me samples of Australian birdsong and reading the *Endeavour* chapter; Mandana Ruane for creating a space (in London and in my head) to which I can escape; Julia Sen and Jessica Bartling for emotional support at all times; Connie Wall and Gerd Hagmeyer for wonderful days in Sweden and translating some Swedish letters; Tineke Walsh for translating a Dutch letter; the late David Wishart for his lecture on the "Fooks-ia," Anne Wigger for our wonderful trip to the gardens at Wörlitz and Potsdam; Brigitte Wulf for always being there and to Herbert Wulf for battling through Swedish eighteenth-century letters and reading all chapters in their many shapes and forms—and thank you for being my chauffeurs in Sweden.

Thank you, Patrick Walsh, my darling agent, friend and reckless race car driver—for always believing in me, working with me and putting me in cabs (when I can't sit on the top deck of the night bus anymore) . . . and for all those evenings that keep me sane.

And thank you to my two fabulous teenage girls, Linnéa and Hannah, who were my umbilical cord to the real world (Linnéa you were named after the delicate and beautiful *Linnaea borealis* not the cantankerous Linnaeus . . . I promise).

As always my most gracious thank you goes to Adam Wishart, for he is still the best storyteller, the most insightful critic (even if it sometimes hurts . . . a lot . . .) and the most wonderful rock to lean against. Thank you for being so brilliant and loving—and for making me laugh. And thank you for doing so much of the weeding in this book and in the garden.

Index

Note: Page numbers in **bold** refer to illustrations, those followed by "n" refer to footnotes